AF682110

Natural Environment and Culture in the Mediterranean Region

Natural Environment and Culture in the Mediterranean Region

Edited by

Recep Efe, Georges Cravins, Munir Ozturk and Ibrahim Atalay

Cambridge Scholars Publishing

Natural Environment and Culture in the Mediterranean Region,
Edited by Recep Efe, Georges Cravins, Munir Ozturk and Ibrahim Atalay

This book first published 2008 by

Cambridge Scholars Publishing

12 Back Chapman Street, Newcastle upon Tyne, NE6 2XX, UK

British Library Cataloguing in Publication Data
A catalogue record for this book is available from the British Library

Copyright © 2008 by Recep Efe, Georges Cravins, Munir Ozturk and Ibrahim Atalay and contributors

All rights for this book reserved. No part of this book may be reproduced, stored in a retrieval system, or transmitted, in any form or by any means, electronic, mechanical, photocopying, recording or otherwise, without the prior permission of the copyright owner.

ISBN (10): 1-84718-658-0, ISBN (13): 9781847186584

TABLE OF CONTENTS

PREFACE

The aim is to examine, for both natural and cultural environment in the Mediterranean Basin. Thirty-one reviewed papers were accepted for publication. They are grouped in two sections in this volume: Natural Environment in the Mediterranean Basin and Cultural Environment in the Mediterranean Basin.

Contributions in each chapter are prepared by experts in the respective fields and mirror the advancement in the approach. This book contains important future tasks of the particular fields and supplies extensive bibliographies at the end of each chapter, as well as tables and figures that illustrate the research findings. All these make this book highly useful and a must read for students, researchers and professionals in geography, geomorphology, ecology, forestry, hydrology, climate soil and environmental sciences.

The Mediterranean Basin is one of the largest archipelagos in the world, surrounding the Mediterranean Sea and stretching west to east from Portugal to Jordan and north to south from northern Italy to Morocco.
The Mediterranean Sea covers about 2,500,000km^2 with an average depth of 1,500 metres. The coastline extends 46,000km running through 22 countries.

The region is known for its particularly mild climate with uniform and moderate temperatures. Rainfall patterns are however, more unpredictable with a high of 1,200mm per year in Antalya (Turkey), Genoa (Italy) to a low of 100mm per year in Djerba (Tunisia).

The mountains as high as 4,500 meters and peninsulas add to its spectacular scenery. A large part of the basin was once covered by evergreen, deciduous and conifer forests, eight thousand years of human settlement and habitat modification have distinctly altered its natural environment and biodiversity. There are now roughly 300 million people living around Mediterranean Sea. The greatest impacts of this demographic outburst have been deforestation, intensive grazing and fires, and infrastructure development, especially on the coast. The area is a meeting place of three continents, which contribute to its high ecodiversity. Its 22,500 endemic plant species are more than four times the number found in all the rest of Europe. Tourism development has placed significant pressure on the region's coastal ecosystems, because it is the biggest large-scale tourist attraction in the world, with 120 million visitors travelling in the area every year. Sensitive dune ecosystems remain as a major threat to coastal areas due to the infrastructural constructions and direct impacts of people. The Mediterranean Basin today is among the four most significantly altered hotspots on Earth. Much of this updated information will be presented in the book.

This book contains a selection of papers of scientists from a range of disciplines interested in the Mediterranean region as one can observe from the contents of the book. It will present a timely and wide-ranging overview of the "Environment and Culture in the Mediterranean Region". The book will examine geomorphology, environmental geography, geographical information science, human-environment interactions, karst landforms, ecology, land degradation, land use, land cover change, sustainability, biogeography, climate and climate change, desertification, ecosystem geography, and eco-tourism. The aim of this book is to link the rapid developments in mountain environments, natural hazards, political geography, population geography, remote sensing, tourism geography and urban geography in the Mediterranean ecosystems.

Finally, I particularly wish to express my thanks to, Prof. Dr. h.c. Ibrahim Atalay, Prof. Dr. Münir Öztürk and Prof. Dr. Georges Cravins, who assisted significantly with the selection and editing of papers for this volume, and to Dr. Andy Nercessian and his team at Cambridge Scholars Press for preparing the book for publication.

Prof. Dr. Recep Efe

PART I

NATURAL ENVIRONMENT IN THE MEDITERRANEAN REGION

CHAPTER ONE

MEDITERRANEAN ECOSYSTEMS OF TURKEY: ECOLOGY OF THE TAURUS MOUNTAINS

IBRAHIM ATALAY, RECEP EFE
AND ABDULLAH SOYKAN

Introduction

Mediterranean ecosystems are limited to five relatively small areas around the world: the region bordering the Mediterranean Sea; Central Chile; the Cape region of South Africa; Southwestern and Southern Australia; and California south to northern Baja California. These ecosystems largely occur along the western edges of continents between the 30° and 40° parallels in both northern and southern hemispheres. The Mediterranean climate is characterized by mild, rainy winters and warm, dry summers (Köppen 1923, Emberger *et al.* 1962, Gottman 1979, Atalay 1993, 2002, Atalay *et al.* 1998, Erinç 1969, Koçman 1993, Naveh & Lieberman 1984, Turkes 1996). This special and unique climate has great influence on natural physical processes, including on soil formation, ecological conditions, landform development, and karst topography as well as on human activities (Atalay, 1987b, 1987c, 1994, 1995, 1999, 2002, 2006; Atalay and Tetik & Yeşilkaya 1997; Di Castri *et al.* 1981; Efe, 1998, 2004a, 2004b, 2005; Efe and Greenwood 2007; Kaniewski *et al.* 2007). Globally, the total area over which Mediterranean climate prevails is only about 2 million square kilometres. About half of this area occurs in the Mediterranean Basin: Although plant species and communities differ between the regions, the dominant vegetation is evergreen woodland with evergreen sclerophyllous shrubs and trees like carob tree (*Ceratonia siliqua*), cork oak (*Quercus suber*), holm oak (*Quercus ilex*), mastic tree (*Pistacia lentiscus*), kermes oak (*Quercus coccifera*) and the fodder shrub (*Medicago arborea*) These species are resistant to summer droughts, have deep rooting systems and resprouting capacity.

Large areas in the Mediterranean countries show impoverished plant cover due to long-term anthropogenic disturbance and harsh environmental conditions. It is widely accepted that vegetation is a major factor in controlling land degradation. The introduction of plants in degraded ecosystems could affect the overall microclimatic and soil conditions. Due to the limited extent and isolation (almost island-like), of each area of Mediterranean biome, there is frequently a high degree of endemism in the flora and fauna (Emberger at al. 1962, Di Castri 1973, Aschmann 1973b, 1973b, Naveh & Lieberman 1984, Thrower & Bradbury 1973, Tyrrel 1982, Walter 1983, Vogl 1982, Smith and Smith 2003, Bailey 1995, Paskoff 1973, Dallman 1998, Atalay 2006b).

Throughout the world, the Mediterranean biome is characterized by evergreen or drought deciduous shrublands. The *chaparral* of southern California is echoed in the old world Mediterranen *maquis*, the Chilean *matorral*, South African *fynbos* and the Australian *mallee scrub* communities.

The landforms of the Mediterranean areas of the world constitute a distinct group, primarily due to the geographical position of the region, which has largely determined the peculiar present morphoclimatic conditions and the marked climatic changes of the recent geological past (Paskof 1973). On the other hand, the use of seedlings has many advantages over planting when reforesting or afforesting bare karstic lands with shallow or medium depth and soil cracks which contain fine soils, creating physiologically deep soil. On serpentine parent material with shallow soil depth, in most cases, the application of seedings is inevitable (Boydak & Ayhan 1990; Boydak 1996, 1997; Atalay 1987a, 1987c, 1993, 1997, 2001; Atalay and Tetik 1997, Atalay *et al.* 1998).

This paper presents an overview of the ecological characteristics of Taurus Mountains in the Southern Turkey in the eastern part of Mediterranean Basin.

Study Area, Materials and Methods

The Taurus Mountains are a wide and long mountain range in the Mediterranean region of Turkey, running approximately 560 kilometres parallel to the Mediterranean coast, and forming the southern border of the Anatolian plateau. it is Turkey's second range of folded mountains, after the Northern Anatolian Mountains rising at the western-most range of the Great Himalayas. The range starts from the western part of Teke Peninsula in the west and extends to the upper side of the Euphrates (Firat) and Tigris (Dicle) rivers in the east after making a long curve and is connected to the Zagros Mountains in Iran. Its northeastern extension across the Seyhan River near Adana is known as Anti-Taurus (Fig. 1-1).

In this study 1/100.000 scaled topographical maps and 1/500.000 scaled geological maps were used together with DEM (Digital Elevation Model). In addition to these, field studies were conducted during 1980 - 2007.

The topographical maps were scanned and transferred to the computer to supply a base for the study. Map sections were coordinated as a standard map projection, UTM (Universal Transverse Mercator–UTM–WGS84). The geological maps and drainage characteristics were also scanned and placed over the base maps, obtained from topographical maps.

For overlapping and processing the scanned maps, ArcGIS Desktop v.9x software was used. Coordinated map sections were transformed to layers by the method of Screen Digitizing.

The Topographical and Geomorphological Properties of the Taurus Mountains

The Taurus Mountains are the largest and most important karstic area both in Turkey and in the Mediterranean Region. It has a 200 km wide zone in the east – west direction along the coast between the Mediterranean Sea and Central Anatolia, composed mainly of the limestone. The limestones are heavily folded and thrusted because of the Alpine orogeny, which was a result of the collision of the Eurasian and African plates. Karstic land forms such as *polje, doline, uvala, lapies* development is influenced by the faults and thrusts and by the border to the surrounding insoluble and impervious schists.

The central Taurus Mountains are mainly composed of Permian metamorphic limestones. Mesozoic limestones and dolomites occur in over 1000 meter thick layers. They are surrounded by various rocks from Cambrian to Tertiary era. The limestone is often surrounded by an ophiolitic melange from the Late Cretaceous forming an impermeable base or cover

These Mountains, lie within the Alpine-Himalayan orogenic belts. The ecological characteristics of the Mediterranean region are mainly determined by the Taurus Mountains. The altitude, exposure, direction, and parent materials of the Taurus range result in the formation of different and special habitat both for plant and animal life as well as human activities. In order to explain the importance of the region, its impact on natural environment, it is neccesary to enlighten some facts on the formation and the evolution of the Taurus Mountains.

The Formation of the Taurus Mountain Range

The Taurus Mountain range was occupied by Tethys Sea during the Mesozoic era, when the African and the Anatolian microplates and the Euro-Siberian plate got dissected by the convectional flow occuring on the upper mantle (asthenosphere) of the Earth. As a result of the sea-floor spreading, ultrabasic magma spread all over the deeper parts of the Tethys Sea. During the Mesozoic era, sediments, mostly composed of carbonates, accumulated within the deeper part of Tethys (Ketin 1983, Atalay 1987b) resulting in a 4000 meters thickness of the carbonates. The pelagic debris materials and flysch, derived from the adjacent lands were swept into the trench, continental slope by turbidity currents in Tethys. The flysch deposits widely occur on ophiolites and pedidotite in the Taurus Mountains.

Tethys started to close up due to the moving of the African plate towards the north direction at the end of Mesozoic era and the area was uplifted and converted into land. First karstification process may have started at the end of the Mesozoic era (Fig. 1-2). During the Alpine orogenic period, which occurred at the end of the Oligocene, Taurus area was subjected to the compressional movements leading to the uptrust and/or overturn and these tectonic activities resulted in the formation of nappe structure in the Taurus. In the mountains extending western and eastern part of Antalya Gulf nappe structure is clearly visible. The oceanic crust composed of peridotite was divided into many stripped and or as slivers of layered mass and trust in the semi horizontal direction toward the upper part of the mountains. The squeezing and stripping of the peridotite lad to the formation of serpantine and the ophiolitic series containing limestone lenses and masses. Ophiolites which occur in the Taurus are important evidence of intensity of plate tectonic.

After the alpine orogenic movement in the neotectonic period in the Taurus Mountains the African plate moving beneath the Anatolian plate resulted in the extensional tectonic regime which lad to the vertical tectonic movements responsible for the block faulting movements in the area. Some parts of the Taurus Mountains were cut by fault lines as in the Antalya- Egirdir line, and some part collapsed and was uplifted. Tectonic basin or corridor within the Taurus Mountains was formed. For example Bucak, Korkuteli, Acıpayam depressions were formed.

Fig. 1-1: Map showing the study area - Taurus Mountains

The depressed areas were occupied by neogene lakes in which clayey, carbonate and sandy sediments were deposited. The uplifting process continued during the Pleistocene time. The movement in the Taurus Mountains accelerated karstification process and fluvial in the karstic land erosional and depositional activities. The surface drainage shifted from the surface to the deeper part. The cave systems which are found in different level in the vertical direction imply the uplifting movements (Atalay 2003).

The poljes found in the Taurus Mountains are of tectonic origin. The enlargement of the tectonic depression is associated with the karstification process. The hums called the remnant hills of the old topography of the karstic lands which are found in the depression indicate the intense karstification. The Antalya travertine deposits can be accepted as a production of the karstic dissolution which occurred in the Taurus Mountains. The rich calcium bicarbonate water evaporated in the Antalya depression so that travertine formation attaining more than 300 m thickness was formed (Efe *et al.* 2007).

Climate in the Turkish Mediterranean Region

The climate of the Mediterranean region of Turkey is somewhat different from other regions, including other regions that are principally characterized by Mediterranean ecosystems. This is mostly related to the topographic effects on the climate. The Taurus Mountains' abrupt rise on the Mediterranean coast contributes to the distribution of precipitation and winds. The precipitation regime shows a highly irregular behaviour in both the spatial and temporal dimensions (Efe 1998, Atalay 2002).

The general atmospheric circulation in the region is under the impact of tropical continental, tropical maritime and polar air masses. During summer is influenced by tropical air masses coming from the Azores (maritime tropical, mT) and northern Africa, notably the Sahara region (continental tropical, cT). In general, precipitation and/or frontal activities do not occur between May and September. The dominant wind during this period blows in the direction of NE-SW, due to a difference in atmospheric pressure in the Aegean Sea. The winds blow during the summer period from high pressure in the Balkan region towards low pressure in the Persian Gulf.

Sometimes temperatures reinforced by regional winds blowing from the Sahara can cause a sudden decrease in relative humidity to 20% and temperatures rise up to 40 °C.

Frontal activities originating from the Mediterranean Sea generally affect only the Mediterranean region of Turkey. The fronts coming from the northern sectors seldom reach to the Mediterranean region due to the Taurus Mountain Range. Fronts from southwest to the northeast are intercepted by these Mountains. Thus, the southwest facing slopes of the Taurus receive abundant rain. The amount of the mean annual rainfall is more than 1000 mm in the vicinity of Antalya and it exceeds 2000 mm in the southwest slope of Geyik Mountains, NE of the Antalya Gulf. The rain shadow areas occurring in the tectonic depression receive less precipitation (Burdur 400mm, Isparta 581mm, Mut 400mm). On the other hand, orographic or relief rainfall occurs on the slopes facing the Mediterranean Sea. The warm maritime air is forced to rise when confronted by a coastal mountain barrier. The Taurus Mountains reduce the water holding capacity of rising air by enforced cooling and increase the amounts of cyclonic rainfall by retarding the speed of depression movement. They also tend to cause air streams to converge and funnel through valleys crossing the Taurus Mountains. The mean annual rainfall varies between 300 - 750mm in other parts of the Mediterranean Basin, being less than 500 mm in the southern part of Spain (Malaga, 447mm), and the other islands of the Mediterranean Sea.

The relative humidity of the Mediterranean coastal belt is generally more than 80 %. This high humidity decreases due to evaporation therefore fog and dew occur during the early morning the decreases in air temperature. These events decrease intensity of drought conditions.

Foehn or föhn events also occur due to the air masses come from the northern sectors in the Mediterranean region. When the air mass crosses the Taurus Mountains and the descending air is compressed and warmed and relative humidity falls to 10 percent, and temperature rises up to 40 °C. This low moisture content and warm air encourages the increase of forest fire. Conversely, the humid air coming from the Mediterranean Sea increases the humidity and decreases the evapotranspiration.

The frontal activities which affect the Mediterranean region change frequently, the amount of the rainfall greatly changes in winter months. For example, the mean annual rainfall of Antalya is about 1000mm. However, from year to year, rainfall varies between 700 and 1400mm.

Soil Formation in the Turkish Mediterranean Region

Climate, parent material, and topographic factors such as altitude, exposure, and slope inclination, determine soil forming processes and soil types in the region. The climate exerts a powerful influence on soil-forming processes. In the wet winter season, rates of weathering and leaching are at maximum levels. Different parent materials are subjected to chemical weathering along cracks and fissures. The weathering processes of hydrolysis and hydration are carried out by rain water charged by carbon dioxide, both from the atmosphere and from soil air, whose higher content of CO_2 comes from the activities of soil fauna and soil micro-organisms. The rainwater with a pH of 5.5, readily attacks soil minerals, and where the parent rock is limestone, cause rapid dissolution by carbonation.

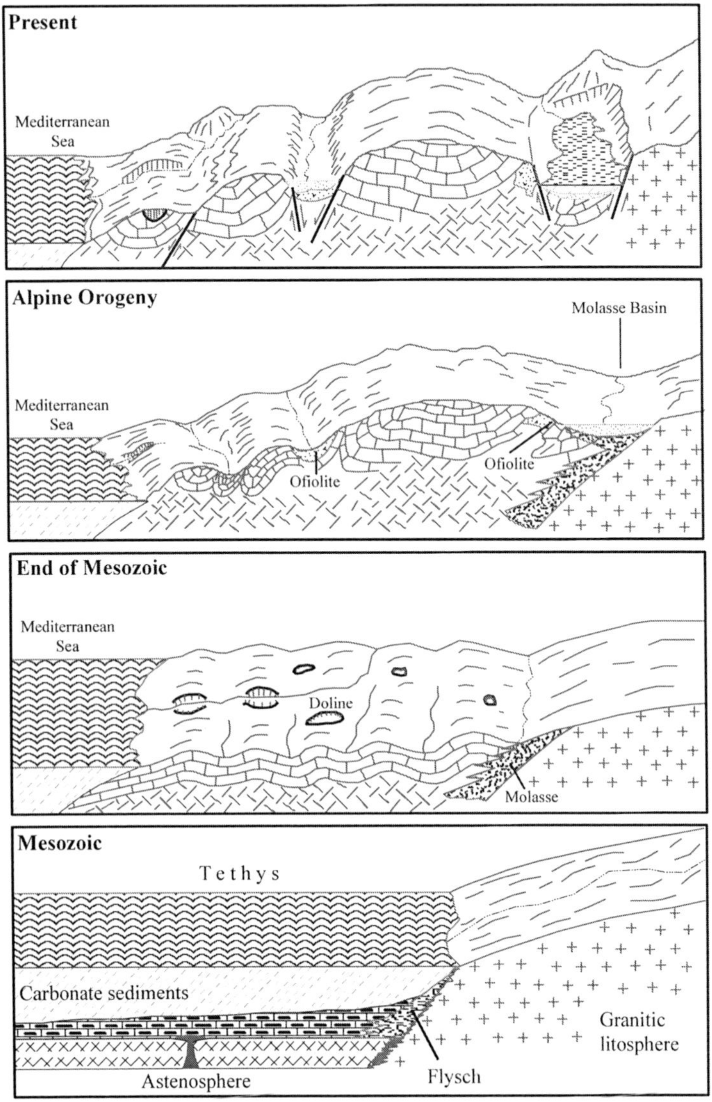

Fig. 1-2: Schematic Formation and Evolution of the Taurus Mountains

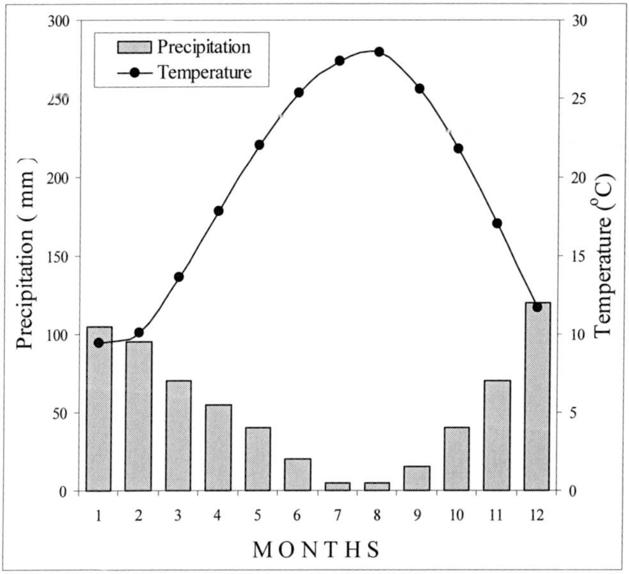

Fig. 1-3: Temperature and Precipitation Diagram of Adana

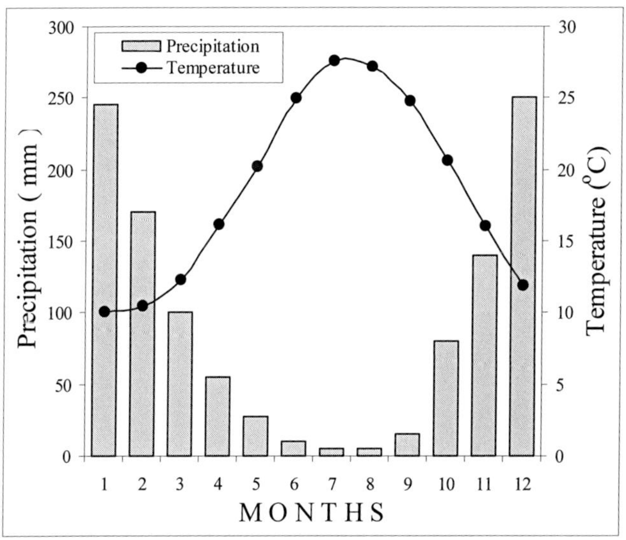

Fig. 1-4: Temperature and precipitation diagram of Antalya

Simultaneously with weathering during the winter leaching also follows, resulting in the removal of weathered product (cation and anions) and any free calcium carbonate (decalcification) from the soil profile. The rates of soil formation and the thickening vary considerably between different parent materials.

In addition to the leaching of ions from the soils, winter precipitation causes the leaching of clay and silt particles from the A into the B horizon to give a clay-enriched or textural B horizon. The blocky structure of B horizon indicates the accumulation of the clay material.

Weathering and leaching occurring during the winter period are responsible for the production of the clay minerals, oxides and hydroxides of iron and aluminium (sesquioxides), and silica. The oxidation of the iron as ironsesquioxide leads to the reddening of the soil, and the development of a red hue will increase with time, and thus the degree of reddening can be used as an indicator of the age of a soil. The reddish soil can be regarded as the climatic soil types of the Mediterranean climate. It is classified as Alfisol, according to 7th Soil approximation, as Chromic Luvisol in the FAO system, and it is termed as terra rossa in the classic soil classification.

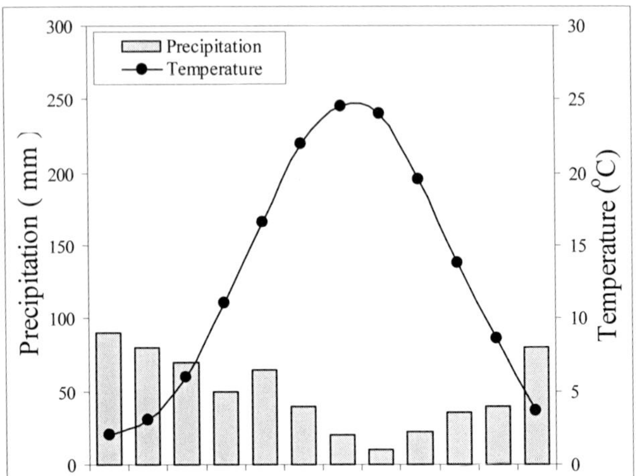

Fig. 1-5: Temperature and precipitation diagram of Isparta

Towards the higher part of the Taurus Mountains the organic matter content of the soil increases due to decrease of biological activities and the colour of the soil is brownish reddish.

Table 1-1: Rainfall changes in the Mediterranean Region.

Stations	Average (mm)	Maximum (mm)	Pos.Deriv. (mm)	Rate (%)	Minimum (mm)	Neg. derivation (mm)	Rate (%)
Antalya	1173	1914	+ 741	+ 63	533	- 640	- 55
Anamur	1032	1415	+ 383	+ 37	540	- 492	- 48
Mersin	617	1035	+ 418	+ 68	278	- 339	- 55
Adana	646	1246	+ 600	+ 93	370	- 276	- 43
Antakya	1173	1550	+ 377	+ 32	651	- 522	- 44
Maraş	723	1133	+ 410	+ 57	309	- 414	- 57
Isparta	619	879	+ 260	+ 42	332	- 287	- 46
Burdur	437	615	+ 178	+ 41	258	- 179	- 41
Beyşehir	447	716	+ 239	+ 50	303	- 174	- 24
Pozantı	703	1172	+ 469	+ 67	304	- 399	- 57
Konya	335	500	+ 165	+ 49	143	- 192	- 57

The soil formation of the Mediterranean region in Turkey can be summarized as fallows:

The climatic soil type of the Mediterranean region is the Reddish Mediterranean soil or Alfisols (Xerallf). This soil occurs on every parent material on slightly undulating and well drained flat lands.

In the coastal belt of the Mediterranean Sea and the forested lands, red Mediterranean soil occurs on the sandstone and gravelly deposits as well as ultrabasic rocks and marly deposits.

Soil profile Characteristics of Red Mediterranean Soils

O horizon, litter is only found in the closed forest canopy, especially pure *pinus brutia* forest. In the lower Mediterranean belt there is no organic horizon due to completely decaying of organic material, but in the upper part the thickness of the organic material increases to 1 cm.

A horizon, reddish, dark reddish, reddish brown, yellowish in colour, neutral and weak alkaline reaction, (pH 6.9-7.8), carbonates completely leached, secondary calcification occurs where the water containing calcium bicarbonate infiltrates on the slope, granular structure and clayey and loamy texture.

B horizon, weak reddish and yellowish in colour, block and coarse blocky structure, calcium carbonate accumulation only occurs where annual rainfall is less than 600–700 mm. It has weak alkaline reaction (pH 7.5-7.8).

C horizon is seen on weathered parent material such as schists, sandstone. There is no C horizon in general on the limestone.

The reddish Mediterranean soil is seen in very limited areas due to intense erosion.

Reddish Mediterranean Soils in the Karstic Lands

Reddish Mediterranean soils are common both on and in the karstic lands of the Taurus Mountains. There is a close relationship between the reddish Mediterranean soil and limestones. No soil cover seen on the steep slopes and/or slope area of karstic lands because atmospheric water (rain and snow waters) can easily penetrate along the cracks and fissures. Water is only held in the thick cracks and the surfaces between the layers. The weathering process for the soil formation occurs at the cracks of the rocks. Therefore, soil appears along the cracks and bedding surfaces of the limestones (Atalay 1995, 1997, 2006; Efe 1998, 2000).

Reddening processes easily take places in the karstic lands because air and water is very well circulated, depending on the cracked structure. The oxidation of iron is as Fe_2O_3, reflecting reddish colour. This situation explains why reddish soils are common in the karstic land. Soil which is derived from the limestones is clayey in texture. The limestone is mostly composed of calcium carbonate and clay; calcium carbonates are dissolved by carbonic acid (H_2CO_3), and HCl, and transported as calcium bicarbonate in the water. This process is realized that clay is remained as the result of the calcium carbonate removals. For this reason, the soil texture of the karstic land is clayey in texture (Atalay 1997, 2006a).

Red Mediterranean soil in karstic area has formed over a long period. The existence of the eolian sediments in the soil and boxite deposits in the big holes explains this situation.

Karstic land also creates special habitat for the growth of shrubs and trees. The plant roots growing within the cracks and fissures easily develop and follow the water infiltrating and seepage along the cracks. During the growing season, comprising 4 and 5 months, the height of the roots attain a length of 1 meter. This is why karstic land is very suitable habitat for the natural regeneration (Atalay 1995, 1997d; Boydak 1990, 1996). Thus, protected karstic land is covered by the dense forest.

On the other hand the structure of the limestones prevents the loss of the water by evaporation, because infiltrated water is held by the soil in the deeper part and reflection is high due to the white colour of the limestone.

Fig. 1-6: Soils usually red (Alfisol) and reddish brown in color on the limestone and occurs in the cracks of limestone.

Fig. 1-7: Soil profile in the Acıpayam Polje in Taurus Mountains.

Fig. 1-8: Many different karstic landforms are found on Taurus Mountains.

In some karstic lands the soil which developed on the cracks is transported in a vertical direction. This is related to the widening of the cracks by dissolution. This situation leads to the vertical transportation of the soils.

The physical composition, layer sequences or layer inclination and the distribution of crack, all determines the formation and the distribution of soil. For example comprehensive and hard limestone produce thin soil due to there fact that it is responsible for fewer thin cracks and a more compact structure. There is a close relationship between layer and the soil appearance. Soil occurs in the bedding surface. In other words soil is found along the layer. On the other hand, thin-layered limestone has much more soil than the thick limestones in weak zones are also favourable for the soil formation due to fact that it is easily weathered.

Karstic holes and/or depressions contribute a somewhat humid habitat for the growth of some plants. Hydrophilic such as *Acer, Fraxinus, Sorbus* grow on the bottom land of the small karstic depressions. The small karstic depressions are ‘U’ and ‘V’ shaped, and deep canyons create suitable habitats for both endemic and relict species and communities. In other words relict and endemic species are sheltered in the karstic depression. *Luquidambar orientalis*, one of the relict and endemic species, occurs in the lower part of Aksu valley, *Sorbus torminalis* and *Quercus vulcanica* are found in the karstic depression of the Davraz and Dedegöl Mountains. The karstic depression of the Taurus, Barla and Dedegöl mountains, the eastern part of Lake Beyşehir, are the spreading area of *Quercus vulcanica*. This endemic species only grows within the karstic depressions, when its branches grow up from the wall of dolines; they dry up due to wind effect (Fig. 1-4).

Vegetation Formations of the Mediterranean Region

Different vegetation formations are distributed in the Mediterranean Region of Turkey under varying climatic conditions and topography.

The shrubby and steppe-like vegetation which is a characteristic of present day wild landscapes in the region has been accepted by biogeographers (Polunin and Huxley 1992, Ajbilou *et al.* 2005, and Thirgood, 1982) as a consequence of human pressures superimposed upon climatic trends. The effects of these human impacts, and their relationships with both degenerative and regenerative trends in vegetation, are common in the region (Atalay 2002; Efe, 2004).

The climax vegetation in the region is evergreen oak woodland which get degenerate into maquis under light exploitation. Maquis formation is typically 1-3 m high, and is more widespread than relict evergreen forests. Many plants that are present in the forest but which prefer more open habitats grow abundantly in maquis (tree heath, buckthorn, holly oak, strawberry tree, myrtle, and juniper). The result is a dense, almost impenetrable shrub community, with plant species varying in different parts of the Mediterranean. Excessive exploitation leads to the formation of a low mixed heath, garrigue, which is a very diverse community of low shrubs and flowers, typically less than 1 m high. The community is colourful and aromatic, with species varying according to the local conditions. The common feature of garrigue formation is the resistance of plants to grazing by sheep and goats due to their poisonous, thorny or "waxy" nature.

Prolonged degeneration can lead to the almost complete disappearance of shrubs leading to the formation of steppe grassland and stony pasture. Such eroded, rocky terrains which support only grasses, annuals and bulbs which indicate the pathways of regeneration should be the cessation of human activity, for example, through the abandonment of agricultural land. However, regeneration of garrigue to maquis and then forest is clearly a much slower process than degeneration, as soil erosion will have reduced soil depth, water-holding capacity and nutrient content. In some cases, extreme degeneration can make regeneration impossible.

A different view of vegetation dynamics has been given lately by several researches (Grove et al. 1991, Grove and Rackham 1993, 2001) which advocates that the use of terms such as "potential climax" and "degradation" is not correct because there is little evidence that maquis, garrigue and steppe can change from one into another (Smithson et al. 2002).

The main characteristics of the Mediterranean plants are their resistance against the summer drought. A range of structural modifications in these plants favour drought tolerance, needle-leaf form, elimination of all leaves and formation of a photosynthesizing stem, sticky, waxy or hairy leaf cuticles; leaf stomata sunk in surface depressions; loss of transpiring leaves in summer (drought-deciduousness); dull coloured leaves and stems to increase reflectivity (higher albedo). A high ratio of below-ground roots to above-ground shoots favours moisture absorption by these plants. A widespread adaptation is the sclerophyllous leaf type.

Although the main characteristics of the Mediterranean vegetation are similar in the all part of the Mediterranean region in the world, The Mediterranean vegetation in Turkey considerably differs ecologically as compared to other Mediterranean regions.

It can be divided into two main ecoregions: 1. Mediterranean biome; 2. Mediterranean orobiome. While the Mediterranean biome extends along the coastal belt of Mediterranean Sea, the second covers mountainous areas above 1000 m.

The vegetation of the Mediterranean biome is divided into three groups: 1) *Pinus burita* forest; 2) Maquis; and 3) Garriques. It begins at the sea level and extends up to 1000/1500 meters (Efe 1998, Atalay 2002).

Ecological Features of Red Pine (*Pinus brutia* Ten.)

Red pine (*Pinus brutia* Ten) is one of the main forest trees of Turkey and is widespread in the Mediterranean, Aegean, southern parts of Marmara, the western and middle section of Black Sea and the western part of the Southeastern Anatolian geographical regionb. These forests constitute one-seventh of the total forest area of Turkey and cover 3 million hectares with an annual production or biomass by such forests is about 5 million m^3 and annual timber and wood production is some 4 million m^3.

This tree has special importance because it is resistant to summer drought and grows on all parent materials and has fast regeneration capability.

The varieties of the Red pine in Turkey are *Pinus brutia Ten.* var. Agrophictii Papaj.; *Pinus brutia Ten.* var. *pyramidalis* Selik; and *Pinus brutia Ten.* var. *densifolia* Yaltırık and Boydak. The formation of the different varieties is related to genetic features depending on various ecological conditions (Kantarcı, 1982; Tetik and Yeşilkaya, 1997, Boydak et al., 2007). The ecological conditions of red pine which is the climax tree of the Mediterranean region are summarized below.

Climatic Features

Red pine grows under the Mediterranean climate as characterized by dry summer, mild and rainy winter seasons, but it also grows in transitional climates such as oceanic climate in the Black Sea region, continental climate of Inner Anatolia, and Southeast Anatolia. The biomass production and physiognomic appearance of red pine is different in accordance with these climatic conditions.

Temperature. The annual temperature of the natural distribution areas of red pine changes between 12°C - 20°C. The mean January temperature ranges between 5 - 10°C. The January temperature of the coastal belt of the Mediterranean is some 10°C and on the upper part of the Taurus Mountains and the Marmara and Black Sea Region it is nearly 4-5°C. The January mean temperature does not fall below the freezing point in red pine areas. The mean July temperature varies from 23 - 28°C. Absolute maximum is over 40°C, minimum about -15°C.

Radiation. Red pine grows in place where direct sun radiation is dominant. In the natural regeneration areas seeds germinate very well under direct sun radiation, but in the sheltered area, germination is poor and plants die. Optimum growth occurs where the intensity of sun radiation is 70% or more. Thus it does not grow in foggy areas. Our studies revealed that red pine seedlings do not grow and the plants die where light intensity is under 50%, while the seedlings easily grow in places where sun radiation is 75% and more.

Fig. 1-9: Red pine stand on limestone in the north of Antalya on the south facing slopes of Taurus Mountains.

Precipitation. Mean annual precipitation varies between 400-2000mm. Although red pine is resistant to summer drought, it grows very well in humid habitats. In other words, red pine is not only a xerophytic tree species but also a hygrophytic tree. The areas which receive more than 900-1000mm of rainfall are favourable habitats for the growth of red pine. Therefore, some of the productive red pine forest stands are found between Anamur and Silifke in the Mediterranean Region and Menteşe Subregion in the southern part of the Aegean Region. In the wet habitats, its annual tree growth ring thickness is more than 1 cm (Atalay et al. 1997).

Relative humidity. The mean annual relative humidity is over 60 % in the naturally distributed areas of red pine. Humidity decreases during the summer and increases during the winter; but summer humidity is 10 % more than in the winter period due to humid wind coming from the Mediterranean Sea. The vapour content of air during summer is much more than the winter period. When relative humidity increases evapotranspiration decreases.

Topographic Factors

The growth and the wide distribution of red pine are determined by exposition, altitude and direction of the mountain ranges.

Exposition. Red pine forests are mostly found on upperland on slopes facing south. We find big clusters at an altitude of 1500m on south-facing slopes of the Taurus Mountains (Efe, 1998).

Slope inclination. Slope factor also determines the growth of red pine in the mountainous areas. The slope inclination increases, the water holding capacity of the soil and parent material decreases so that transporting of soil and parent materials increases. Except for the karstic lands, the less productive and degenerated red pine cluster are found on the steep slopes due to eroded soil and weathered and exposed parent material. The steep slopes composed of ultrabasic rocks such as serpantine-peridotite are, in general, especially devoid of climax vegetation.

Altitude. Red pine grows at considerably high temperatures. In the mountainous areas the altitude determines the wider distribution of these forests. Its upper boundary is about 800-1200 meters in the Taurus Mountains. The upper line is about 800 meters in the Aegean region and 400 meters in the Marmara Region and 1000 m in the backward area of Black Sea Region.

The biomass productivity of red pine increases from the coastal belt to the higher elevations, because respiration and consumption of biomass decreases during the night in the upper part, due to the low temperatures. On the other hand, an increase in the precipitation in the higher areas produces a wet habitat where biomass productivity increases up to 600 meters in the Mediterranean Region in Turkey (Fig. 1-12).

The direction of the mountain range: The direction of the mountains as a barrier is responsible for the amount of precipitation. In almost all regions, mountain ranges prevent the moving of the fronts coming from the northern and southern sectors. Therefore the fronts coming from the Middle part of the Mediterranean Sea are prevented by the Taurus Mountains so that the slopes facing southwest receive much more precipitation than other slopes. The leeward facing areas receive less precipitation. For example, Mut Basin, which is found in the middle part of the Mediterranean region, is the aridest part of the region due to its leeward position relative to precipitation processes (Atalay, 2002; Efe, 1998).

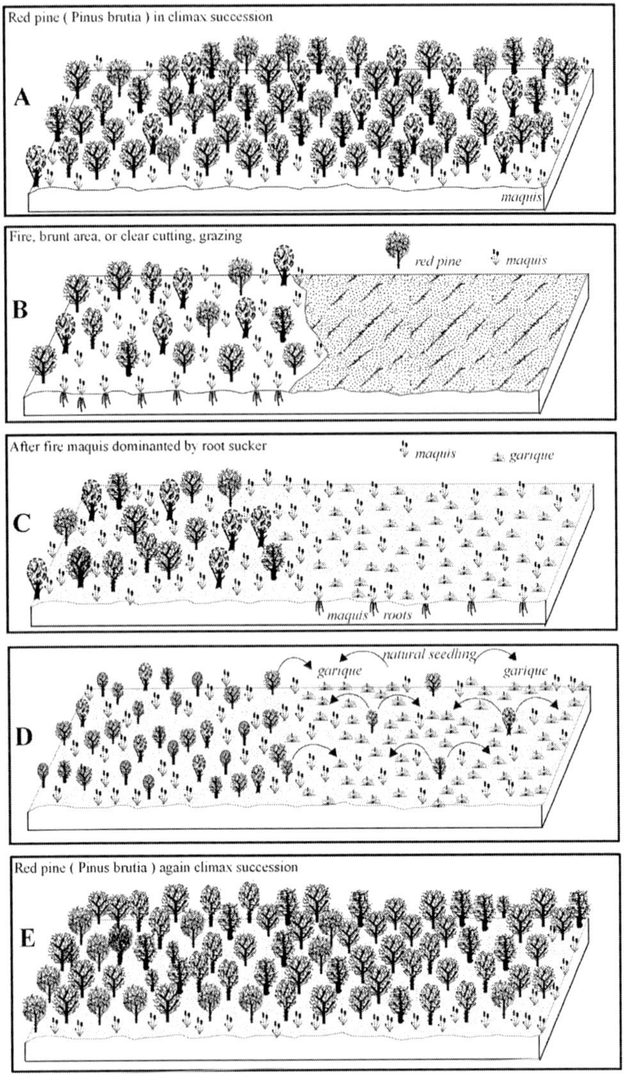

Fig. 1-10: Regressive and progressive succession of Mediterranean vegetation. A-Climax red pine stand. B-Degradation of red pine by fire. C- The development of first seral or pioneer vegetation composed of maquis and garrique. D-The growth of red pine and maquis and elimination of garrique. E- Transformation to climax

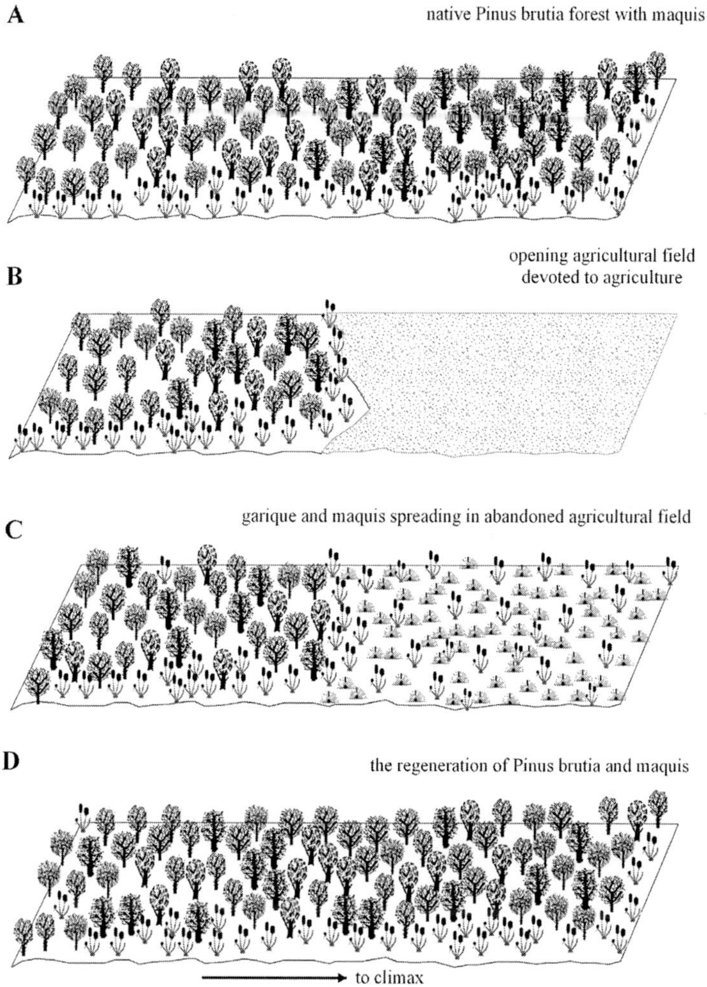

Fig. 1-11: Regressive and progressive succession of pinus brutia in agricultural field and abandoned areas.

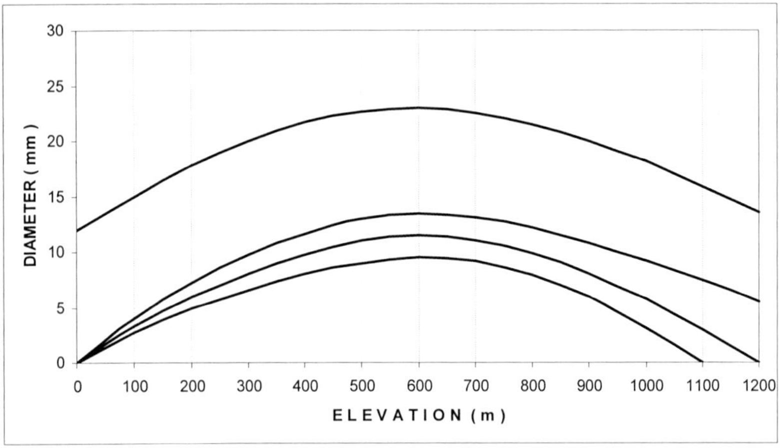

Fig. 1-12: Relationship between altitude and increase in diameter for *Pinus brutia* in the Mediterranean ecosystem in Turkey.

The amount of annual precipitation is less than 400mm in the bottom of this basin; wheras the total annual precipitation is more than 600mm on the mountains encircling the Mut basin. The transitional region extending between Inner Anatolia and the Mediterranean region has been determined to be a separate region due to less precipitation.

Soil and Parent Material

In the mountainous areas the physical and chemical properties of the parent material are an important factor in the growth of vegetation (Atalay 1987c, 2006b, Atalay et al. 1998, Efe 2004b). Entisol and inceptisol are dominant in the natural areas of red pine in Turkey due to rugged topography. This situation is clearly seen where soil and parent material erosion continues. Rendzinas are common on soft limestones and marly deposits. Red Mediterranean soils are common in karstic areas and on flat lands. But Red Mediterranean soils are widespread in the karstic land due the fact that soil erosion does not occur. Red Mediterranean soils have formed along the cracks and bedding surfaces between the layers. Because water is held in this land weathering processes have taken place. In addition to this, run off does not occur in the karstic lands because the water infiltration capacity is high. Most of the Red Mediterranean soils show neutral reaction and are in a clayey texture.

The site quality of red pine is mostly affected by parent material. Good quality stands are found on the cracked limestone, colluvial, marly and flysch deposits alternating sandy, silty and limey layers. Poor quality sites are common on unweathered serpantine-peridotite rocks. But well weathered ultrabasic rocks and clayey schists produce good quality for red pine sites. Volcanic tuffs and siliceous materials are mostly responsible for poor quality sites.

Most of the natural areas of red pine sites are found in the karstic lands due to the fact that natural regeneration occurs rapidly and cation exchange capacity is about 40me/100 g soil (CEC). It is about 8-12 me/100g in the sandy soils derived from gneiss and mica schists.

Biotic Factors

The factors determining the present occurrence of red are related to the fire, grazing, excessive cutting and competition between maquis and red pine.

Red pine is associated with the maquis elements which are climax plants of the Mediterranean region in most areas. Natural and good quality sites and/or dense forests are in pure stands and there is some shrub layer, but in degenerated areas under story is composed of maquis vegetation. Maquis elements are abundant in the open red pine areas. The seeds of the pine do not germinate in the dense maquis area. Here the maquis community is found as stabilized climax community. Overgrazing and highly exploited areas are especially covered by maquis. Maquis vegetation also regenerates by the root suckers. Thus even if maquis are cleared form an area, it is again is occupied by maquis within a short period.

Some maquis members are also indicators of the habitat as they found out the stands and divide into red pine region. *Arbutus andrachne, Arbutub unedo*, *Laurus nobilis* and *Mrytus communis* and *Styrax officinalis* are indicators of wet habitats notably on the higher elevations. The red pine stands which are associated with these maquis elements imply the good habitat for the growth of red pine. In the dry habitat and the upper part of the Mediterranean region kermes oak (*Quercus coccifera*) communities are common. *Quercus coccifera-Pinus brutia* community reflects the existence of dry habitat. For example pure *Quercus coccifera* community and *Quercus coccifera-Pinus brutia* community are dominant in the inner section of the dry forest areas. Garrique vegetation also indicates both the destroyed sclerophyllous woody and dry and sunny areas.

In the Mediterranean region, forest fires occur in the summer period. Red pine is a fire-resistant species. Fire in red pine areas produces post-fire regeneration. It can be stated that fire is the regeneration factor for the older red pine stands. *Pinus brutia* starts to hold cones at very early ages. A four-year-old juvenile may bear several cones, and the number of cones increases quite rapidly as the age increases. Cones held four years are well below the average fire frequency determined for low density fires. This implies that the juveniles, by producing seeds, can secure their survival even before the occurrence of the first low density fire. In addition, red pine produces a high number of cones every year.

In terms of fire, these features can be considered as fire adaptive traits, because the tree, starting from the fourth year of age, produces seeds which will ensure its reproduction after any fire, every year till the end of its natural life. On the other hand, the resin of unopened cones melts during the fire and after the fire the seeds stored in-cone fall down. Remarkable germination of seeds in recently burnt forests can usually be attributed to the shedding of seed stored in these cones. In addition seeds both stored in cones and the fallen seeds can be stored in the soil for few years without losing their germination ability. These seeds too contribute to the remarkable germination observed generally soon after fires. Thus, fire cannot be regarded as only a detrimental factor, but its beneficial effects on the growth and reproduction of red pine must be understood comprehensively.

Human impact also determines the spreading of red pine. The Mediterranean region is also the cradle of civilizations. One of the first Paleolithic and Neolithic settlements was established on the coastal belt of the Mediterranean Sea and its backward section. First vegetation destruction activities were began in the vicinity of Beyşehir which is named "Beyşehir Occupation Phase" 3000-4000 yrs ago. The maquis vegetation is widespread as a result of red pine forest destruction. The Garrigue vegetation which is found in red pine areas are seen in the destroyed and abandoned field areas (Atalay 1992, van Zeist et al. 1975, Zohary 1973).

It also grows in the southern part of Greece, Italy, Spain and some parts of the Eastern Mediterranean coastal belt, in Jordan, Palestine, Israel and in the northern part of the Black Sea, especially in the Crimea, extending as far as Tbilisi in Georgia, Madrid in Spain and the southwest part of Africa.

Maquis Vegetation

As mentioned before, maquis is the natural Mediterranean vegetation, but constitutes the shrub layer and/or under story of the red pine in their natural environments. The maquis vegetation is more widespread than the red pine forest in the Mediterranean region. This is related to the destruction of the red pine forests.It can be clearly stated that maquis forms the secondary and/or regressive succession stage in the Mediterranean region in Turkey, as maquis spread all part of the lower Mediterranean region where *Pinus brutia* forests are cleared. Almost all maquis elements regenerate with root sucker and becomes stabilized in karst areas. This is related to the natural generation of the maquis vegetation. Indeed, in the brunt areas, the roots of the maquis in the karstic cracks regenerate by the root suckers. It is impossible to remove the maquis in the karstic lands due to its deep root system. For example, after fires, the root sucker begins to grow fast. Four or five years later, burnt areas are again covered by the young maquis vegetation. Maquis recovers quickly, and kermes oak and strawberry tree can reach 1 m height after two years.

Clear-cutting also leads to fast growth of the maquis. Dense maquis areas prevent the growth of Pinus brutia, because the seeds of the *Pinus brutia* fall on the maquis canopy and the lower layer do not germinate. Conversely, the sparse maquis areas are replaced by red pine forest, because red pine trees grow fast, become dominant, and the maquis vegetation remains in the lower layer of the red pine. Natural competition between maquis and red pine is common in the maquis and red pine mixed stands.

The elements of maquis differ, depending on to habitat features. *Spartium junceum, Witex agnus castus, Nerium oleander* generally grow in wet habitats, like along the creeks, and the edges of the slopes getting water from the upper slopes.

Arbutus unedo, A. andrachne, Myrtus communis, Ceratonia siliqua, Laurus nobilis appear in the areas receiving more than 800mm rainfall. In other words these maquis elements are indicators of the sub-humid part of the Mediterranean region. *Phillyrea latifolia, Ceratonia siliqua* grow a somewhat *Quercus coccifera, Paliurus spina christi*, *Calicotome villosa* are common on the arid part and degenerated area.

The most common maquis element is *Quercus coccifera* because it is very resistant to drought, overgrazing, and flourishes on poor habitats. Some maquis elements like *Quercus coccifera, Arbutus andrachne, Ceretonia siliqua, Olea* sp. reach a height of more than 10 meters and become trees in the protected areas.

Garrigue Vegetation

Garrigue, which is a regressive stage of succession or early or first stage in burnt areas, occurs in abandoned fields and poor habitats. For example, abandoned agricultural areas are first covered by garrigues due to the fact that very small seeds can easily be transported by wind. The water requirements of the garrigues are very low, and their seeds germinate on sunny, bare land. During the summer drought, some garrigue elements shed their leaves and the dimensions of the leaves become narrower and dry up.

The main elements of garrique vegetation are *Cistus salviifolius, Cistus creticus, Cistus parviflorus, Coridothymus capitatus, Erica verticillata, Fumana arabica, Fumana thymifolia, Genista acanthocloda, Micromeria nervosa, Majorana syriaca, Phlomis viscosa, Phlomis cretica, Salvia triloba, Sarcopoterium spinosum, Satureja thymbra, Thymbra spicata, Teucrium divaricatum, and Teucrium creticum.*

The leading garrigue element is *Poterium spinosum* it is common on abandoned, destroyed and poor habitats. The occurrence areas of garrigue in the vicinity of Antalya are less than other areas due to high humidity.

Ecology of Oromediterranean Biome

The Taurus Mountains rise up to 3000 meters, create an orobiome after the lower Mediterranean region. Mixed and pure cedar and black pine stands are widespread in this belt. The mountains having orobiomes, such as Anti-Lebanon, Atlas Mountain (in Northwest Africa), are also accupied by cedar forest, but the majority of the cedar has been destroyed; and only small clusters and stands are seen.

Ecological Features of the Cedar Growing Areas

Cedar is generally common between 1000 - 2000 meters in the Taurus Mountains. It begins at 800 meters and rises up to 2000 - 2100 meters in the western Taurus. Cedar forests start at between 500 and 550 meters in the eastern part of the Amanos Mountains in the Eastern Taurus. The upper boundary of cedar is determined by the low temperature so that tall cedar gets shorter towards 2000 meters (Atalay 1987c).

Climatic Features of Cedar

Temperature. The mean annual temperatures on the elevations between 1000-2000 meters in the Taurus Mountains change between 6°C-12/14°C. The mean annual temperature of optimum growth area of the cedar is about 7-8°C. July temperatures are over 20°C, and it descends to 18°C in the upper boundary. Maximum temperature is recorded as 35°C. The January temperatures are about between -2.5 and 3°C, while minimum temperatures are below -15°C. The temperatures below freezing days are more than two months.

Precipitation. The amount of mean annual rainfall ranges between 650-1200 mm. The majority of annual rainfall occurs between October and April. This makes up only 5 % in the summer period. The snow cover duration is a few day in the lower boundary and three months in the upper parts. Snow fall starts at the end of November and continues till the end of the March.

Relative humidity. The mean annual relative humidity changes between 60 - 80 %. It is more than 80% along the coastal belt due to maritime air mass, but decreases up to 40 % in the inland section. The distribution of relative humidity also determines the natural occurrence areas of cedar. Low relative humidity limits the spreading of the cedar. Cedar does not grow where relative humidity is under 50 percent. In the depression areas of the inland part of the Taurus, the relative humidity decreases as low as 10%; occasionally, it decreases to 1%.

Radiation. Cedar prefers sunny and semi-shade habitats. During its young period, cedar grows in the semi sunny habitat under canopy and grows fast in open areas. Semi-closed canopy creates an optimum habitat for its seed germination, but a sunny habitat also permits the germination of seeds.

Vegetation composition. Oromediterranean forest including cedar forms both pure and mixed forests. In the oromediterranean belt, pure cedar and black pine forests are common. Pure Taurus fir forests only occur on the slopes facing north. In the remaining area, Taurus fir (*Abies cilicica*) grows with *Pinus nigra* and cedar forest on the slopes facing north.

Because cedar and black pine form a semi–shade environment for the growth of the Taurus fir, the pure Taurus fir forests occur where *Pinus nigra* and cedar trees have been completely destroyed. Indeed, Taurus fir trees are not preferred for building material because its timber decays in a short time.

Fig. 1-13: *Quercus coccifera* reaches to 20 meters hight in the protected areas.

Fig. 1-14: Cedar forest on the Taurus Mountains in the West of Antalya

In the lower boundary of the cedar and black pine forest associated with *Pinus brutia*, some maquis elements such as *Phillyera latifolia, Quercus coccifera, Arbutus andrache, A. unedo, Paliurus spina-christi, Styrax officinalıs, Juniperus excelsa, and Quercus cerris* are common, but in the humid habitats *Ostrea carpinifolia, Sorbus torminalis, Fraxinus sp. Populus tremula, Acer* sp. are found within the cedar and black pine forest. In the upper part, juniperus species become dominant tree in the cedar forest (Atalay 1987c, Efe 1998, Eliçin 1987).

Parent material. Cedar grows on all parent materials, ranging from metamorphic siliciferous schist to utrabasic volcanic parent material (peridotite-serpantine), from limestone to sandstone and conglomerate, but *Pinus nigra* can grow very well on soft parent material, especially weathered schists, dejection cones, colluvial deposits and ultrabasic parent materials.

Exposure. Cedar grows very well on the slopes facing north in the inland part of the Mediterranean region, and spreads well on these slopes in the Mediterranean sector. This situation is mostly related to air humidity. Humid and cool air comes from the northern direction in the backward section of the Taurus Mountains, depending on etesian winds. Conversely, the humid and cool air originates from the Mediterranean Sea in the southern slopes of the Taurus Mountains.

Biotic factors. It is impossible to establish the exact natural occurrence areas of cedar, due to the fact that cedar forest has been completely cleared or destroyed for during the last 5000 years. The cedar was used for construction of ships and palaces in the ancient times. Thus, the natural occurrence areas of cedar are getting progressively narrower, but nowadays, the Ministry of Forestry started reforestation and afforestation activities meant to enlarge of the cedar forest areas. In accordance with these activities, ecological and silvicultural activities are taken into consideration.

One of the main reasons for the destruction of cedar is transhumance activities and or nomadic herding and grazing in the Taurus Mountains. The Yörük (nomads) socicty engaged in the animal husbandry graze their animal in the forest areas and they cut the cedar trees to meet their fuel wood needs during the summer period. Additionally, needs for construction material by local villages in the oromediterranean region have been met by local forests. The upper limits of the cedar area were either converted into bare land or juniper communities due to intense destruction. Indeed, the main reason for the widespread dispersion of juniper communities is the clearing of cedar. In the destroyed areas, the juniper seeds eaten by the birds are subjected to fermentation in the stomach and the seeds fall down with birds' feces and germinate very well. For this reason, destroyed cedar areas are covered by juniper communities. It can be said that the juniper forest appearing in the upper limits of the forest is related to bird plantation areas. The cedar in Turkey can be regarded as a regressive or secondary succession (Atalay 1987c).

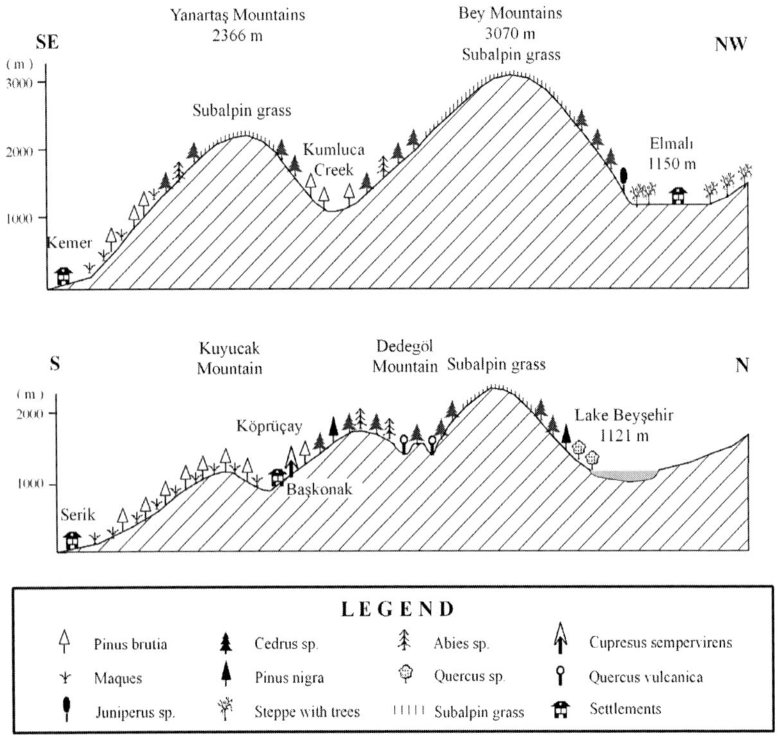

Fig. 1-15: Vegetation profiles in the Taurus Mountains

Conclusions

Some of the important conclusions we get from this study on the Taurus Mountains reveals that;

First, Tethys Ocean or Sea volcanic and sedimentary rocks are common in the Taurus Mountains. The formation and the structure of the Taurus Mountains were determined by Alpine and Neotectonic movements. After compressional movements occurred during the Alpine orogeny, vertical tectonic movements started, leading to uplifting and dissecting of the Taurus Mountain. The present-day topography of Taurus Mountains is the production of neotectonic vertical movements. The main poljes and corridors and/or basins were formed in tectonic depressions.

Second, the abrupt rise of the Taurus along the coast of Mediterranean sea forms a barrier between the inner part of Anatolian and Mediterranean Sea. The upper level of the Taurus exceeding 2000 meters has led to the orobiomes or Oro-Mediterranean biomes on which cedar; black pine and Taurus fir grow.

Third, the Mediterranean region which is classified under a Mediterranean-type is characterized by mild and rainy winters, dry and hot summers, and this type of climate prevails on the lower belt of Taurus Mountains. On the other hand the natural courses of the fronts are prevented by the Taurus Mountains. The areas where fronts can easily reach receive abundant precipitation.

Fourth, the altitude of Taurus Mountains leads to the different altitudinal vegetation belts, like the Eu or lower Mediterranean belt extending from 0 to 1000 meters, forest with black pine, cedar and the Taurus fir orobiome, which is found between 1000-2000/2100 meters, and a subalpine biome after the natural timberline.

Fifth, climax vegetation of a lower belt is composed of *Pinus brutia* forest with maquis under story. Maquis being a secondary succession are widespread where *Pinus brutia* forest has been completely destroyed. Maquis vegetation becomes stabilized on the karstic lands due to root suckers. Garrigues are the pioneer succession, and are only found on the abandoned fields and poor habitats.

Sixth, the Oro-Mediterranean biome is the growth environment for cedar, black pine and Taurus fir. Good stands of cedar forest are a monopol for Taurus Mountains. Seventh, Mesozoic limestone and karstic landforms contributes a special habitat for the increase of biodiversity and reddish Mediterranean soils. Relict and endemic species are generally found within the dolines in the upper part of Taurus.

References

Ajbilou, R., T. Marañón, and J. Arroyo. 2005. "Ecological and Biogeographical Analyses of Mediterranean Forests of Northern Morocco." *Acta Oecologica*. Volume 29, Issue 1, January-February 2006, 104:113

Aschmann, H. 1973a. "Man's Impact on the Several Regions With Mediterranean Climates." In F. di Castri and H.A. Mooney, eds. *Mediterranean Type Ecosystems, Origin And Structures.* Volume 7, pp. 363:371. Berlin, Heidelberg, and New York: Springler-Verlag

—. 1973b. "Distribution and Peculiarity of Mediterranean Ecosystems." In F. di Castri and H. A. Mooney, eds. *Mediterranean Type Ecosystems, Origin and Structures.* Berlin, Heidelberg, and New York: Springler Verlag. Volume 7, pp. 11-19.

Atalay I. 1987a. "Vegetation Levels of the Taurus Mountains of the Mediterranean Region in Turkey." *Aegean Geographical Journal.* Izmir: Turkey. Volume 4. pp. 88-122

—. 1987b. "Introduction to Geomorphology of Turkey." Izmir: Ege University, Faculty of Letters Publication Number 9. 451 pp.

—. 1987c. *General Ecological Properties of the Natural Occurrence Areas of Cedar (Cedrus Libani A. Rich) and Regioning of Seed Transfer of Cedar in Turkey.* Ankara: Orman Genel Müd. Yay. Number 663, 167 pp.

—. 1992. *The Paleogeography of the Near East (From Late Pleistocene to Early Holocene) and Human Impact.* Izmir: Ege University Press.

—. 1993. "Ecological Conditions and Natural Occurrence Areas of Pinus brutia in Turkey." *Papers, International Symposium on Pinus Brutia Ten, 18-23 October, 1993.* pp.57:66. Marmaris, Turkey: Ministry of Forestry.

—. 1994. *Vegetation Geography of Turkey.* Izmir: Ege Üniversitesi Basımevi.

—. 1995. "Pedogenesis and Ecology of Karstic Lands in Turkey." *Acta Carsologica.* Volume XXIV. 53:67.

—. 1997. *Red Mediterranean Soils in Some Karstic Regions of Taurus Mountains*. Turkey, Catena. 247:260-28.

Atalay, I., and M. Tetik. 1997d. "The Importance of the Karstic Lands in Terms of Forestry in the Taurus Mountains." Antalya: *Proceedings of the Xth World Forestry Congress, 1997*. p. 208.

Atalay, I., L. I. Sezer. ve H. Çukur. 1998. *The Ecological Properties of Red Pine (Pinus brutia Ten.) Forests and Their Regioning in terms of Seed Transfer.* Ankara, Ministry of Forestry Publication Number 6.

Atalay, I. 1999. "Land Use in the Karstic Lands in the Mediterranean Region." *International Journal of Speleology*. 28 B (1/4). 111:118.

—. 2001. Ecology of Forests in Turkey. Silva Balcanica, 1/2001,s. 25:34.

—. 2002. *Ecoregions of Turkey.* Ankara: Ministry of Forestry Publications.

—. 2003. “Effects of the Tectonic movements on the karstification in Anatolia, Turkey.” *Acta Carstologia.* Volume 32, Number 2. pp. 195-204.

—. 2006a. Soil formation, its classification and geography. Izmir: Meta Publishing Company. 584 pp.

—. 2006b. “The Effects of Mountainous Areas on Biodiversity: A Case Study from the Northern Anatolian Mountains and the Taurus Mountains. 2006.” Grazer Schriften der Geographie und Raumforschung. Band 41 / 2006

Bailey, R. G. 1995. *Ecosystem Geography.* New York, Berlin, Hong Kong, London, Paris, and Tokyo: Springer.

Boydak, M. and A. Ayhan. 1990. “Cedar Forestations of Barren Karstic Lands Around Anamur.” *Journal of Institute of Forestry Research.* Ankara: Number 36 (1) ?No? 71 pp. 7:21.

Boydak, M. 1996. *Ecology and Silviculture of Cedar of Lebanon (Cedrus Libani A. Rich.) and Conservation of its Natural Forests.* Ankara: Ministry of Forestry Publications Number 012, 68 pp.

Boydak, M., M. Doğru. 1997. “The Exchange of Experıence and State of the Art in Sustainable Forest Management (SFM) by Ecoregion: Medıterranean Forests”. ?s.?146-158.? *XI World Forestry Congress, 22 October 1997.* Antalya, Turkey: Volume 6, pp. 146:158-13.

Boydak, M., H. Dirik, and M. Çalikoğlu. 2007. *Biology and Silviculture of Turkish Red Pine (Pinus Brutia Ten).* Ankara: OGEM-VAK Publication.

Colak, A. H. andI an D. Rotherham., 2006. A Review of The Forest Vegetation of Turkey: Its Status Past And Present And its Future Conservation. Biology And Environment: Proceedings of The Royal Irish Academy, Vol. 106b, No. 3, 343- 354 (2006).

Dallman, R., 1998. Plant Life in the World’s Mediterranean Climates. California Native Plant Society, University of California Prcss. California, USA. Pp:257.

Davis, P.H. 1965-1988. Flora of Turkey and the East Aegean Islands. Edinburgh. Edinburgh University Press.

Di Castri, F. 1973. “Climatographical comparisons between Chile and the western coast of North America.” In F. di Castri and H.A.Mooney, ed. *Mediterranean Type Ecosystems, Origin and Structures.* Berlin, Heidelberg, and New York: Springer-Verlag. Volume 7, pp.21-36,

Di Castri, F., D.W. Goodall, and R. L. SPECHT. 1981. *Ecosystems of the World 11. Mediterranean-Type Shrublands.* Amsterdam, Oxford, and New York: Elsevier Scientific Publication Company.

Efe, R. 1998. *Ermenek River Basin: Natural Environmental Properties.* Istanbul, Number 1. pp 210. ISBN 975-303-001-0

—. 2000. *Interrelationships between Karst Geomorphology, Soils and Vegetation: A Case Study in the Central Taurus Mountains (Southern Turkey).* Karst'2000 International Symposium and Field Seminar on Present state& Future trends of karst studies. September 17-26, 2000, Marmaris, Turkey. Abstracts page 76.

—. 2004. Anthropogenic Degradation of Natural Vegetation in Karst Ecosystems in the Southern Turkey. *European Geosciences Union, 1st General Assembly, 25-30 April 2004. Nice, France, EGU-Geophysical Research Abstracts.* Volume 6, 01779.

—. 2004b "Biogeography: Principals, Vegetation and Zoogeography." *Cantay Publication, ISBN 975 7206 86 5.* Rec. Istanbul

—. 2005. "Land Degradation in Taurus Mountains (Southern Turkey)." *European Geosciences Union, 2nd General Assembly, 24-29 April 2005, Wien, Austria. EGU-Geophysical Research Abstracts.* Volume 7, 00922.

Efe, R., and M. Greenwood. 2007. Vegetation Zonation Patterns on the Goksu Delta (Southern Turkey). *Journal of Applied Sciences.* Volume 7 (16) p. 2277-2284.

Emberger-Gausen-Kassas-de Phillippis. 1962. Bioclimatic Map of the Mediterranean Region. Published by UNESCO-FAO.

Erinc, S., 1969. Climatology and Methods. 2nd Edition. Instanbul: University of Istanbul, Institute of Geography Publication, no. 35. Rec. İstanbul.

Gokklah, J. 1979. *A Geography of Europe.* New York, Holt. Rinehart and Wiston. 588pp.

Grove, A. T., J. Moody, and O. Rackham. 1991. *Stability and Change in the Cretan Landscape.* Cambridge: Corpus Christie College. (Petromarula 1).

Grove, A. T., and O. Rackham. 1993. "Threatened landscapes in the Mediterranean: Examples from Crete." *Landscape and Urban Planning.* 279:292-24.

Grove, A. T., and O. Rackhan. 2001. *The Nature of Mediterranean Europe: An Ecological History.* London: Yale University Press.

Kaniewski, D., V. De Laet, E. Paulissen and M. Waelkens. 2007. "Long-term effects of human impact on mountainous ecosystems, western Taurus Mountains, Turkey." *Journal of Biogeography* (Online Early Articles). doi:10.1111/j.1365-2699.2007.01753.x

Kantarci, D., 1982. *The relationships of the Regional Growing Properties of the Natural Tree and Bush Species in the Metiterranean Region.*

University of İstanbul, Faculty of Forestry Publications Number 330. Rec. Istanbul.

Ketin, I., 1983. *Introduction to the Geology of Turkey*. Istanbul Technical University Publications. Rec. Istanbul.

Koçman, A., 1993. Climate of Turkey. Ege Universitiy, Faculty of Letters Publication Number 72. Rec. Izmir.

Köppen, W. 1923. *Die Klimate der Erde*. De Gruyter, Berlin, Leipzig.

Naveh, Z., and A. S. Lieberman. 1984. *Landscape Ecology, Theory and Application.* New York, Berlin, Heidelberg, and Tokyo: Springer-Verlag. 356p.

Ozturk, M., Celik, A., Yarci, C.,Aksoy, A., Feoli, E., 2002. *An overview of plant diversity, land use and degradation in the Mediterranean region of Turkey*. Environmental Management & Health, 13,5,442-449.

Ozturk, M.A., 1995.*Recovery and rehabilitation of mediterranean type ecosystem- A case study from Turkish maquis*. In: Evaluating and NATO-ARW (Eds. D.Rapport, C. L. Gaudet, P.Calow), Springer-Verlag, Berlin, pp: 319-332.

Ozturk, M., Guvensen, A., Gork Ç., Gork, G., 2006. *An overview of coastal zone plant diversity and management strategies in the mediterranean region of Turkey*. In: Biosaline Agriculture& Salinity Tolerance in Plants (Eds.Ozturk et al.). 89-100, Birkhauser Verlag (Springer Science), Basel.

Paskoff, R. P. 1973. "Geomorphological Processes and Characteristic Landforms in the Mediterranean Regions of the World." In: F.di Castri and H.A.Mooney, eds. *Mediterranean Type Ecosystems, Origin and Structures.* Berlin, Heidelberg, and New York: Sprigler-Verlag. Volume 7. pp.53-60.

Polunin, O., and A. Huxley. 1992. *Flowers of the Mediterranean.* Trowbridge, Wiltshire: Printed in Great Britain by Redwood Books. 260p.

Quezel, P.1985. *Definition of the Mediterranean Region and the Origin of its Flora*. In: Gomez Campo,C.(ed.), Plant Conservation in the Mediterranean Area 9-24, W.Junk, Doderecht.

Sakcali,M.S.,Ozturk,M.,2003.*Eco-physiological behaviour of some Mediterranean plants as suitable candidates for reclamation of degraded areas*. Journal of Arid Environments , 57, 141-153.

Smithson, P., K. Addison, and K. Atkinson. 2002. *Fundamentals of the Physical Environment.* London and New York: Routledge.

Smith, R. L., and T. M. Smith. 2003. *Elements of Ecology*. San Francisco, London, Madrid, and Tokyo: Benjamin Cummings.

Tetik, M., and Y. Yeşilkaya. 1997. *The Parent Material-Soil Depthness and Bonitet Relations in Red Pine Forests in Antalya Vicinity*. Antalya: Technical Bulletin Number 6. Western Mediterranean Forestry Research Institute Publications.

Thirgood, J. V. 1982. "Man and the Mediterranean Forest: A History of Resource Depletion." *The Journal of Applied Ecology*, Volume 19, Number 3. December, 1982. pp. 980:981

Thrower, N. W., and D. E. Bradbury. 1973. "The Physiography of the Mediterranean Lands With Special Emphasis on California and Chile." In F. di Castri and H. A. Mooney, ed. *Mediterranean type ecosystems, origin and Structures.* Berlin, Heidelberg, and New York: Springer-Verlag. Volume7, pp.37:52.

Turkes, M. 1996. "Spatial and temporal analysis of annual rainfall variations in Turkey." *International Journal of Climatology.* 1057:1076-16

Tyrrel, R. R. 1982. "Chaparral in Southern California." *Proceedings of the Symposium on Dynamics and Management of Mediterranean-Type Ecosystems June 22-26, 1981, San Diego, California.* USDA Forest Service, General Technical Report, PSW-58. p. 56:59.

Walter, H. 1983. *Vegetation of the Earth and Ecological Systems of the Geo-Biosphere.* Berlin, Heidelberg, New York, and Tokyo: Springer-Verlag.

Van Zeist, W., H. Woldring, and D. Stapert. 1975. "Late Quaternary vegetation and climate of southwestern Turkey." *Paleohistoria.* 17 .53:143.

Vogl, R. J. 1982. "Chaparral Successions." *Proceedings Of The Symposium On Dynamics And Management of Mediterranean-Type Ecosystems, June 22–26, 1981, San Diego, California.* USDA Forest Service, General Technical Report, PSW-58. p. 81:85

Zohary, M. 1973. *Geobotanical Foundations of the Middle East*. Vol. 2. Stuttgart: Gustav Fischer Verlag.

http://www.allaboutturkey.com/taurus.htm

CHAPTER TWO

DETERMINATION OF THE MORPHOLOGIC-TECTONIC EVOLUTION OF THE BOLKAR MOUNTAIN THROUGH DRAINAGE PATTERN, MIDDLE TAURUS

TÜRKAN ALTIN

Introduction

The aim of the present study is investigate the relationship between the formation of the drainage system and regional tectonism on the south slope of the Bolkar Mountains. In this study, numerical data on 51 rivers, which are feeding from the Bolkar Mountains and are located in a basin between the Goksu River (Silifke) and the Tarsus River (Tarsus), were evaluated by correlation. Hierarchical structures of rivers were taken into consideration for this evaluation. Every main river and its tributaries were evaluated (Fig. 2-2, 2-3). River density, river frequency, bifurcation ratio and valley profiles of each basin were numerically correlated. The distribution and common characteristics of outcomes were examined.

Method

The study was conducted on numerical maps of the Bolkar Mountains with 1:25.000 and 1:100.000 scales using the softwares Mapinfo7.5 and Surfer 8. Since all the rivers in the study area do not directly flow into the Mediterranean Sea, three different sub-basins were identified (Fig. 2-3).

The Goksu River basin is separated from the western watershed line extending roughly in a direction of NW-SE in the west of the area. The middle section, which is called the Mediterranean Basin, covers all parts of the slopes facing the Mediterranean Sea. This area was deeply dissected by running water flowing into Mediterranean Sea. Rivers that split the wide plateau of the Bolkar Mountains and then reach to Mediterranean Sea are located in this section.

The Third basin is called Tarsus River basin which is found in the east, and is also deeply dissected by the streams of the Cehennem, Kadıncık and Gülek Rivers which are the main tributaries of Tarsus River. Each main river and their sub-basins containing the tributaries of the main basin were recognized (Fig. 2-3, 2-5). Moreover, a main river and its tributaries were classified in the sub-basins. A main river is distinguished by a hierarchical classification as increasing segments such as the main river first, second and third tributaries, etc., similar to the Scheidegger (segmantation formation) expansion. By using the appropriate software, the length, frequency, density of rivers, valley profiles and river directions were obtained by raw numerical values on numerical topographical maps which are used to determine morphographic characteristics. Additionally, several tectonic belts were defined, depending on the length and bifurcation numbers and river length among certain altitude levels (Fig. 2-3, 2-6). Thus, the relationship between the development of the river basin and tectonics could be evaluated.

Geological and geomorphologic properties

The Bolkar Mountains (Middle Taurus) lie in a WSW-ENE direction. Its altitude increases from a westward to an eastward direction. Medetsiz Hill, at 3524 meters, is the highest summit. Peridotites and ophiolite melange series are widely exposed between Guzeloluk and Camliyayla in the middle of the basin. Permo-Mesozoic crystalline rocks appearing on the higher parts of the Bolkar Mountan are found on these series (Akay and Uysal, 1988). Between Arslankoy-Camliyayla, Oligocene detritics and Jura-Cretaceous limestone are outcropped (Ozgul, 1983).

Miocene marine sediments form the most common unit of extending Creteceous series among the ophiolitic series. Between Erdemli and Tarsus, a cemented travertine series extends as a parallel to the Mediterranean coast line. Coasts without steep cliffs and delta areas form recent alluvial sediments.

The eastern part of the Bolkar Mountains is limited by the left lateral strike-slip fault attaining 80 kilometres in length (Ozgul, 1976, 1983, Sengor and Yilmaz, 1981). The uplifting of the Bolkar Mountains is related to the extentional tectonic regime that occurred between the Miocene and the Quaternary Period (Dhont et al., 1999).

Erosion-sedimentation surfaces deeply dissected by rivers on the Bolkar Mountains form wide plateaus and decline towards Mediterranean

Sea by cascading with tectonic thresholds. Pleistocene glacial features are located over 3000 meters and periglacial forms are seen over 3100 meters (Altin, 2001; 2002; Altin and Bayer Altin; 2005a, b, Bayer Altin, 2007).

Fig. 2-1: Map showing the southern part of the Bolkar Mountains.

The karstic land forms in the different dimensions are found on the limestones belonging to different geological periods. Karstification processes are responsible for the development and deterioration of the drainage system. The streams which are drained by sinkholes recharge as a resources in the deep valley. Capture events are also related to tectonic-karstic processes.

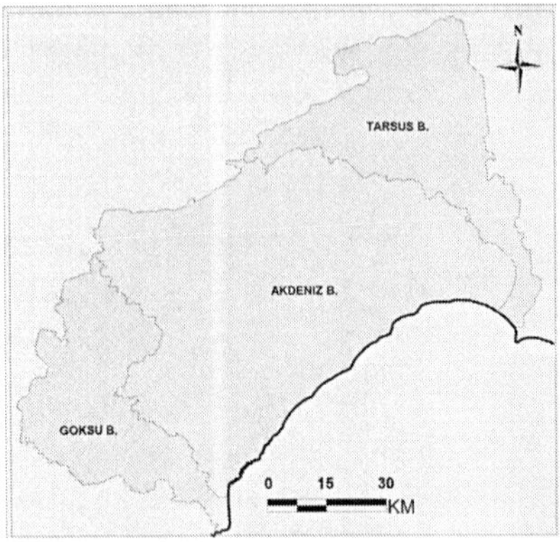

Fig. 2-2: Main Basin classification

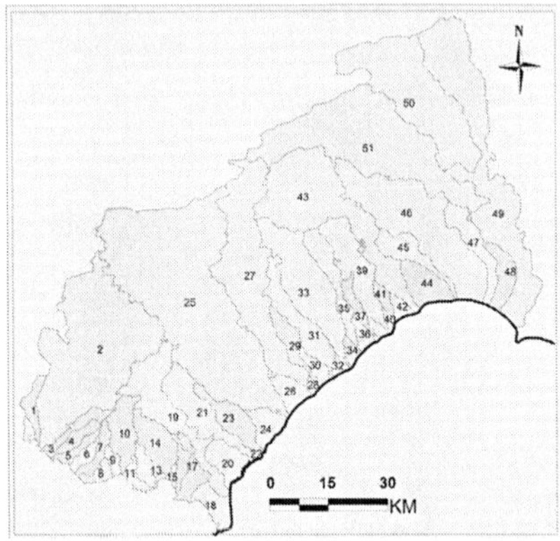

Fig. 2-3: Distribution of three sub-basins in the main basin

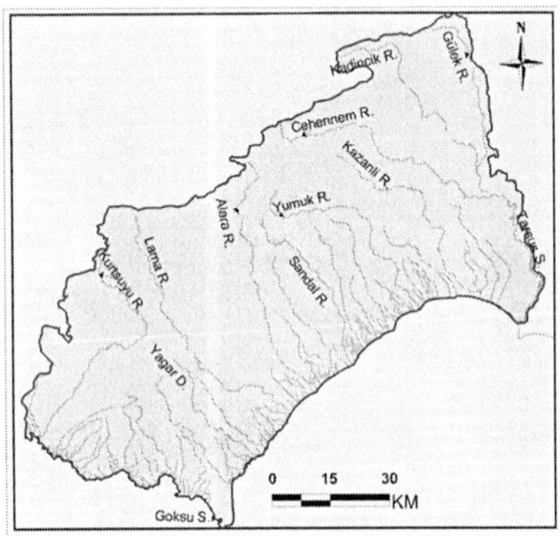

Fig. 2-4: Distribution of the main rivers

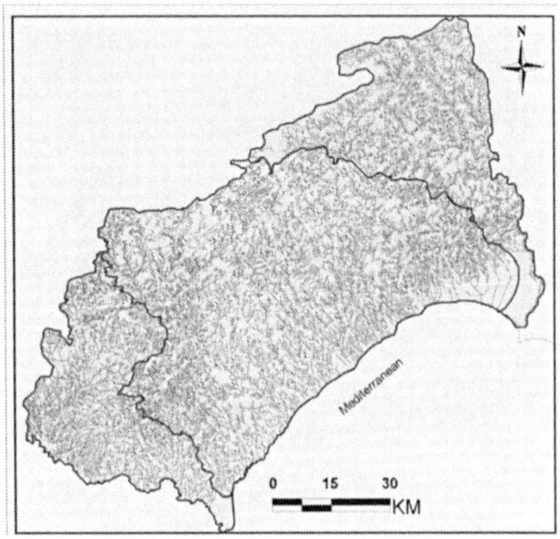

Fig. 2-5: River drainage pattern

Morphodynamic-morphoclimatic factors and processes which formed the Bolkar Mountains are deeply dissected by parallel and sub parallel-extended rivers. Tectonic movements have great importance in drainage formations, which are directed by structure and lithology. Traces of continuing seismic activity can be observed, especially in coastal zones, developing river valleys, and river and marine terraces. The sea started to rise in the early Holocene and then reached today's maximum level in the Middle-Holocene Period (Oner *et al.*, 2005).

Drainage Characteristics in the Bolkar Mountains

Although the river drainage of the Bolkar Mountains has similar characteristics in many places, exceptional cases can be also seen. There is no doubt that tectonic and karstic formations are shaping drainage in the region. River drainage on homogenous limestone is different than that of heterogeneous areas where limestone and ophiolitic series are present. Structural extensions, tectonic thresholds, lithological changes, and slope and height factors of the Bolkar Mountains had an important role in the formation of river drainage. Therefore, the difference in drainage pattern or network of the Tarsus and Göksu Rivers is related to tectonic properties. Namely, the Tarsus River was established on the weak zone which is controlled by fault lines.

Most of the rivers that are located in the Mediterranean Basin of the Bolkar Mountains have a parallel and sub-parallel drainage. In addition to elevation increase, drainage characteristics also change from east to west, from coast to inlands because of their structure and the direct effect of the lithology. In the plain areas with low slope, rivers show either dantritic or deteriorated drainage characteristics due to karst formation. In time, the effect of karstification on the river drainage formation is seen on the plateaus located among the Karayelek, Karagüney and Kilimli Hills. Throughout the places where tectonic linearity occurs, rivers have parallel extensions. All the river formations on plateaus have dantritic drainage and all of those in coastal areas (0 to 100 meters) have a parallel drainage network. Also, in local areas, rivers collecting their water from a single hill, which were formed by different abrasion, have a radial drainage pattern.

Bifurcation Ratio

Bifurcation of a river is an indication of the development stages after formation of a main river. Increased numbers of segments towards mountainous region indicates backward erosion of rivers and the dissecting degree of the basin with fluvial processes (Scheidegger, 1961; Strahler, 1975; Atalay, 1986).

The appearance of river segments, depending on certain levels, indicates a possible tectonic linearity or contact lines in the lithological sequences. Except for very short slits, all of the rivers collecting their water from the seashore, which is 0 to 50 meters in elevation, and on the east side of the Goksu River valley, 50 to 100 meters in elevation, are young rivers without secondary and tertiary segments or on an order of 2nd and 3rd. However, parallel to these rivers, old rivers with high segment numbers range to the interior (Table 2-1, Fig. 2- 7).

Likewise, when rivers from 200 meters and 600 meters in elevation are evaluated, there should be a reason for river rejuvenation. This situation is common for all the rivers which are parallel to the Mediterranean Sea, and collect their water from 50-100-200-600 meters elevation, and lead us to think of eustatic or diastrophic movements in the Mediterranean Basin.

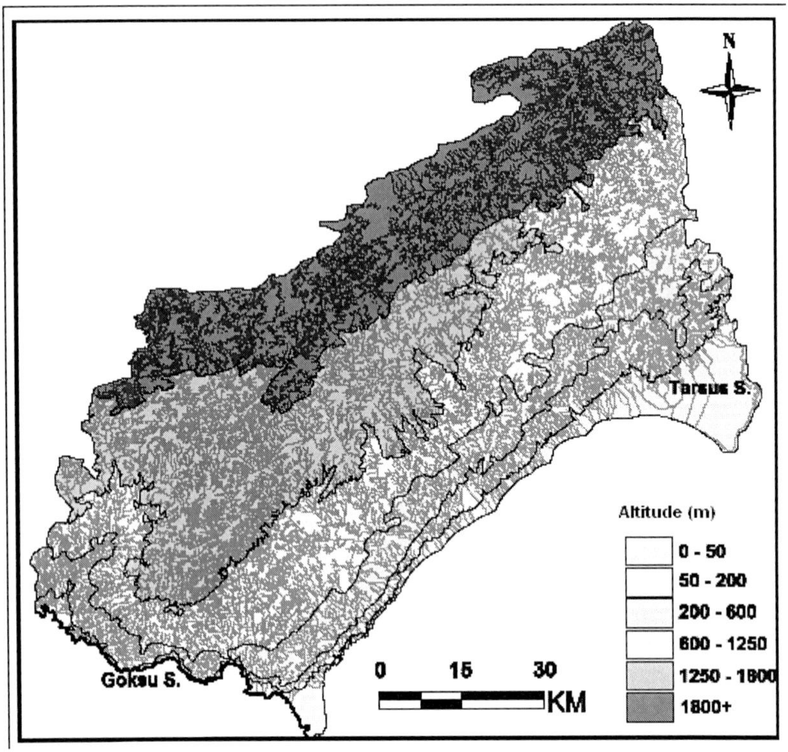

Fig. 2-6: Certain elevation steps resulted from river segmentation numbers.

Table 2-1: Distribution of river bifurcation numbers at certain elevations

Altitude (m)	1.	2.	3.	4.	5.	6.	Total
0 – 50	111	156	93	16	1	-	377
50 - 200	206	600	560	297	66	7	1736
200 - 600	124	852	907	480	99	4	2466
600 - 1250	43	1369	2205	1150	216	17	5000
1250 - 800	18	751	1895	1814	1021	301	5800
1800+	9	541	1509	1494	1256	138	4947

Oner et al., (2005), suggest a sea level change between the last glacial period and the Late Holocene periods. During the last glacial period (whose maximum phase occurred 20,000 years before the present), the sea level was lowered by 125 meters. Therefore, a change of 50 meters and 100 meters in sea level at the coast caused new river formations. Depending on the dense karstic formations on the plains located between the Lama and Alara Streams, river segmentation is greater on karstic basins. Further towards the east, the dissection with fluvial processes appears to be due to the deterioration in lithological homogeneity.

Drainage Frequency

Bifurcation numbers increase, depending on the developmental stages of the streams. The tributary numbers of the rivers per unit area are in line with the development of the main river, beginning from its formation and increasing segmentation numbers. This change varies for the smaller basins which are located among east, mid and west basins, with respect to area dimensions. There is a great contribution of lithologic resistance against erosion and slope. Values obtained by summation of each river flow channel reflect the toughness, depending on dissecting situation and the extension of the basins that are draining (Strahler, 1975; Atalay, 1986). The number of rivers in the basins is compared with river frequency and river density analysis in Table 2-2.

Table 2-2: Density and frequency rate distributions of the total number and total length of rivers according to both their own basins and through the Bolkar Mountain

Basin Name	Square (km^2)	Length (km)	Total River Number	Density		Channel Frequency	
				Basin	Bolkar Mts.	Basin	Bolkar Mts.
Göksu R. B.	1475	3567.114	3939	2.42	0.4	4.06	0.14
Akdeniz B.	4693	9977.63	10910	2.13	1.25	3.14	1.09
Tarsus R.B.	1804	3391.745	3217	1.88	0.43	2.45	0.13
TOTAL	7972	16936.49	18066	2.12	2.12	3.13	3.13

The Goksu River Basin is based at a high rate of river frequency (4.06) location of rivers in Miocene sediments, with heavy effect due to karstic formation. Rivers in the west of the Bolkar Mountains developed their drainage networks in a narrow region. Incised of the Tarsus River's tributaries into different lithologic units and high slope values took the river frequency (2.45) under control. Since the Mediterranean Basin's rivers are located on a wider region, the river frequency is the highest through the Bolkar Mountains (1.09).

Since the rivers are in their maturation stage, the river frequency yields high values. River frequencies are higher in their basins, at the edges of basin, on the Goksu and Tarsus Rivers, but lower through the Bolkar Mountains. When considering total length of rivers and total segment number, the middle section of Mediterranean Basin Rivers have higher values than do the edge basins.

Drainage Density

Factors that affect the drainage density of the Bolkar Mountains' drainage and river frequency are almost the same. River length per unit area is calculated and mapped (Fig. 2-9). River lengths change base on homogeneous structure and lithology. They change parallel with the bifurcation ratio of rivers.

The Cehennem and Kadincik Streams, which feed from the upper part of the Bolkar Mountains, have dense drainage networks due to their water from high hills, and thus drain the region densely. The main reason for the deeply dissected situation and dense drainage networks is related to the increase of precipitation in the higher parts of mountains, high slope inclination and the change of the lithology within a short distance. Drainage density is different through three basins, based on river development. The high number of short rivers depends on karstification processes on the plateau, slope erosion and high drainage density in the middle and western parts of the basin. The dark grey areas in Fig. 2-10 indicates the distribution area the shortest rivers and severe karstification.

Cross-Section and Longitudinal Profiles

Bifurcation of rivers at certain elevations and parallel development of levels at certain isohip belts are correlated with the scarp of vertical faults. Normal faults also reflect valley profiles. The vertical tectonic movements of the big main rivers and some of their long segments have knickpoints (rejuvenation heads). The segmentation rates of elevation-depended rivers segmentation rates depend on important change has close parallelism with river valley profiles.

Depending on the relationship of isohips to main river lengths and segmentation numbers, from seashore to mountainous region, five belt zones (from elevations of 0-50 meters, 50-200 meters, 200-600 meters, 600-1250 meters, and 1250-1800 meters) are distinguished.

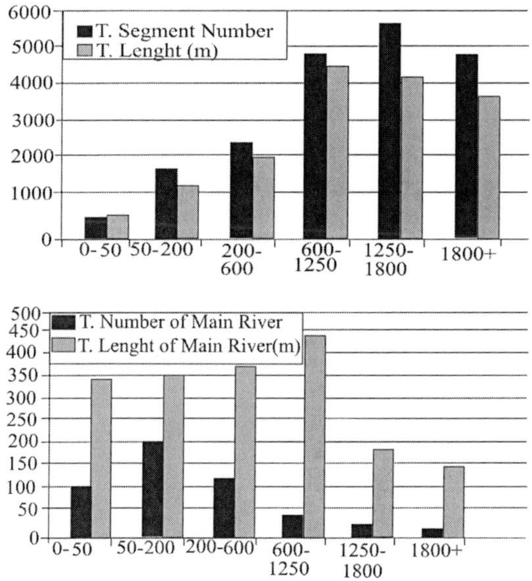

Fig. 2-7ab: Density analysis according to river tributary number

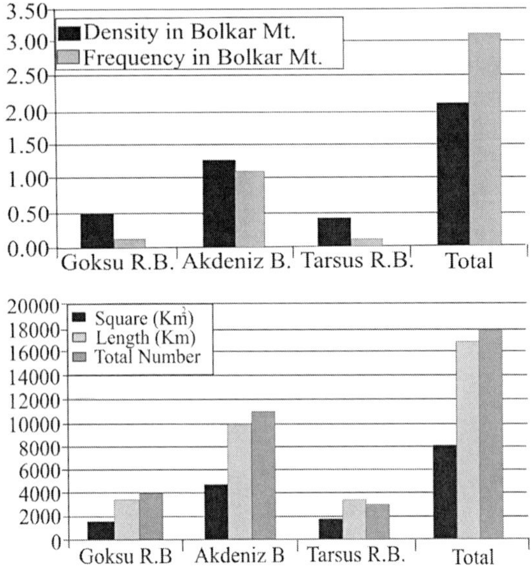

Fig. 2-8ab: Density analysis according to length of river branches

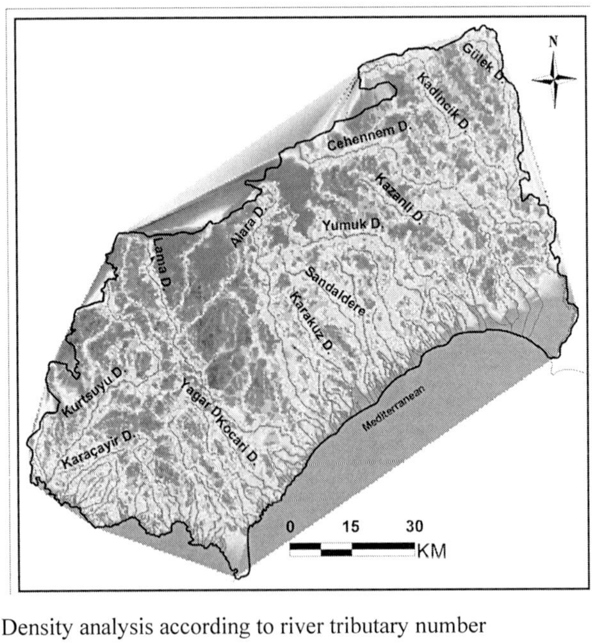

Fig. 2-9: Density analysis according to river tributary number

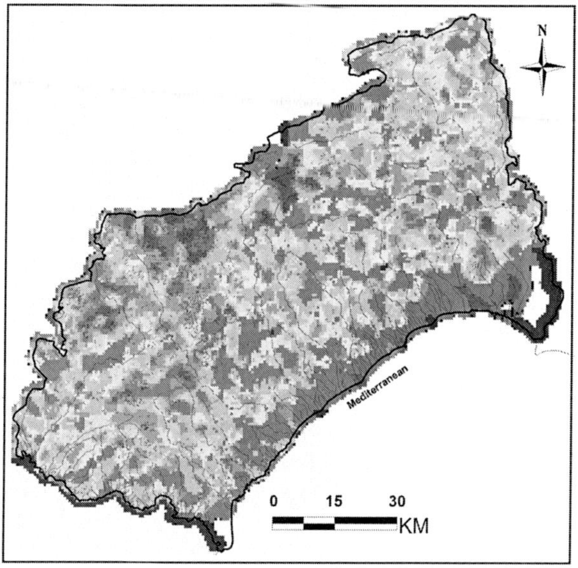

Fig. 2-10: Density analysis according to length of river branches

Table 2-3a. Data on rivers located in basins

Basin Name	Basin No	Square km²	Length (km)	Total Number	Density		Frequency	
					Basin	Bolkar Mts	Basin	Bolkar Mts
Acidere	1	42.47	131.8	131	3.10	0.017	6.68	0.00019
Gokboyun	1	12.22	35.8	58	2.93	0.004	5.96	0.00001
Kurtsoyu	2	789.5	1872.	2256	2.37	0.235	3.90	0.03829
Karacay	3	44.25	108.8	122	2.46	0.014	4.20	0.00013
Agandere	4	34.2	87.5	88	2.56	0.011	4.54	0.00008
Asmali	5	27.22	78.7	87	2.89	0.010	5.79	0.00007
Kapizdere	6	24.27	63.0	46	2.59	0.008	4.67	0.00004
Domuzca	7	37.67	101.6	92	2.70	0.013	5.05	0.00011
Keben	8	24.47	55.6	58	2.27	0.007	3.58	0.00003
Devedere	9	17.73	36.7	28	2.07	0.005	2.98	0.00001
Agildere	10	96.36	259.1	317	2.69	0.033	5.02	0.00073
Eksiler	11	24.02	133.4	53	5.55	0.017	21.40	0.00019
Kayaini	12	34.34	89.2	95	2.60	0.011	4.68	0.00009
Akdam	13	9.17	16.9	19	1.84	0.002	2.36	0.00000
Incindere	14	101.02	287.0	323	2.84	0.036	5.60	0.00090
Eyceli	15	7.07	11.7	10	1.65	0.001	1.89	0.00000
Bebek	16	28.47	57.8	48	2.03	0.007	2.86	0.00004
Akarda	17	57.2	106.7	82	1.87	0.013	2.41	0.00012

Esenbel	18	62.23	33.8	26	0.54	0.004	0.21	0.00001
Topbogazi	19	237.97	693.3	898	2.91	0.087	5.89	0.00525
Cehennem	20	74.71	151.9	128	2.03	0.019	2.87	0.00025
Yagar	21	152.48	452.4	598	2.97	0.057	6.11	0.00224
Çakal	22	4.66	6.7	6	1.45	0.001	1.45	0.00000
Kocari	23	90.43	169.5	145	1.87	0.021	2.44	0.00031
Ayas	24	61.28	90.4	54	1.48	0.011	1.51	0.00009
Mergin	25	1227.6	2798.0	3718	2.28	0.351	3.60	0.08553
Kocaman	26	56.32	87.4	55	1.55	0.011	1.67	0.00008
Sorgun	27	488.14	1026.5	1154	2.10	0.129	3.07	0.01151
Tabya	28	8.47	15.5	7	1.83	0.002	2.33	0.00000
Kargicak	29	61.73	128.7	108	2.09	0.016	3.02	0.00018
Çalilikoyak	30	29.94	61.3	37	2.05	0.008	2.91	0.00004
Karakuz	31	136.63	281.9	254	2.06	0.035	2.95	0.00087
Kargipinari	32	15.37	8.1	3	0.53	0.001	0.19	0.00000
Sandaldere	33	296.3	564.5	483	1.91	0.071	2.52	0.00348
Seymenli	34	18.78	44.0	29	2.34	0.006	3.80	0.00002
Tece	35	102.56	195.6	173	1.91	0.025	2.52	0.00042
Kovanli	36	24.74	58.4	40	2.36	0.007	3.86	0.00004
Cevikdere	37	36.53	80.6	71	2.21	0.010	3.38	0.00007
Cesmeli	38	13.25	34.0	23	2.57	0.004	4.57	0.00001

Table 2-3b. Data on rivers located in basins

Basin Name	Basin No	Square km²	Length (km)	Total Number	Density		Frequency	
					Basin	Bolkar Mts	Basin	Bolkar Mts
Mezitli	39	169.56	344.4	299	2.03	0.043	2.86	0.00130
Kale	40	9.6	11.8	11	1.23	0.001	1.06	0.00000
Merdiven	41	23.13	63.7	59	2.76	0.008	5.27	0.00004
Pozcu	42	22.11	53.9	32	2.44	0.007	4.12	0.00003
Yumuk	43	493.92	887.8	961	1.80	0.111	2.24	0.00861
Tirmil	44	92.87	186.0	109	2.00	0.023	2.78	0.00038
Icmedere	45	87.82	196.7	176	2.24	0.025	3.48	0.00042
Kazanli	46	487.76	999.0	965	2.05	0.125	2.91	0.01090
Karabucak	47	97.96	230.7	272	2.36	0.029	3.85	0.00058
Aynaz	48	69.66	54.8	42	0.79	0.007	0.43	0.00003
Belen	49	609.09	1089.5	1017	1.79	0.137	2.22	0.01297
Kadincik	50	499.14	1073.8	1051	2.15	0.135	3.21	0.01260
Cehennem	51	695.9	1228.4	1149	1.77	0.154	2.16	0.01649

A sudden tightening in the isohips of the upper and lower limits of these zones corresponds to steep slopes. These steep slopes reflect the existence of knickpoints along river valley profiles.

The valley profiles change from the local base level of the Göksu and Tarsus Rivers to the upper parts. That is, the base level of the tributaries is the altitude of the Göksu and Tarsus Rivers. From the Goksu valley to the inner parts, valley profiles become convex. This results from the Bolkar Mountains, showing a convex structure at west side and less depth split. However, in eastern section, the river profiles show a concave prospect. This indicates that the abrasion at eastern part is at a further stage of development. A general profile extending in a SW-NE direction taken from Silifke-Medetsiz Hill, shows that the height of the Bolkar Mountains increases from west to east direction and tectonic movements are more effective in the east.

Conclusions

Numerical characteristics of the rivers originating from the Bolkar Mountain and flowing into the Mediterranean Sea change, depending on morphological and morphoclimatic factors and processes. It is difficult to evaluate river basins based on structure, lithology, slope, height, precipitation, feeding regime etc., which are effective in river formation. However, in light of geologic, tectonic, geomorphologic and climatic studies carried out in this region, it is possible to present some comments.

Changes from east to west, through seashore to inner lands can be observed by grouping the numerical values of main basins and sub-basins that are located in the main basins, according to the table and graphs, which were obtained by the correlation of numerical values for rivers. The data obtained for the river basin and a number of tributaries explain the morphological evolution of the study area.

In the light of all the evaluations, it can be said that a polycyclic topography has developed in the Mediterranean Basin of the Bolkar Mountains where parallel and sub-parallel rivers exist. This topography has been greatly affected by tectonic movements the occurred in the Neotectonic Period. Levels that formed under the control of vertical slip faults based on young tectonic movements have affected the backward erosion of rivers. Young rivers (between 0 and 250 meters) observed around seashore indicate that young tectonic movements occurred during the Quaternary Period and as a result of sea level changes in the Mediterranean basin.

References

Akay, E., Uysal, Ş. 1988. Orta Toroslar'ın Post-Eosen Tektoniği. Post Eocene Tectonics of the Middle Taurus. Mineral Research and Explaration Inst. Pub 108, pp 57-68, Ankara.

Altın, B. N. 2002. Bolkar Dağları'nın Morfolojik Yapısı ve Arazi Kullanımı. Morphologic structure and land use of Boklar Mountain. Ministery of Environment and Forestry. National Parks and Hunter-Wild Life General Directorate, East Mediterranean Region Directorate, Its important, problems and solution proposals with regard to the eco-tourism of Bolkar Mountain Panel, June 5, 2002, Niğde.

—. (2001) Ecemiş Fay Kuşağı Doğu Bloğunda Karstlaşmanın Esasları . The principles of karstification in the eastern block of the Ecemiş Fault Zone. Workshop I, Proceedings, pp 163-170, Nigde.

Altın, B. N., Bayer Altın, T. 2005a. Bolkar Dağlar'ında Buzul Morfolojisinin Etkisi ve Dağılışı. The effect of glacial morphology and its distribution on Bolkar Mountain TURQUA Türkiye Kuvaterneri Sempozyumu 5, Eurasia Earth Science Institute. Book of abstract, page 258.

Altın, B. N. 2005. Bolkar Dağları'nda Doğal Ortam Koşulları ve Arazi Kullanımı. Natural environment condition and land use on Boklar Mountain. TURQUA Türkiye Kuvaterneri Sempozyumu 5, Eurasia Earth Science Inst. Book of abstract, page 272.

Altın, B. N., Bayer Altın, T. 2005b. 'Dağlar', Doğal Ortam Koşulları ve Arazi Kullanımı Sürdürülebilir Dağ Alanları Yönetimi (DAY) Açısından Yaklaşımlar. 'Mountains' Natural condition and land use, Sustainable Mountain Areas Management. Hacettepe University, Sport Science and Technology Faculty. 3th Natural Sports and Science Symposium November 26-27, 2005, Ankara. Book of abstract, page 42, Ankara.

Atalay, İ. 1986. Uygulamalı Hidrografya. Applied Hydrography. Ege Üniv. Education Fak. Pub. No: 38, İzmir.

Bayer Altın, T. (2007) Orta Toroslar'da (Aladağlar ve Bolkar Dağları) Görülen Periglasiyal Şekiller, Periglacial forms on Middle Taurus (Aladag and Boklar Mountain) Turk Geography Journal, 46: 105-122, Istanbul.

Dhont. D., Chorowicz.J., Yürür. T. 1999. The Bolkar Mountains (Central Taurides, Turkey): a Neogene Extansional Thermal Uplift? Turkey Geology Jurnal, 42: 69-87, Ankara.

Öncr, E., Hocaoğlu, B., Uncu, L. 2005. Tarsus Ovası'nın jeomorfolojik gelişimi ve Gözlükule Höyüğü. The geomorphologic development of

Tarsus plain and Gözlükule tumulus TURQUA Türkiye Kuvaterneri Sempozyumu 5, Eurasia Earth Science Inst. Book of abstract, pp 82-89, Istanbul.

Özgül, N. 1976. Toroslar'ın Bazı Temel Jeolojik Özellikleri. Some basic geological features of Taurus. Bulletin of the Turkey geology society, 19: 65-78, Ankara.

—. (1983). (Eds) Geology of the Central Taurus Between Hadim and Alanya. Int.Symp.on the Geology of the Taurus belt. Field Guide book for Excursion II, page 16, Ankara.

Scheidegger, A.E. .1961. Theoretical Geomorphology, Springer-Verlang, Berlin, Göttingen, Heidelberg.

Strahler, A. N. 1975. Pysical Geography (4. Edition): John Wiley and sons Inc., New York, London, Sydney, Toronto.

Şengör, A.M.C., and Yılmaz, Y. 1981. Tethyan Evolution of Turkey; A Plate Tectonic Approach. Tectonophysics, 75 (3-4), pp 181-241.

CHAPTER THREE

HYDROLOGIC PROCESSES WITHIN THE BALACITA PIEDMONT

SANDU BOENGIU, ALINA VLADUT, GHEORGHE CURCAN AND EMIL MARINESCU

Introduction

The Bălăciţa Piedmont represents a distinct relief subunit within the Getic Piedmont; it is a lower step, with a general aspect of a plateau fragmented in large interfluves and a West-East directed hydrographical net. The Bălăciţa Piedmont, also called The High Plain of Bălăciţa, due to its generally high plain, is a transition subunit between the Coşuştei Hills, in the North, and the Oltenia Plain, in the South, the main differences being related to altitude, its unique river system, general relief configuration, drainage density, and pedological and climatic features.

The eastern limit within the piedmont is made through a vast depression couloir directed NW-SE, which is cut by the Jiu River in a monocline structure that inclines in the same direction. In the North, the Motru, the Huşniţa, and the Ghelmegeoaia rivers, which are consequent valleys, separate it from the Motru Piedmont. In the West, the limit with Mehedinţi Plateau continues through the Severin Depression to the Danube. In the South, it neighbours with the Oltenia Plain, which, in this sector, is separated into the Terraces Plain of Oltenia and the Sălcuţa Filed (a piedmont plain) on the alignment Drobeta Turnu-Severin-Vânju Mare-Pleniţa-Podari (Fig. 3-1).

General Geographical Features

The geological conditions represent the starting point and conditions of all the future geomorphologic processes through paleogeographic, lithostratigraphic, and paleoclimatic features. From the lithological point of view, the oldest Quaternary deposits can be divided into three horizons made up of sands, gravels, blocks, and some intercalations of grey clays and lignite beds; the newest deposits belong to Holocene and they represent accumulations within the alluvial plains, dune deposits, and deluviums located on the slopes of the valleys.

Dilivium deposits display a high development within the analysed region and they cover most of the slopes of the valleys and the taluses of different morphological steps; these deposits are made up of sandy clays and clayish sands and, sometimes, gravels.

The general morphological elements, due to their features, reflect the presence of three distinct sectors within the Bălăciţa Piedmont. The first sector includes the catchment areas of the Blahniţa and of the Drincea, inclusively the slope of the Danube between Şimian and Batoţi and it is characterized by the presence of the rivers with subsequent valleys, which deeply cut a NW-SE directed monocline structure, made up of soft rock deposits, such as the sandy ones (predominantly). The most representative cuestas developed on the left slopes of the Danube, of the Corlăţel, and of the Blahniţa Rivers.

The second sector, represented by the catchment area of the Desnăţui, is completely different; the interfluves are large and horizontal, the relief intensity is low, the valleys are consequent along the upper course and subsequent in the area they go out of the piedmont, where there also appear cuestas. The third sector is characterized by the symmetry of the main valleys and the consequent course of the Jiu River's tributaries. The interfluves are narrow and sometimes formed by intersecting hills, while the valleys preserve a relatively high width. The slopes are generally affected by settled or active landslides and large glacises at the contact with the flood plain, while certain tributaries of the Motru, of the Argetoaia, and of the Raznic Rivers display, on the left, slopes under the shape of cuestas along their lower course.

The important climatic parameters are included in the present study are mean annual amounts of precipitation, mean monthly amount of precipitation, maximum monthly amounts, maximum amounts of precipitation in 24 hours, and the intensity and duration of showers.

The mean annual amounts of precipitation increases from south to north and decreases from west to east, due to the altitude, on the one hand, and to the diminution of south-western influences, on the other (568.8 millimetres at Craiova, 577.3 millimetres at Vânju Mare, 670.7 millimetres at Drobeta Turnu-Severin, and 603.1 millimetres at Bâcleş).

The mean monthly amounts of precipitation increase from January-March, the period of minimum registered rainfall, May-June, when the maximum is registered. In August, September, and October, the amounts are low and these months correspond to the secondary minimum, while in November and December, there is a new increase, which corresponds to the secondary maximum.

The maximum monthly amounts correspond to October 1972 for three of the stations. Thus, at Vânju Mare, there were registered 197.4 millimetres, meaning a 493 per cent higher than normal amount; at Craiova, the maximum amount reached 238.3 millimetres, which represents an increase of 658 per cent as compared to the mean; at Bâcleş, the amount of 229.3 millimetres represents an increase of 573 per cent. At Drobeta Turnu-Severin, the maximum value was registered in *July 1999* and it reached 331.4 millimetres, which means an increase of 500 per cent and a positive deviation of 268.8 millimetres. Such extreme monthly amounts have a deep impact on relief as they are one of the main causes that induce the present geomorphologic processes.

The maximum amounts of precipitation registered in 24 hours display quite similar or higher values than the monthly means and follow the same annual regime, meaning they are lower during the cold season, when cold dry air masses predominate, while during the warm season, especially in May-June, they increase due to the intensification of frontal activity and thermal convection. Thus, at Craiova, the maximum amount in 24 hours reached 91.5 millimetres (the 6th of September 1968), a similar value being registered at Vânju Mare, 91.3 millimetres (the 13th of July 1982). At Drobeta Turnu-Severin, the amount was almost double, 171.7 millimetres (the 30th of July 1969).

With regard to the mean intensity of the showers within the Bălăciţa Piedmont, values higher than 5 millimetres/min are locally registered in the South-west, but most of the analysed area displays a mean intensity of 3-4 millimetres/min (Carmen Dragotă, D. Bălteanu, 2000). Generally, there predominate short local showers (at Craiova, the mean of the highest four values of the maximum intensity is of 4.5 millimetres/min, while the mean duration is of 1.6). Showers occur irregularly and the analysed area presents an intermediary vulnerability from this point of view.

The river system within the Bălăciţa Piedmont belongs to two catchment areas: the Danube and the Jiu. The territory form the North-eastern half is directly drained by the Jiu or its tributary, the Motru, while the South-western half is directly drained by the Danube. The main watershed that separates the catchment area of the Jiu from the catchment area of the streams that flows into the Danube is generally directed NW-SE and it lowers from 307.9 m in Culmea Bâcleş to 164.5 m in Dealul Livezile. It separates the piedmont into two approximately equal and symmetrical parts. The geological conditions, even if they seem to be simple, represent an original mark when it comes to the organization and evolution of the river system within the Bălăcița Piedmont, due to the general divergent character and the predominance of autochthonous rivers, except for the Motru and the Jiu.

Erosional Processes

The action of the water within the Bălăcița Piedmont can be divided into two types: sheet erosion and concentrated drainage.

Sheet erosion is manifested through ablation; most of the times, it acts in association with other degradation processes, but there also appears to be certain surfaces that are affected almost exclusively by ablation, as it is the case of the right slope of the Blahnița and of the Drincei (Fig. 3-2), where the sandy deposits, the dry periods, and the removal of the vegetation favour this phenomenon.

Fig. 3-1: Slope affected by rain wash at Scorila

Within the Bălăcița Piedmont, the intensity of this process reaches a maximum and extended value during spring or autumn rain showers, when the slowly inclined slopes with cereals crops are tilled; a high value characterized the slopes without vegetation, as well. The soil material is loose due to the tillage and, thus, the rain drops scatter it easily. The removed material is carried as a water pellicle charged with fine particles that flow towards the base of the slope.

In July 2005, a heavy rain shower induced a strong denudation on Popicului Hill, with a slope of $2^{o}/_{oo}$. It generated a pellicle of fine dusty sand within the yards of the households in the settlement of Breasta, located at the foot of the hill.

The concentrated drainage of the water along the slopes appears as streams and it is represented by incipient erosion forms, such as rills, erosion ditches, furrows, as well as by well-evolved forms, such as gullies and torrents in different evolution stages. The rain wash represents the rapid flow of the rain or snow thawing on the slopes. There appear rills, furrows, and gullies.

The rills and *the gullies* are the simplest forms, which appear due to the rain wash. They are quite frequent at the spring of the Blahnița, West of the settlement of Valea Izvorului, but the greatest spreading is within the following valleys: the Poroinița, the Orevița, the Motru catchment areas, the Huşnița on the left slope, at Scorila (Fig. 3-2).

The furrows mainly develop in the spring area of the rivers and their tributaries. The furrows evolve within deforested areas with mean and inclined slopes, along carts paths that are transformed and enlarged as rills. Most of the furrows are located in the West and North-west of the Bălăcița Piedmont, especially on the slope of the Danube between Erghevița and Hinova, within the catchment areas of the Blahnița and of the Drincea.

Many of the furrows also formed along slope roads that go up on the slopes of the Argetoaia and of the Raznic (Fig. 3-4). The rapid development of this form of concentrated drainage is favoured by the friable rocks (sands, clays, marls).

The gullies from the Bălăcița Piedmont evolve according to the realized model (Bălteanu D., Taloescu 1978). The incipient stage of discontinuous erosion, which starts when the vertical erosion does no longer modify its path from a rain shower to another and the gullies cannot be filled up by ploughing and unlikely the rills, there appear a succession of crossings along the longitudinal profile; the stage of the formation of the discontinuous gullies, which corresponds to the beginning of a catchment area and contributes to the accentuated concentration of the discharge; the stage of the integration of the discontinuous gullies in continuous gullies. The types of gullies noticed within the Bălăcița Piedmont emphasize a wide range of their shapes (Fig. 3-5).

The configuration of the gullies is mainly dominated by the type of surface drainage, by the presence of human activities on the slopes, and by the existence of small cracks in the soil that favours the development of pipping processes.

Fig. 3-2: The Evolution of a Rill (1) to a Furrow (2) and then to a Gully (3) at the Spring of the Raznic

- Bifurcated (confluence) gully on the Bistrița Valley at its spring;
- Dendritic gully in Odăilor Valley at Hinova;
- Composed (linear, bifurcated) gully at Poroina Mare;
- Composed gully at Valea Lungă (linear, dendritic);
- Unilaterally latticed gully at the spring of the Corlățel upstream Izvorălu de Jos;
- Composed (dendritic, bulb-like shaped) gully on the Orevița Valley at its origin;
- Poorly dendritic gully on the western slope of the Jiu, at Sfârcea below Mănești Hill;
- Bulb-like gully spatulate at its apex, on the left of the Argetoaia Valley at Dumbrava de Jos;
- Linear gullies evolved through the development of certain rills, on the right of the Valea Rugilor stream.

The most spectacular gullies as length, depth, and shape appear on the left slope of the Danube, upstream of Hinova (Fig. 3-5), and on the right slope of the Jiu between Bralostița and Sfîrcea, within the catchment area of the Drincea and on the Slătinic Valley in the proximity of the settlement of Plopi. The proportion of the gullies from the right slope of the Danube and from the left slope of the Jiu is due to the altitude of more than 100 meters, to the inclination of these valleys' slopes, as well as to the great thickness of the sandy deposits. As for the gullies from the catchment area of the Drincea, which develop in a slowly-inclined relief, we mention the deepening of the Corlățel course and the sandy deposits.

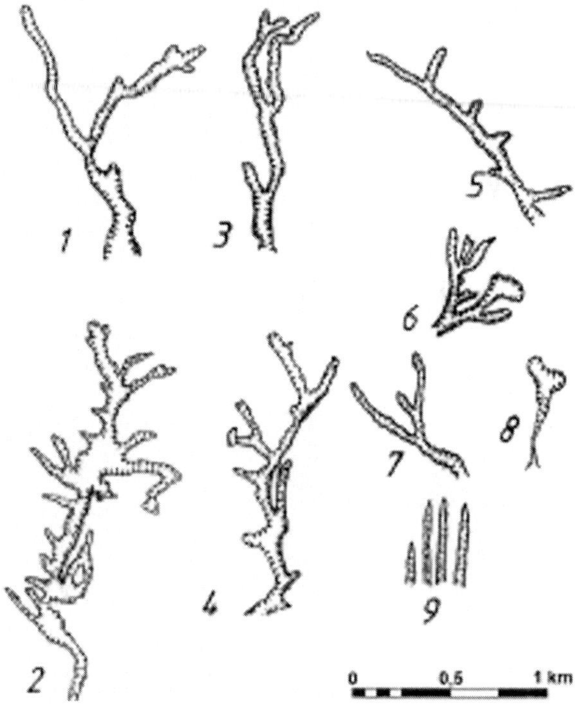

Fig. 3-3: Types of Gullies within the Bălăciţa Piedmont

The torrential valleys are relief forms which result from the associations of the furrows and gullies, concentrated towards a discharge channel along which the volume of water and eroded materials are carried. Their appearance within the Bălăciţa Piedmont is due to the less hard rock, to the positive neotectonic movements, and to the deepening of the permanent river system.

In the North of the Bălăciţa Piedmont, at the origin of the rivers, there can be noticed numerous torrents in different evolutionary stages (Fig. 3-6). Through the union of many torrents, there has developed complex torrential valleys, as it is the case of the Branişteea Valley, a tributary of the Desnăţui.

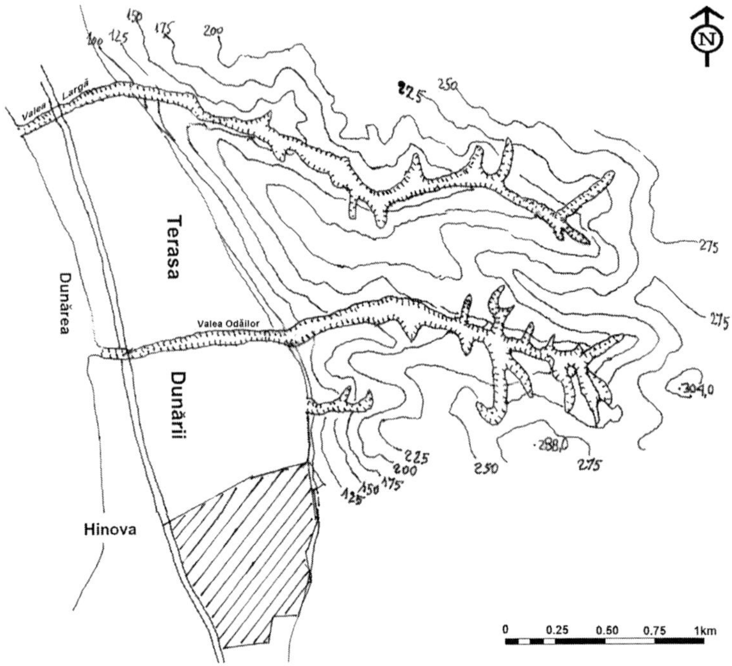

Fig. 3-4: Gully erosion on the Left Slope of the Danube, Upstream of Hinova

The analysis of the frequency of the torrents on the slopes of the valleys: In describe hierarchical system of the torrential valleys that accompany the water courses within the Bălăcița Piedmont, we have chosen the criterion of number, of length, as well as their location on one side or another of the water course. We have taken into account the tributaries of the Jiu within the piedmont area, the Argetoaia and the Raznic and their tributaries: the Țânțar, the Mașcot, and the Merețel, as well as the tributaries of the Motru: the Slătinic and the Breznița de Motru (the Tălăpan).

For comparison, we have also analysed the Blahnița, the Drincea, and the Danube along its piedmont sector (there were taken into account only the slopes on the left, which form cuestas, as the right ones present large valleys, slowly inclined slopes, where permanent streams predominate).

By centralizing the data regarding the number and length of the torrential valleys from the both side of the rivers, the results are the following situation (Table 3-1 and 3-2).

Fig. 3-5: Active Discharge Channel at Gropanele

According to the above tables, on the left slope, which is generally more inclined, the torrents are mainly short, up to one (1) kilometre.

The Slătinic displays the most numerous torrents with lengths below 0.5 kilometres on the left (64 percent); the Breznița de Motru, the Țânțarul, and the Raznic present the highest rates of torrential valleys with lengths between 1 and 5 kilometres (46.1 percent the Breznița de Motru, 70.8 percent the Țânțarul, and 73.1 percent the Raznic) on the same side. The Maşcot displays such torrents at a rate of 80 percents on both slopes.

Table 3-2: The Number and Rate of Torrential Valleys on the Right Slope

VALLEY	RIGHT SLOPE								A	B	C
	< 0.5 km		0.5 – 1 km		1 – 5 km		> 5 km				
	Nr.	%	Nr.	%	Nr.	%	Nr.	%		km	No./ km
Slătinicul	6	31.5	10	88.9	3	11.1	-	-	19	22	0.86
Breznița de Motru	-	-	6	60.0	4	40.0	-	-	10	18	0.55
Argetoaia	11	55.0	1	5.0	5	25.0	3	15.0	20	50	0.40
Țânțarul	-	-	8	88.9	-	-	1	11.1	9	22	0.41
Raznicul	-	-	52	8.1	6	10.2	1	11.7	59	58	1.01
Maşcotul	8	88.9	1	11.1	-	-	-	-	9	20	9.54
Merețelul	13	72.2	2	11.1	2	11.1	1	5.6	18	42	0.43

A: No of torrents B : Total Length C: No of torrents / Total length of valley

As compared to the tributaries of the Jiu and of the Motru, which flow West-East and are consequent, in the case of the Blahnița and the Corlățel, which flow subsequently southwards, the rate of the torrential valleys with lengths under 0.5 kilometre (33.3 percent the Blahnița and 54 percent the Corlățel) is close to that with lengths between 0.5 and 1 kilometres (30.1 percent the Blahnița and 35.5 percent the Corlățel).

The 0-1 kilometre torrents appear only in the case of the Slătinic, of the Breznița de Motru, and of the Tălăpanul (all tributaries of the Motru) and they present rates higher than 60 percent on the right. The 1 – 5 kilometre torrents display rates below 40 percent on both slopes for all the rivers.

With regard to all the torrential valleys from both slopes for the tributaries of the Jiu and of the Motru, the situation is rendered in Table 3-3.

In the case of the Maşcot and of the Merețel, which have small catchment areas, the short torrents from both slopes with lengths below 0.5 kilometres present a high rate (84.2 percent for the Maşcot and 71.1 percent for the Merțel) due to the other streams parallel with them.

The Slătinic, on both slopes, has a high rate of the torrents with lengths below 0.5 kilometre (64.0 percent), but the Țânțar and the Raznic display a similar rate for the torrent with length of 0.5-1 kilometre (70.8 percent the Țânțar and 73.1 percent the Raznic).

Table 3-3: The Number and Rate of Torrential Valleys from Both Slopes of the Rivers

VALLEY	BOTH SLOPES								A	B	C
	< 0.5 km		0.5 – 1 km		1 – 5 km		> 5 km				
	Nr.	%	Nr.	%	Nr.	%	Nr.	%		km	No./ km
Slătinicul	16	64.0	6	24.0	3	12.0	-	-	25	22	1.14
Breznița de Motru	2	15.4	6	46.1	5	38.5	-	-	13	18	0.72
Argetoaia	16	42.1	14	36.8	2	5.3	6	15.8	38	50	0.76
Țânțarul	-	-	17	70.8	5	20.8	2	8.4	24	22	1.09
Raznicul	9	11.5	57	73.1	9	11.5	3	3.0	78	58	1.34
Maşcotul	16	84.2	1	5.3	2	10.5	-	-	19	20	0.95
Merțelul	32	71.1	6	13.3	6	13.3	1	2.3	45	42	1.07

A: No of torrents B : Total Length C: No of torrents / Total length of valley

The Argetoaia displays similar rates on both slopes, 42.1 percent torrential valleys with a length below 0.5 kilometre and 36 percent torrents with a length of 0.5-1 kilometres.

It is worth mentioning that all the autochthonous streams within the Bălăcița Piedmont do not start from permanent springs, but gradually build their course through two or more torrents.

Conclusions

Erosion and depositing processes induced by the action of the water courses within the Bălăcița Piedmont acquire a high importance, especially during floods. This is especially the case if we take into account the size of the catchment areas and the multitude of tributaries, as well as the climate characterized by rain showers from spring till autumn. These processes are less intense within the Bălăcița Piedmont as compared to other sectors of the Getic Piedmont, due to the reduced drainage density and relief intensity, to the lack of strong springs able to supply the river system, and to the important amount of the rain water in the discharge of the rivers within the inner part of the piedmont.

Generally, the minor bed of the autochthonous rivers within the Bălăcița Piedmont presents deep thalwegs, 1-5 metres below the level of the flood plain. The greatest depths of the stream channels can be noticed on the Drincea, the Raznic, and especially on the Desnățui, downstream from Dragoia.

The undercutting of the high slopes during floods leads to their collapse, to landslides, to the acceleration of the head erosion and to the alluviation and of the minor bed and its transformation into a swamp. The high humanization degree within the Bălăcița Piedmont inevitably affected the environment from early times. The deforestation of the secularly forests led to terrains degradation and strengthened relief drainage density. The diversity of economic activities and the intensification of agriculture modified many of the natural aspects of the region, such as the vegetation and the soils.

Man-induced modifications, which significantly influenced the environment and especially relief, are obvious throughout: deforestations on vast surfaces, extension of crops, significant increase of the number of animals' (cows, goats), which require intensive grazing, construction of dams and other schemes, construction of an increasingly number of roads, extension of the precincts of the settlements and development of new ones, construction of economic and industrial units.

The building of roads required embankments or cuttings along certain sectors, which modified the runoff regime of the pluvial water on the slopes and the profile of the slopes. Road building required the excavation of certain sandy materials for the embankments, from what the constructors call "borrowing pits" that facilitated the appearance of gullies in their proximity.

Many roads of access to the settlements located on the slopes or even on the watersheds have been affected by rills and gullies. It is the case of the road that links the settlements of Bistrița and Valea Izvorului, which was unusable after floods each spring until it was modernized, and of the road that goes from Livezile to Izvorălul de Jos, which climbs below the "La Gâlmele" summit and then crosses the Ostraşeva stream "at the bridge of Bârca" and which is still affected by rain wash after showers. We must also mention the road towards a sand and gravel pit situated on the Buciumul valley, from Corzu, recently affected by rills.

The building of dwellings, which intensified during the last century, and the maintenance of the local roads also required materials exploited from pits, as well as of clays for bricks taken from the flood plains. These activities led to the occurrence of landslides or of swamps in the holes made for clay exploitation.

Due to the frequent floods that occurred within the flood plain, where the cereals crops and the settlements extended, the Brabova stream was harnessed from Brabova to Sârsca, a distance of 12 kilometres, and the Pleşoiul stream, a distance of 2.8 kilometres, in the proximity of the homonymous settlements.

The construction of dams especially in the valleys located in the south of the Piedmont, at Verbița and Vârtop on the Baboia, at Cornu at the confluence between the Caraula and the Cornu streams, at Caraula on the stream with the same name, Fântânele basin on the Desnățui, had as its main purpose pisciculture and irrigation activities but the basins also regulate the discharges during floods.

References

Badea, L. 1973. Piemontul Getic – Vol. Piemonturile. Centrul de multiplicare al Universității din Bucureşti.

Bălteanu, D., Taloiescu Iuliana (1978) Asupra evoluției ravenelor. Exemplificări în dealurile şi podişurile de la exteriorul Carpaților. SCGGG- Geografie. Vol. XXXV. Bucureşti.

Boengiu, S. 2002. Observaţii hidrogeografice în Piemontul Bălăciţei. Revista Geografică T. VIII –2000, Institutul de Geografie, Academia Română, Bucureşti.

—. 2005. Caracteristici morfometrice ale versanţilor din Piemontul Bălăciţei, Editura Universităţii din Bucureşti, Revista de Geomorfologie, vol. 4-5 / 2002-2003, Bucureşti.

Dragotă, Carmen, D. Bălteanu, D. 2000. Intensitatea precipitaţiilor extreme pe teritoriul României, Revista Geografică, t. VI, Bucureşti.

Ielenicz, M. 2004. Geomorfologie, Editura Universitară, Bucureşti.

Vlăduţ, Alina. 2002. Caracteristicile ploilor torenţiale în Oltenia, Analele Universităţii din Craiova, Seria Geografie, vol. V, Serie Nouă, Editura Universitaria, Craiova.

Stroe, R. 2003. Piemontul Bălăciţei. Studiu geomorfologic, Edit. MondoRo, Bucureşti.

CHAPTER FOUR

ACTUAL PROCESSES OF ELEVATION MODELING ON THE VALLEY OF LLAP AND THEIR IMPACT ON THE ENVIROMENT

SHPEJTIM BULLIQI, GANI GASHI,
FLORIM ISUFI AND IBRAHIM RAMADANI

Introduction

Development of an area elevation is a multidimensional process and it is expressed by indicators that have a complex dependency between internal forces and external geomorphologic factors.

General study of an area elevation makes possible the knowing of one of the most complex components of the environment. Considering the fact that the major part of Kosovo territory has been studied relatively little bit on a morphological aspect, in this project it was seen needed to make efforts for a partial approach related to the process of recent elevation modeling of the valley of Llap, and also to the other created processes and forms.

Methods

To fulfill this work there have been used the following main methods: cartographical, statistically-mathematical and field activities. Through cartographical and in combination with statistical-mathematical method it has been determined area of study. From calculation it has been finalized horizontal and vertical fraction. Horizontal fraction performs report of water flows on proportion with surface. From erosion maps we got surface of poignancy, surface of erosion and have been determined most risk surfaces. On geographical studies especially on geomorphologic and environmental, work methods in the field remains most needed. Observing directly processes and occurrences in the field made the problem more imminent and understandable .In the field has been concluded soil slides depend from geological construction and slope of terrain. Meantime also have been verified damages foresees for construction dwellings on these areas from unaware inhabitant for risks.

Actual processes of relief modeling

Today's sight of the elevation of Llap (Fig. 4-1) valley is a result of activity of geomorphologic processes in space and time.

Elevation modeling begun after the valley remained without water of the lake that once existed and it is continuing until today's day. Some factors that they have influenced on elevation modeling are: natural factors, natural-anthropogenic and the anthropogenic factor.

First we should take into consideration all those processes that came to be as result of the activity of endogenous and exogenous forces.

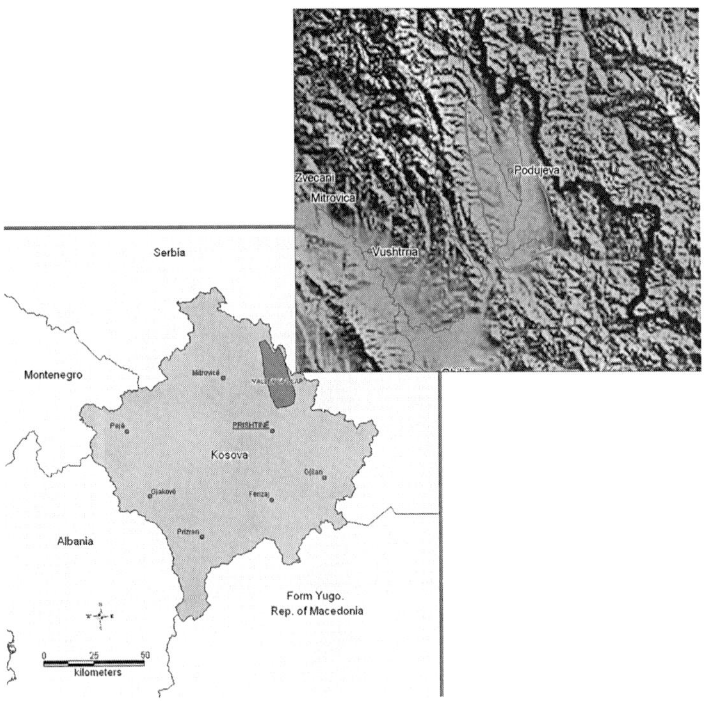

Fig. 4-1: Geographical position of valley Llap

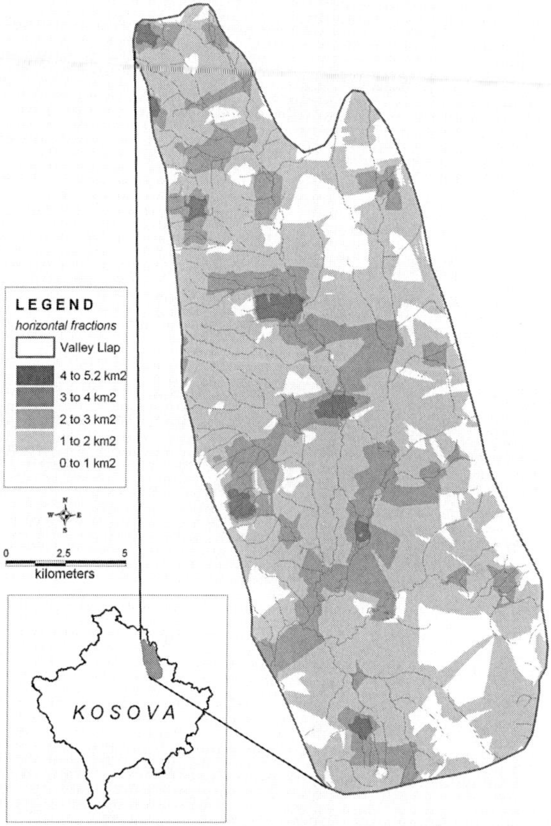

Fig. 4-2: Map of horizontal fractions in valley Llap

Secondly we should take into consideration all the actions of a human being that in a way initiate or intensify these processes, while in anthropogenic processes we should take into account all kinds of activities of a human being (agriculture, construction, mining etc), that have or would have affect in the modeling of elevation (Bognar, 1990). Recent modeling of an elevation was conditioned by some factors where the main role had the tectonics of a terrain, litho-logic forming, climate, water and plants etc., which are directing these processes (Markovič, 1983). The Geographical position of Llapi valley conditioned this area to receive for many years a medium rainfall of 691.1mm and to have a medium temperature of 9.6 °C. As a consequence of the aforementioned factor and litho-logic forming of the terrain, there came to a medium development of temporary and permanent surface flows, and also to the development of various plants. The level and intensity of erosion activity of flowing waters and slope processes, apart from climate conditions, is closely connected also with the medium horizontal fractioning (from 1-4 km/km^2) and also with the medium slope of the terrain (from 4-12 °) (Bulliqi, 2000) (Fig. 4-2).

All these together conditioned that in this valley to appear many deepened forms of river beds, ridges, alluvial depositions, stream runways, collapsing, sliding etc (Bulliqi, 2006). In the_territories made from flysch, the torrential river processes have larger intensity. Occurrence of collapsing and various fallings that are characteristics for unstable banks, for sure in some cases are also initiated from anthropogenic factor are noticed within this area. Such occurrences are characteristic for tectonic escarpments and litho-logic ones of bordering banks of this valley, which are forming the main morphologic element of the elevation. They appear at certain parts along the main river beds. The main cause of equilibrium disorder of these banks is the side-erosion action of respective flows.

In the complex of slope occurrences, an important role in elevation modeling of this part plays the sliding of the ground of highly different sizes, which in some territories are falling into a powerful collapsing, especially during the tectonic escarpments.

The occurrences have the widest spreading along the bordering banks of Llapi valley, especially on the complex escarpment of western bank between Pakashtica of the northern side and Llapashtica on the southern side. Over-satiation of ply-ocenic lake argils with water, during their contact with dibasic rocks, and regressive erosion of branches of Llapi River, created conditions for displacing of fractional material masses placed over these argils.

The size and the level of sliding development are introduced in very different ways, since the plan of sliding in this sector is passing along the contact of highly tectonic dibasic rocks with ply-ocenic argils. Thus the whole complex escarpment part on the height of 700 – 900m is noticed for some active sliding, such as those along the river of Pakashtica, Dobratin, Obrançe, Llapastica etc.

A relatively dense spreading of the sliding is also along the eastern banks of Llap pond, which is noticeable also due to high level of tectonic activity of magma rocks and the flysch rocks of mioplyocen molasses from tectonic separation of Merdari. As result of this tectonic occurrence is the formation of a wide generation of destruction, during which were developed sequences of sliding, which is expressed also with the over fragmentary forms of complex escarpment and an over degraded elevation, especially between Shakovica on the southern side and Merdari cool on the northern side.

In relatively smaller sizes, the sliding also meet each other at the incline southerner of this pond, between embouchure of Batllava river and Terrnava river, which are connected with the placing of conglomerates of the lake between Dyzi and Terrnave over the core of conveyance of Cretaceous flysch. The last one, by getting satiated with water from the water-keeping conglomeratic size, which is also provoked from the regressive erosion of aforementioned streams, leads to an unstable condition of the last ones and their continuous displacement (Fig. 4-3).

Fig. 4-3: Sliding on the left side of Llapi valley–Bellopojë (Bulliqi, 2006)

Other processes of degradation of the terrain such as surface erosion, sliding, and other occurrences also get an important part. If we base at the erosion map (Institute, Jeroslav Černi, 1983), we could come to the conclusion that within the valley of Llap are represented the surfaces on which is presented high erosion that includes around 9.4% of the entire surface. (Fig.4-4)

Than the surfaces on which dominates the medium erosion processes that includes around 22.2% of the entire surface of the valley. The surfaces that are dominated by the weak and very weak erosion are represented on 53.8% of the surface; and the territories on which dominates the accumulation of the material and that includes around 14.6% of the surface (Table 4-1). The surfaces on which are dominating the high erosion processes are mainly spread on the south-western parts, south and south-eastern parts.

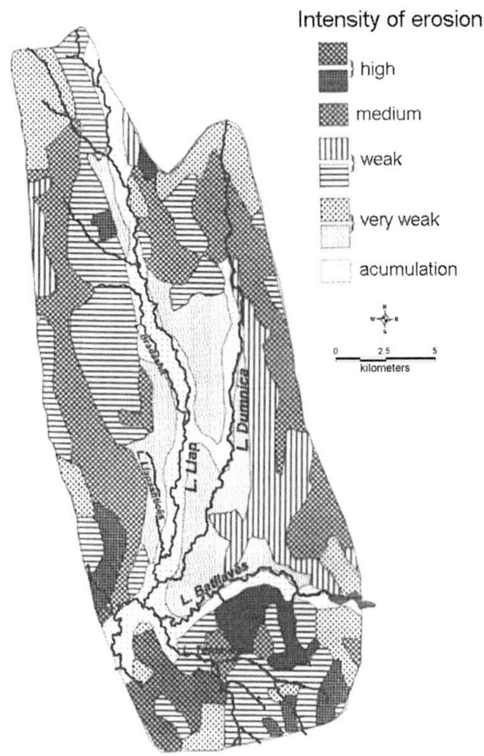

Fig. 4-4: Map of intensity of erosion in valley Llap

Results and Discussions

On these terrains, within a year in every km^2 is eroded an amount of material from 1200–3000m^3 /km$^{2/}$ a year, while the total amount of eroded material is somewhere around 35400m^3/a year.

Terrains which are dominated by medium erosion occupy around 22.2% of Llapi valley. These terrains are spread on the eastern and western part of the valley. Within a year, in every km^2 is eroded an amount of material from 800–1200 m^3/km^2/ a year, respectively the total amount of eroded material per year reaches around 56000 m^3/ a year. This erosion participates with 40.7% of the entire erosion that is taking part within this valley. Terrains that are dominated by weak erosion are spread on the south-eastern parts, eastern, and western part of the valley. The amount of eroded material in every km^2 is somewhere around 400-800 m^3/km^2/a year, respectively around 38400 m^3/a year. These territories are spread over the Cretaceous of flysch. Territories with very weak erosion are spread on the lower parts, central parts of the valley, but the same ones are also met on the south eastern past of the valley. In every km^2 during a year are eroded 100-400 m^3/km^2/ a year, while the total amount of eroded material reaches around 7715.5 m^3/a year. These terrains are mainly built from the laky sediments, while on the south-eastern part over the thick-grain arenitic units. While the territories that are dominated by the accumulation of the material are occupying around 14.6% of the surface of the valley and they are mainly spread across the main water-flows.

Table 4-1: Participation of the surface according to the intensity of erosion processes on Llapi valley.

Category	Intensity of erosion process	Erosion coefficient	Amount of eroded material	Surface		Product of eroded material	
		Z	m^3 / km^2 / a year	km^2	%	km^3 / a year	%
II 4 5	High Erosion	0,86 - 1,00 0,71 - 0,85	1200 - 3000	29,5	9,4	35400,0	25,7
III 7	Medium erosion	0,41 - 0,55	800 - 1200	70,0	22,2	56000,0	40,7
IV 8 9	Weak erosion	0,31 - 0,40 0,21 - 0,30	400 - 800	96,0	30,5	38400,0	28,0
V 10 11	Very weak erosion	0,71 - 0,20 0,01 - 0,10	100 - 400	73,5	23,3	7717,5	5,6
2	Accumulation of material	-	-	46,0	14,6	-	-
Total	-	-	-	315,0	100,0	137517,5	100,0

Source: Erosion map of SR Serbia, Beograd, 1983.

Conclusions

These processes are reflected in the environment in two aspects. The first aspect is that through sliding are being damaged the plants, especially forests. The second aspect is that during the wet period of the year the melting rocks and the easy-digested rocks are polluting the rivers. This is a natural pollution which in this territory is hard to prevent. The polluting effect is also expressed in shallow underground waters that may be polluted by organic matter that is washed from the surface of the earth.

Erosion is important for studying as morphological factor, it is also important on protection of environment. On natural balance destruction in this case, human factor contributes on forest cutting and increase of erosion from poignancy and also on spread surface .Sliding as process on this hollow brings damages not only to environment, or even on changing on water quality but in some cases also house devastations. Geomorphologic studies in some branches are applicable on spatial planning on conservation and use of space.

References

Bognar, A. 1990. Geomorfologija Baranje. Zagreb. pp. 275:278.

Bulliqi, Sh. 2000. Physic geographical characteristics of Llapi valley, (Project of Magistracy in handwriting). Pristina. pp.84:93.

—. 2006. Morphological features of NE Kosovo, (Project of Doctor's degree in handwriting). Pristina. pp. 71:74.

Erosion map of SR Serbia. 1983. Institut Jeroslav Černi. Beograd.

Markovič, M. 1983. Osnovi primjene geomorfologije Beograd. pp.23:37.

CHAPTER FIVE

THE ASSESSMENT OF EROSION PROCESSES AND THE POTENTIAL FOR DESERTIFICATION: A CASE STUDY OF ANTALYA PROVINCE, TURKEY

UFUK COŞGUN AND A. CEM AYDIN

Introduction

The "Turkey Action Plan on Combating Desertification" was published by the Turkish Ministry of Environment and Forestry in 2005. Soil erosion is the primary factor which turns potentially useful land to desert. The classification of land according to is soil characteristics, as well as its potential for erosion, can provide planners and political authorities with a means to productively use the land and its of resources. Learning about the dimension of the erosion is only possible by determining the land use ability. With this land ability classification, the slope rate of the land and soil depth distribution also becomes necessary, as these are critical to assessing erosion.

The focus of the present study is an assessment of erosion of agricultural land, pasture land, and forest land in Antalya Province. The sum rate of erosion for every town in the Antalya region would present the total erosion situation. For this study, first of all, the distribution of the land according to ability classes, slope and soil depth, etc., will be determined. Next, identification of land used for agriculture, pasture and forest and their distribution by town, will be determined. By using a ranking method, the sensitivity of towns to erosion is established by specialists. After ranking the towns, short, medium and long-term strategies will be developed for combating erosion.

Material and Method

For the present study, twenty-six specialists helped to develop nine main criteria for judging erosion. The method of analysis was determined by ranking and the linear equations methods which are used in analytic hierarchy process in order to classify land by significance levels. According to these data, a multiple criteria decision making process approach was estabilished in order to develop an erosion working plan.

The Analytic Hierarchy Process (AHP) is a commonly used solution technique for solving any multicriteria decision-making problem. This method is one of the most flexible and easily implemented multicriteria decision making techniques. It is a theory of measurement concerned with dominance priorities from pairwise comparison of clements with respect to a common criterion. The AHP allows the consideration of both objective and subjective factors in selecting the best alternative. Despite the widespread use of the AHP in diverse decision problems, this multi-attribute approach has not been without criticism. In spite of these criticsms, it is one of the most popular multicriteria decision making methodologies available today. Thus, the areas in wich the AHP is applied are diverse and numerous. The popularity of the AHP is due to its simplicity, flexibilty, ease of use and interpretation etc., in analysing complex decision problems.

The Ranking Method is one of the multiple criteria decision making methods used to rank decision units (Yılmaz, 2006). Decision makers weigh the relative criteria by significance levels in this ranking method as it is shown in Figure 5-1.

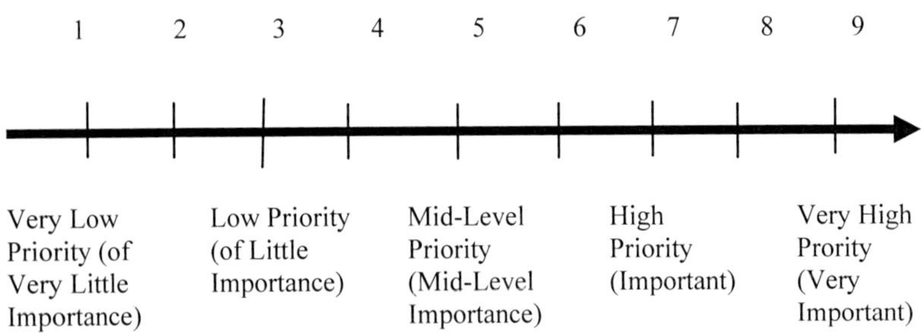

Fig. 5-1: Ranking Method Scale

$$W_{ji} = \frac{\sum_k r_{jki}}{\sum \sum_k r_{jki}} \qquad (i=1,2,...m)$$

The scale on Figure 5-1: 1. Very Low Priority (of Very Little Importance); 3. Low Priority (of Little Importance); 5. Mid-Level Priority (Mid-Level Importance); 7. High Priority (Important); 9. Very High Prority (Very Important). 2,4,6,8 are used as intermediate values of the scale. The relative significance levels or weighs of the criteria ranked in order to given points. For example a "k" decision maker ranked the subcriteria of the "j" criterion like

r_{jk1}, r_{jk2}, r_{jk3},...., r_{jkm}.. The weigh value (W_{ji})of the subcriteria of "i" criterion can be estimated as it shown in the priority determination matrix. A = Priority Values Matrix, a_{ij} = the upper hierarchy level unit's significance to other, the significance of "i" unit to the"j" unit.

$$A = \begin{bmatrix} a_{11} & a_{12} & . & . & . & a_{1n} \\ a_{21} & a_{22} & . & . & . & a_{2n} \\ . & . & . & . & . & . \\ . & . & . & . & . & . \\ . & . & . & . & . & . \\ a_{n1} & a_{n2} & . & . & . & a_{nn} \end{bmatrix}$$

The Lineer Equation Method. In this method, the criterion reel value should be fixed to the same scale by muliplying with its criterion weigh. To obtain the ranking of each town and village, the sum of the weighed criterion reel values is necessary. With this aim, the Lineer Equation was developed (Yılmaz, 2005) as;

$$TAP = \sum_{i=1}^{n} A_i P_i$$

TAP= Total weighed criterion reel values ,

A_i= Criterion weigh,

P_i= Reel value of the criterion.

Short (3 years), middle (5 years) and long term (8 years) application roadmap approaches aimed to be planed for applicators with this ranking. Towns were divided into 3 groups in order to express erosion sensitivity by using the ArcGIS 8.3 program.

Determining the Basic Factors Which Lead to Erosion

There are nine possible situations and factors leading to erosion in Antalya:

1) The Distribution of Agricultural Lands;
2) The Distribution of the Forest Areas;
3) The Proportion of Degenerated Forest Areas;
4) Pasture Areas and Their Usage Amount;
5) Land Use Capability Classes;
6) Distribution of Soil by Slope Groups;
7) Water Erosion;
8) Distribution of Soil Depth;
9) The Aridity and Senility of the Soil.

Distribution of Agricultural Land

Land distribution in Antalya is as follows: 56.37% forested; 22.00% agricultural; 18.66% pastures; 1.57% settlements; and 0.99% other (Table 5-1).

Table 5-1: Land Distribution of Antalya Province, by Town

Towns	Total Agricultural Lands	Pasture Lands	Forest Lands	Settlements	Other Lands	Total
Akseki	20267,15	51742,74	174985,01	900,67	541,97	248437,54
Alanya	35011,66	40123,67	115995,72	1768,30	817,62	193716,97
Antalya	66806,81	5973,73	153664,67	17329,82	4791,11	248566,14
Elmalı	54778,60	57083,77	56286,79	1338,78	914,70	170402,64
Finike	10417,03	15322,81	44926,98	600,08	160,09	71426,99
Gazipaşa	17849,43	29043,56	60882,76	895,47	511,52	109182,74
Gündoğmuş	10038,94	53443,39	42875,36	487,20	447,77	107292,66
Kaş	42254,75	18556,38	148331,71	1622,89	5351,38	216117,11
Korkuteli	84935,77	73159,57	93256,10	1865,69	671,40	253888,53
Kumluca	20423,59	18751,04	67183,00	713,58	1406,45	108477,66
Manavgat	51895,40	22506,51	149134,80	2640,35	3701,48	229878,54
Serik	51785,71	2709,67	65725,23	2466,43	1362,75	124049,79

Agricultural lands are also classified as watery absolute agricultural lands, dry absolute agricultural lands, planted citrus trees, vineyards and garden lands, dry marginal agricultural lands (Tarım İl Müdürlüğü, 2006) (Table 5-2).

It is an important data for erosion threat that 40.03 % of the total agricultural lands in the Antalya region are dry marginal agricultural lands. The "Dry Marginal Agricultural Lands" class is one of the most important classes of the land.

Table 5-2: Agricultural Land Character Distribution in Antalya Province, by Town

Town Name	Watery Absolute Agriculture	Dry Absolute Agriculture	Planted Fruit Tree	Citrus Trees	Planted Olive	Planted Vineyards	Dry Marginal	Watery Special Products	Total Agricultural Lands
	WAA	DAA	PFT	CT	PO	PV	DM	WSP	TAL
Akseki	0	0	114	0	0	1.386	18.767	0	20.267
Alanya	1.949	7.082	4.077	949	2	592	20.360	0	35.012
Elmalı	6.343	19.490	5.323	0	0	325	23.298	0	54.779
Finike	1.994	337	353	3.220	22	24	4.467	0	10.417
Gazipaşa	787	2.350	587	0	9	571	13.545	0	17.849
Gündoğmuş	0	217	523	0	0	37	9.262	0	10.039
Kaş	3.959	4.062	3.270	329	3.962	483	26.190	0	42.255
Kemer	512	997	47	1.352	0	0	1.448	0	4.357
Korkuteli	6.400	32.756	5.009	0	1	155	40.615	0	84.936
Kumluca	3.739	894	284	3.196	43	0	12.267	0	20.424
Manavgat	13.614	18.793	441	301	409	4	18.334	0	51.895
Antalya	18.323	18.072	1.019	3.477	4.471	0	12.584	4.504	62.450
Serik	27.983	14.017	593	180	89	13	8.910	0	51.786
TOTAL	85.603	119.069	21.642	13.005	9.008	3.589	210.046	4.504	466.465

These kinds of lands are in the fourth or lower classes in Land Ability Classification and they are commonly sloping, low quality lands. Therefore, these lands are very sensitive to the threat of erosion. In this sense, Dry Marginal Agricultural Lands are quite important for determining the erosion sensitivity of the province. We can also determine the erosion threat of the towns by comparing the amount of the Dry Marginal Agricultural Lands (Aydemir, 1973).

Just over forty-five percent (45.03 %) of the total agricultural land of Antalya Province is Marginal Agricultural Lands. Akseki comes first as having the biggest ratio (92.60%) of the Marginal Agricultural Lands to the total by towns. Gündoğmuş (92.26%), Gazipaşa (75.89%), Kaş (61.98%) and Kumluca (60.06%) are the following towns of the Antalya province.

Table 5-3: The Effects of the Land Usage Types to Surface Flow and Soil Transport.

Land Usage Types	Slope (%)	Surfaced Flow Amount of Total Precipitation (%)	Transported Soil Amount (ton/ha/yr)
Corn	15	27.8	6.687
Wheat		27.0	5.294
Fallow		33.6	6.493
Filbert		10.2	1.296
Forest		0.5	0
Corn	28	34 .7	10.225
Wheat		34.3	7.394
Fallow		39.4	7.507
Filbert		15.2	2.053
Forest		1.3	0
Corn	45	42.5	12.079
Wheat		42.1	9.843
Follow		47.4	10.018
Filbert		20.4	3.182
Forest		2.9	0

Agricultural Lands to the Total Agricultural Lands is also a significant criterion for erosion potential. For example, Korkuteli comes first of all with 19, a 34% ratio by towns. Kaş (12.47%), Elmalı (11.09%), Alanya (9.69%) and Akseki (8.93%) are the following towns in Antalya. The usage types of the Marginal Agricultural Lands are also significant. Some usage types and soil erosion amounts are given at Table 5-3. Considering the annual soil losses per hectare, on fallow and forest lands with different slopes, the Marginal Agricultural Lands become much more important. Therefore, these lands must be supported by alternative plantation systems which will lead to sustainable usage.

Distribution of Forest Lands

Antalya is a rich area in terms of forest lands. It is generally considered that a land which is rich with forests must contain at least 30 % forest lands. The rate of forest lands (1,125,698, 9 hectares) to the total is 54.55 % for Antalya (BALBI, A. N. 1958). The type of the plantation (tree, bush or shrub) is also important for erosion (Table 5-4).

Table 5-4: The Effects of Different Land Types on Erosion, Surface Flow and Infiltration.

Trial	Precipitation	Amount of Infiltration			Amount Of Surfaced Flow			Erosion
Parcel	mm	mm	m^3/ha	%	mm	m^3/ha	%	ton/ha
Uncovered (Agricultural)	1336	591,7	5920	44	774,5	7440	56	10,014
Lawn		855,6	8550	64	481,6	4810	36	1,36
Forest (marsh)		1094,8	10950	82	214,4	2410	18	-

The quality and distribution of forest land is potentially very important to understanding potential erosion processes. The extent to which forest land experiences erosion depends on whether these lands are high forest or coppice forest. The rate of forest lands in a town indicates the erosion amount. The rate of the forest lands to the total is 71.04% in Akseki. It is 69.12% for Kaş, 64.78% for Finike, 64.78% for Manavgat, and 28% for Alanya. The towns which have the lowest rates are 28.10% Elmalı, 37.04% for Korkuteli, and 40.74% for Gündoğmuş. On the other hand, the degenerated forest amount is also significant to the erosion threat. The size of these lands is significant as deforestation threats, which leads to erosion. Akseki has the biggest rate of degenerated forest lands to the total lands (40.19%), while Kaş has a rate of 39, or 88%, Finike 28, 0.04%) and Elmalı, 27, or 14 %.

Usage Density of the Pasture Lands

The amount of pasture lands and grazing density are factors that directly affect the erosion level. Therefore, determining the usage density and amount of pasture lands becomes important. By this aim, the whole animal subsistence of the towns is converted to cattle units (CU) and all pasture lands are assumed to have middle quality. A cattle unit assumed 200 kilo's weigh. Animals can graze 1/10 of their weighs (Aşk, 1987). A cattle unit can graze 20 kilo's green grass per hectare. Animals can graze 1000 kg's green grass on an average pasture land. So, (1000/20) only 50 cattle units can graze per hectare. Assuming that average grazing takes 180 days a year, only 3, 6 Cattle Unit could be graze per hectare. According to these assumptions, the animal subsistence and actual pasture lands are shown in Table 5-5 (Cosgun, 2006).

It is obvious that pasture lands are being overgrazed, in view of the actual usage and pasture capacity. This overgrazing of the pasture lands leads to degeneration and erosion threat. The negative amount of the comparison of actual usage and pasture capacity shows the amount of erosion. We can see intensive pasture usage at the central towns of Korkuteli and Manavgat

Table 5-5: The Distribution of Grazing Pressure by Towns

Towns	Pasture Lands (ha)	Green Grass Production Amount (thousand) (kg/ha)	Heavy Livestock Amount	Small Animals Amount	Working Animals	Cattle Unit	The Grazing Capacity of the Pasture lands (3,6 ha)	Actual Situation-Usage Potential
Akseki	51.743	51.742	4.189	56.380	1.219	15.008	54.030	-2.288
Alanya	40.124	40.123	17.345	43.220	1.579	20.630	74.268	-34.144
Antalya	5.974	5.973	34.857	84.980	527	38.437	138.374	-132.400
Elmalı	57.084	57.083	9.147	71.835	335	20.190	72.685	-15.601
Finike	15.323	15.322	1.200	20.000	450	5.170	18.612	-3.289
Gazipaşa	29.044	29.043	9.327	46.800	1.100	16.056	57.802	-28.759
Gündoğmuş	53.443	53.443	3.135	19.000	640	6.321	22.756	30.688
Kaş	18.556	18.556	5.517	53.500	610	14.620	52.633	-34.076
Korkuteli	73.160	73.159	15.885	200.000	450	49.981	179.932	-106.772
Kumluca	18.751	18.751	3.030	19.100	1.070	6.708	24.149	-5.398
Manavgat	22.507	22.506	15.774	117.210	1.875	34.781	125.213	-102.707
Serik	2.710	2.709	16.549	31.000	840	16.969	61.090	-58.380
TOTAL	388.417	388.416	135.955	763.025	10.695	244.873	881.543	-493.126

Land Use Ability Classes

The land amount by land use ability classes of the towns is an important desertification and erosion factor. According to this classification, I-IV land classes are classified as agricultural usage. Class V lands can be used for agricultural purposes, provided some precautions are taken. The other land classes are not suitable for agricultural use. Some serious problems could be expected in terms of erosion and desertification, when we consider the land distribution of land use ability classes (Table 5-6) for Antalya Province. I-IV class lands are 20.33% of the total lands of Antalya. Class V is only 1, 74 % of the total. Serik has the biggest proportion (39.98%) on the distribution of I-IV classes. Other towns and their values include Korkuteli, 37.87%, Central Town, 33.70%, and Elmalı, 31.18%. The land of Gündoğmuş town that was in VI and VII classes was 97, or 42 %. That means that Gündoğmuş is vulnerable to the threat of erosion. The additional values-based on VI and VII classes are Gazipaşa, 93.26%, Akseki, 92.78%, and Alanya, 91.25%.

Table 5-6: The Distribution of Land Use Ability Classes by Town

Towns	Land Use Ability Classes (ha)							General
	I	II	III	IV	V	VI	VII	Total
Antalya	19811	30157	14590	4191	23828	12856	98596	204029
Akseki	2206	7004	2077	2344	0	9186	165925	188742
Alanya	1944	4217	2678	2267	0	20529	95239	126874
Elmalı	11223	18876	4204	5636	128	7143	80899	128109
Finike	1598	1169	2624	834	110	2544	51604	60483
G.paşa	127	3488	1201	1041	81	9573	72571	88082
G.doğmuş	91	1168	211	709	103	13618	72520	88420
Kaş	5930	4072	7227	6809	887	15056	152354	192335
Korkuteli	28560	19348	17734	12877	733	11883	116210	207345
Kumluca	2329	2624	2512	1593	331	12341	76541	98271
Manavgat	5485	13204	9567	12073	1122	36791	122842	201084
Serik	10608	16439	12357	6234	2128	8111	58287	114164
Total	89912	121766	76982	56608	29451	159631	1,00E+06	1697938

Table 5-7: Slope Group Distribution by Town

Towns	Flat	Slight	Middle	Steep	Very Steep	Rugged	Total
Antalya	73430	24083	8412	9844	42874	45386	204029
Akseki	2969	8102	3585	10445	11775	151866	188742
Alanya	3757	3084	7021	24120	6231	82697	126910
Elmalı	25575	9441	8202	5975	14171	64745	128109
Finike	4891	663	1135	2190	5337	46267	60483
G.paşa	1635	2660	2827	8389	7573	64998	88082
G.doğmuş	275	1682	2486	11735	14012	58230	88420
Kaş	10583	6572	8046	19496	20661	126977	192335
Korkuteli	36233	18990	30193	13242	70352	38355	207365
Kumluca	6818	863	1739	14911	10561	63379	98271
Manavgat	17324	11103	15951	34874	27049	94783	201084
Serik	29251	13128	7966	7124	12149	44546	114164
Total	212741	100371	97563	162345	242745	882229	1697994

Slope Groups

The slope of towns in Antalya Province is an important factor in an analysis of erosion. Any increase in the angle of slope of the land increases the surface flow amount.

We can group the land of the towns into categories such as flat, slight, medium, steep, very steep and rugged. Towns listed in order by flat slope group lands include Central Town, 35.99%, Serik, 25.62%, and Elmalı, 19.96%). Towns listed in order by slight slope group lands include Central Town, 11.80%, and Serik, 11.50%.

Steep and very steep land groups are very sensitive to erosion. Towns that are listed in an order by steep slope group lands include Alanya, 19.01 %, Manavgat, 17.34%, Kumluca, 15.17%. Towns listed in order by very steep slope group lands include Korkuteli, 33.93%, Central Town 21.01%, and Gündoğmuş 15.85%).

Table 5-8: The Distribution of Water Erosion Classes by Towns

TOWNS	Water Erosion				TOTAL
	Class I	Class II	Class III	Class IV	
Antalya	74321	19133	59444	51131	204029
Akseki	2969	10499	23408	151866	188742
Alanya	3455	5965	32726	84728	126874
Elmalı	25690	9988	28884	63547	128109
Finike	5001	1186	8029	46267	60483
Gazipaşa	1635	4500	56130	25817	88082
Gündoğmuş	275	1682	29569	56894	88420
Kaş	12090	11892	92794	75559	192335
Korkuteli	36952	31080	100978	38335	207345
Kumluca	6818	1913	22775	66765	98271
Manavgat	16311	17063	72927	94783	201084
Serik	29251	16264	24103	44546	114164
Total	214768	131165	551767	800238	1697938

Water Erosion

We can see III and IV water erosion classes are very common, when we evaluate the towns of Antalya by water erosion.

Highest rated towns of class I water erosion are Central Town, 36.43%), Serik, 25.62 % and Elmalı 20.05%). For towns class II water erosion are Korkuteli, 14.99%, Serik, 14.25%), Elmalı, 9.38 %. Highest rated towns of class III water erosion are Gazipaşa, 63.72%, Korkuteli, 48.70%), and Kaş, 48.25%. Highest rated towns of class IV water erosion are Akseki, 80.46%, Finike, 76.50%, Kumluca, 67.94%, and Alanya, 66.78%. The towns which have the biggest amount of water erosion are Korkuteli, 12.21%, Central Town, 12.02%), Manavgat, 11.84%), Kaş, 11.33% and Akseki, 11.12%).

Soil Depth

Soil depth is another factor in the erosion threat. There are two main classifications for soil depth distribution (Table 5-9).

Deep and middle deep classes pose erosion threat. Shallow and very shallow soil depth groups are areas that face a significant erosion threat. These groups of lands aren't suitable for agricultural usage because plants are not able to hold onto the soil. The towns which have the largest amount of land containing deeper soils are Serik 25.74%, Korkuteli, 22.27% and Elmalı, 22.18 %. The towns which have the largest amount of lands with the middle soil group are Central Town 9.10%, and Serik, (9.03%).

The most important land facing an erosion threat is shallow and very shallow soil depth groups. The towns which have the most land within the shallow soil group are Alanya, 18.49%, Gazipaşa, 16.17%, Gündoğmuş, 15.60 %, and Kumluca, 15.25%. The towns which have the largest amount of land within the very shallow soil group are Akseki (87.91%), Finike (85.35%), Gündoğmuş (82.39%) and Kaş (79. 11%). It is very meaningful that the proportions of very shallow soil depth lands are quite high. These types of lands are about two times larger than the deep soil group's land. On these shallow lands, there must be vegetation cover. Agriculture is the most important livelihood for rural areas. It is obvious that some of the agricultural methods used are very dangerous for these shallow lands.

The Senility and Aridity of the Soil

The senility and aridity of the soil is also another factor leading to erosion. Serik and Kaş are very important for the aridity factor, having the highest level of arid spaces (Table 5-10). Serik and Elmalı are very important for the senility factor, along with Manavgat and Central town as well.

Table 5-9: Soil Depth Distribution by Town

Towns	Deep	Middle	Shallo	Very	Total
Antalya	35022	18573	27822	122612	204029
Akseki	4931	8262	9624	165925	188742
Alanya	4213	3238	23454	95969	126874
Elmalı	28412	4225	13836	81636	128109
Finike	4956	622	3282	51623	60483
Gazipaşa	2225	2086	14247	69524	88082
Gündoğmu	755	1023	13793	72849	88420
Kaş	12153	7686	20338	152158	192335
Korkuteli	46172	10104	27744	123325	207345
Kumluca	7041	996	14982	75252	98271
Manavgat	18454	9586	20145	152899	201084
Serik	29384	10310	10940	63530	114164
Total	19371	76711	200207	1227302	169793

Importance Level of Factors Leading to Erosion and Desertification in Antalya Province

Erosion and desertification factors were determined for towns in Antalya. Each town has different erosion sensitivity for each criterion. An "Erosion and Desertification Combat Plan" should consider these kinds of differences because towns have different sensitivity to those factors (Coşgun, 2006).

This makes it very difficult to choose which factor to combat. The importance of each criterion or factor might be subjective. So, some of the factors should be combined in order to reflect the sensitivity level. The matrix below explains the major terms and concepts used.

Soil Depth Factor	The proportion of "very shallow" land to total lands of the towns
Slope Factor	The proportion of "very steep and rugged" land to total lands of the towns

Water Erosion Factor	The proportion of "Class IV water erosion" land to total lands of the towns
Senility and Aridity Factor	The proportion of "Class V-VIII senility and Class V-VIII arid" land to total lands of the towns
Misused Lands Factor	The proportion of "Class I-VI" land to total lands of the towns
Marginal Agricultural Lands Factor	The proportion of "dry marginal agricultural" land to total lands of the towns
Grazing Pressure Factor	Pasture grazing frequency of the towns
Forest Substance Factor	The proportion of "forest" land to total lands of the towns
Degenerated Forest Factor	The proportion of "degenerated forest" land to total lands of the towns

Table 5-10: Senility and Aridity of the Soil by Town

TOWNS	Senility		TOTAL	Aridity		Total	General Sum.
	I-IV Class	V-VIII Class		I-IV Class	V-VIII Class		
Antalya	7546	376	7922	1118	34	1152	9074
Akseki	594	307	901				901
Alanya	1109	69	1178				1178
Elmalı	14224	128	14352				14352
Finike	2764		2764		91	91	2855
G.paşa	603	197	800				800
G.doğmuş		487	487				487
Kaş	2399	238	2637	996	940	1936	4573
Korkuteli	5083	733	5816	482		482	6298
Kumluca	3496	331	3827	327		327	4154
Manavgat	8932		8932	1545		1545	10477
Serik	12589	323	12912	1824	1158	2982	15894
Total	59339	3189	62528	6292	2223	8515	71043

Table 5-11: Erosion and Desertification Sensitivity by Town

TOWNS	Soil Depth	Slope	Water Erosion	Senility and Aridity	Misused Lands	Marginal Agricultural Lands	Grazing Freq.	Forest Substance	Degenerated Forest	TOTAL
Antalya	189	759	796	26	378	65	326	527	798	3864
Akseki	277	1521	1364	3	81	93	6	661	941	4947
Alanya	238	1229	1360	6	98	151	84	552	693	4411
Elmalı	201	1080	1060	66	350	171	39	262	314	3543
Finike	269	1497	1319	28	116	69	9	603	725	4633
G.paşa	249	1445	1367	5	75	144	70	517	778	4650
G.doğmş	260	1433	1436	3	28	98	-75	379	846	4408
Kaş	249	1346	1286	14	140	128	84	643	959	4850
Korkuteli	187	920	987	18	425	184	264	344	1142	4471
Kumluca	241	1320	1338	25	104	117	13	522	537	4218
Manavgat	240	1063	1225	31	225	86	253	589	635	4346
Serik	175	871	883	82	449	73	143	470	492	3640

Prior Arrangement of the Towns by Factors Leading to Erosion

The main factors leading to erosion and desertification are determined in nine sections for Antalya. Twenty-six (26) erosion and desertification specialists graded these 9 factors.

The average of total points of each factor is accepted as the coefficient of that factor. The coefficients and the proportion values of each factor are multiplied and the result is accepted as the erosion and desertification values of the town. The erosion and desertification sensitivity is shown by town at Table 5-11.

Conclusion

Antalya is classified into four erosion and desertification sensitivity classes. These are:

- Towns that need urgent precaution and care;
- Towns that need short-term prccaution;
- Towns that need mid-term precaution;

- Towns that need long-term precaution.

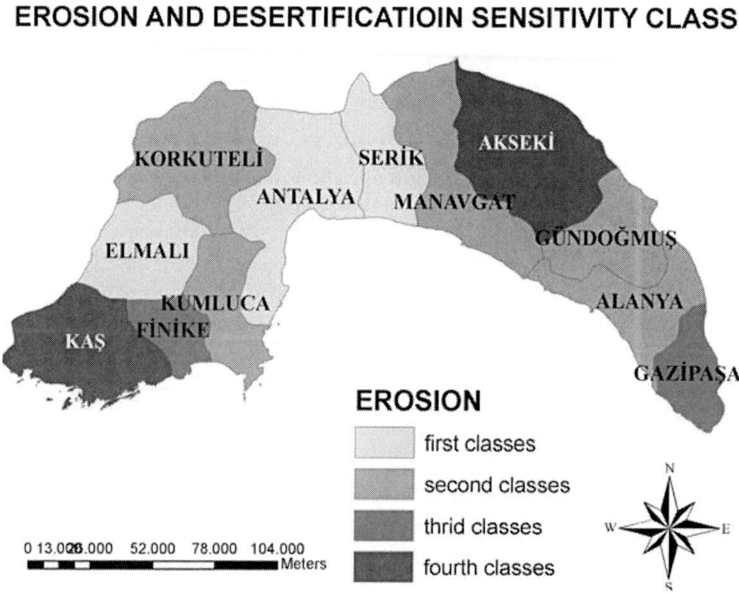

Fig.5-2: Erosion and Desertification Sensitivity Classes

Towns of in Class 1 are Kaş and Akseki; Towns in Class 2 are Finike and Gazipaşa; Towns in Class 3 are Alanya, Gündoğmuş, Korkuteli, Kumluca, and Manavgat; in Class 4, Central Town, Elmalı and Serik. Any future precautions should be considered under this classification in order to plan logical combat against desertification.

References

K.H.G.M., 1993. Antalya İli Arazi Varlığı, T.C. Tarım Orman ve Köyişleri Bakanlığı, Köy Hizmetleri Genel Müdürlüğü Yayınları. Ankara.

Aydemir, H. 1973. Yüzeysel Akışla Su Kaybı ve Toprak Erozyonunun ulusal Ekonomiye Etkisi, İ.Ü. Orman Fakültesi, Yay. No: 1643, O.F. Yay. No:167, Istanbul

Balbı, A. N. 1958. Elmalı Barajının Siltasyondan Korunması İmkanları ve Vejetasyon-Su Düzeni Münasebetleri Üzerine Araştırmalar, İ.Ü. Orman Fakültesi, Istanbul.

Coşgun, U., 2006. Antalya Çölleşme Eylem Planı, Tarım, Orman ve Mera Çalışma Grubu Raporu, Antalya.
Tarım İl Müdürlüğü, 2006. Antalya Tarım İl Müdürlüğü, İstatistik Şube Müdürlüğü, Antalya.
Yılmaz, E., 2005. Bir Arazi Kullanım Planlaması Modeli: Cehennemdere Vadisi Örneği. T.C. Çevre ve Orman Bakanlığı, Doğu Akdeniz Ormancılık Araştırma Müdürlüğü, Çevre ve Orman Bakanlığı Yayın No: 253, DOA Yayın No: 37, Çeşitli Yayın No: 3, Tarsus, 131 sayfa.
—. 2006. R'WOT Tekniği: Arıcılık Sektöründe Katılımcı Yaklaşım ile Örnek Bir Uygulaması. T.C. Çevre ve Orman Bakanlığı, Doğu Akdeniz Ormancılık Araştırma Müdürlüğü, Çevre ve Orman Bakanlığı Yayın No: 274, DOA Yayın No: 40, Çeşitli Yayın No: 6, 93 sayfa, Tarsus.

CHAPTER SIX

THE FORMATION OF ANTALYA TRAVERTINE DEPOSIT AND KARSTIC GROUND WATER SYSTEMS

RECEP EFE, IBRAHIM ATALAY, ABDULLAH SOYKAN, ISA CÜREBAL AND CEMALI SARI

Introduction

Karstic landforms that have formed from both the dissolution of limestone and the accumulation of calcium bicarbonate as the result of evaporation in the water have a special topographic form and habitat. Travertine deposits of calcium carbonate can dominate channel geomorphology in streams where travertine deposition creates a distinct morphology characterized by travertine terraces, steep waterfalls, and large pools. Algae and microorganisms can facilitate travertine deposition, but how travertine affects material and energy flow in stream ecosystems is less well understood (Marks *et al*, 2006).

Travertines are largely physico-chemical settlements formed by thermal and hydrothermal sourced spring waters and microbial agents are frequently found inside. These are usually characterized by frequently thin laminated and bush-like bacterial growth structures of solid crystal (Chafetz and Folk, 1984; Guo and Riding, 1998; Özkul *et al*, 2002).

The Pliocene and Early Quaternary travertines are exposed in the vast area around the Antalya region. Travertine is the major sub grade formation in this area. For geological and geotechnical purposes the Antalya travertine can be divided into three different groups as follows: spongy, weak, and massive travertine. Geomorphologically, the travertine is dissected into four major plateaus, namely the Dösemealti, the Varsak, the Düden, and the Arapsuyu (Kılıç and Yavuz, 1994).

Antalya travertine deposits have produced the flat land lying on the terrace levels. Ground water, rivers, and karstic springs supply drinking and irrigation waters. In addition to this, the rivers flowing on the upper terrace level produce hydroelectric energy (Kepez hydroelectric power plant). Most parts of the Antalya city settlement areas have been set up on travertine deposits.

The travertine formation in the study area is controlled mainly by the paleoclimatic fluctuations, and the periods of growth are connected with the improving climatic phases.

The introduction to and study of Antalya travertine/tufa formations have been the focus of interest by various researchers (Darkot and Erinç 1948; Alagöz 1943; Atalay 1987, 1996; Aydar and Dumond 1979; Güldalı 1970, 1981; Robertson and Woodcock 1982) In these studies, it was determined that chemical corrosion, which has caused karstic sedimentation in Taurus, was isochronal while Antalya travertines and the age of the travertines may be younger than upper Pliocene. Additionally, a relationship has been established between sea level changes and the formation of travertine plateaus. Through observation of vegetal evidence, Planhol (1956) has shown that these carbonates were formed by cold spring waters. İnan's (1985) study has been established over-formation mechanisms and explained travertine formations by four basic facieses distinction.

Study area methods and materials

The Antalya Travertine Plateau, bordered by the Aksu basin in the east, the southern feet of the Taurus mountains in the north, the Bey Mountains in the west, and the Gulf of Antalya in the south, has an area of approximately 630km^2 (Fig. 6-1). It is 21 kilometer from Aksu Stream in the east and the Bey Mountains in the west, and 30 kilometers from the shore in the South and Kırkgöz Springs in the north.

In this study to determine the properties of Antalya travertine deposits and karstic ground water, topography and geology maps and DEM (Digital Elevation Model) have been used. In addition to these, field studies were conducted in May and June of 2007. 1/25000 scaled topographical maps and 1/50000 scaled geological maps were used as cartographic materials.

In the first level of the study, data from already completed studies was collected about Antalya travertines, which are an interesting locality as far its geomorphology is concerned. A draft text was then composed by forming acquired notes. Next, visual material was prepared for the study. First of all, topographical maps of the area were scanned and transferred to the computer in order to supply a base for the study. These map sections were coordinated as a standard map projection, UTM (European Datum 1950-UTM Zone 36N). Also, maps of geological and drainage properties were scanned and placed over the bases which were formed with topographical maps.

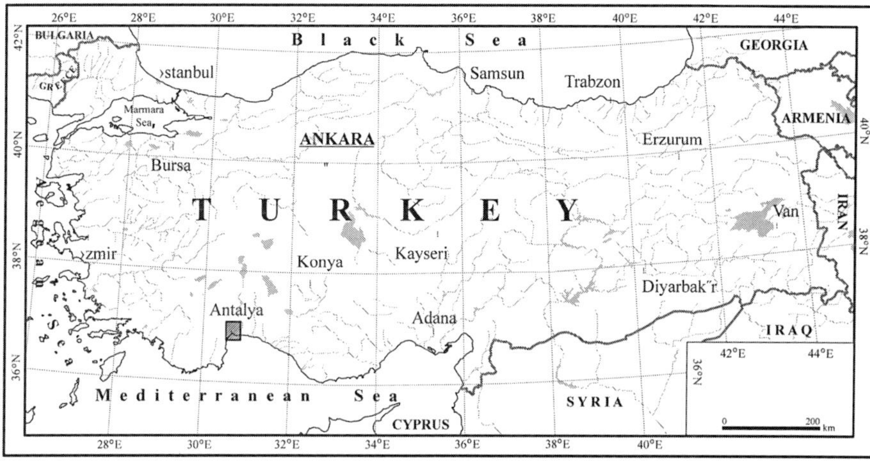

Fig. 6-1: Map showing the study area-Antalya Travertines

For overlapping and processing the scanned maps, ArcGIS Desktop v.9x software was used. For overlapping the data of the field, ground control points–GCPs were selected and Root Mean Square Error (RMS Error) was reduced to under 1 pixel.

Coordinated map sections were transformed to layers by the method of Screen Digitizing. Specifically, harmony between travertines and slope fatigues – travertine terraces were controlled.

To show the travertine terraces and levels, previously obtained data from digitizing the topographical maps was transformed to Digital Elevation Model (DEM) with intervals of 20 meters by ArcScene v9.2 software. After this transformation, block diagrams and sections generally expressing travertine deposits, terraces, and the terrain were formed.

In the last phase of the study, data from previous studies and properties obtained in this study and data obtained in field observations were brought together and analyzed, and a conclusion text was prepared.

Results and Discussion

Jurassic, Cretaceous, Paleocene, Miocene, Pliocene, and Quaternary rocks occur in the research area as autochthonous units. Jurassic–Cretaceous limestones cover a wide area on the north and west of the study area as a comprehensive series. Paleocene argillaceous limestone fits well on the comprehensive series. Miocene aged alterations of sandstone, marls, clay stones, and limestone, surfacing on the north of the study area indicates a malaise property (Poisson, 1977; Lombardo et al., 2003).

Pliocene aged conglomerates cover Miocene units concordantly. Still forming travertines are Plio - Quaternary aged and are wide spread in the study area. Debris in the slope of the hill and alluviums are quaternary aged (Fig. 6-2).

Formation Model of Travertines

Travertines and tufas are rocks which were formed by the precipitation of $CaCO_3$ in terrestrial environments starting from the spring waters which were enriched with calcium bicarbonate through the dissolution of the carbonated rocks by the effects of atmospheric factors and underground waters. Although travertine and tufa terms are generally being used together or with the same meaning, they actually express differences, especially in their formation conditions (Pedley 1990; Atalay 1996; Ford and Pedley 1996; Basar & Ekmekçi, 2000; Özgül & Arpat, 1978).

Carbon dioxide arriving at the internal waters of Jura–Cretaceous autochthonous limestone from the atmosphere or other sources through rain, forms carbonic acid and increases the melting effect of the water (1), this water melts and captures calcium carbonate and becomes intensive for calcium bicarbonate while passing through the limestone (2), this formation, which is the result of the secondary precipitation, forms the travertines. While these waters, which come out of the surface through flooding, flow over the older topography in various temperature and pressure conditions. They materialize the secondary calcium carbonate precipitation by releasing the internal carbon dioxide into the atmosphere (3).

(1) $H_2O + CO_2 \rightarrow H_2CO_3$
(2) $H_2CO_3 + CaCO_3 \rightarrow Ca(HCO_3)_2$
(3) $Ca(HCO_3)_2 \rightarrow CaCO_3 + H_2O + CO_2$

Antalya Travertines

Travertines, the subject of the study, have been formed by carbonates which have been precipitated by intensive karstic waters rich in bicarbonate, flowing through little inclined and partially permeable layers which are sourced from Kırkgöz, 30 kilometers north of the Antalya district and other springs. Travertines which are the product of terrestrial and very shallow swamp environments are divided into two terraces as a lower and upper plateau. The Plain of Döşemealtı (between 200-340 meters in elevation) is located in the upper plateau and the Plain of Düden and Varsak are located between 40 – 100 meters in the lower plateau (Fig. 6-3, 6-4, 6-5). A third plateau continuing below sea level ends with a 50 meter fault or cliff 2.5 kilometers north of the coast.

Antalya travertines have formed step-shaped terrace structures. The thickness of a travertine varies between 160–240 meters in the upper plateau and 60–120 meters in the lower plateau (Fig. 6-2). These two main terrace systems combine in the NE section where the total thickness of travertines decreases to 20-30 meters (Özüş, 1992). The largest known freshwater carbonate precipitation area in the world, the Antalya travertine plateau consists of 12 small and large plateau systems. Three (3) of these 12 plateaus are upper plateaus (Döşemealtı Plateau), Lower Plateau (Düden Plateau) and main plateaus under the sea and the other 9 plateaus are located on the upper and lower plateaus mainly on the Düden Plateau (Fig. 6-3, 6-4, 6-5) (Koşun et al., 2005; Denizman, 1989).

According to the fossils found in the drilling core samples from the Plain of Düden, the age of the travertines is Pliocene. Travertines which continue to form today show four different types as follows: thick textured, massive vegetative textured, spongy and oolitic, according to the changes in the shape of the former topography, temperature in the precipitation environment, depth, carbonate intensity, flora and fauna. Thick textured massive travertine: Porosity and permeability of this type, which is the product of slow precipitation, is very little. It is generally quite difficult to recognize them with their massive look and off-white color which is similar to limestone. Limestones, however, are formed of calcite crystal and are a result of marine or lacustrine environments and they consist of fossils from the environment.

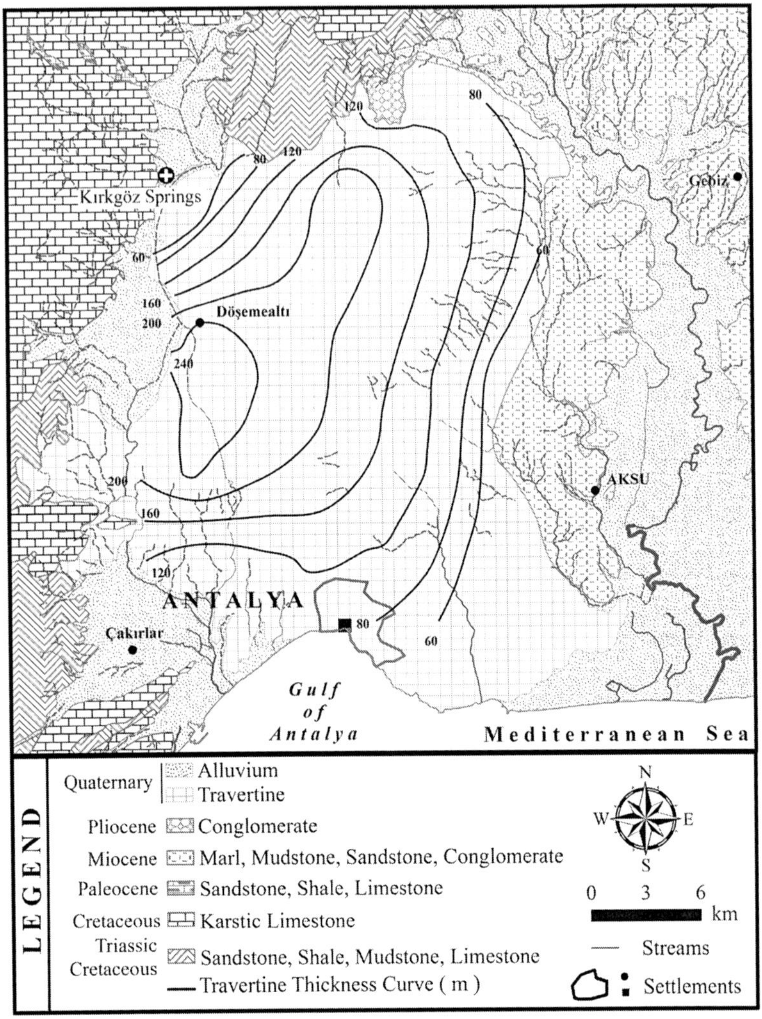

Fig. 6-2: Geological – Travertine Thickness Map of the Vicinity of Antalya

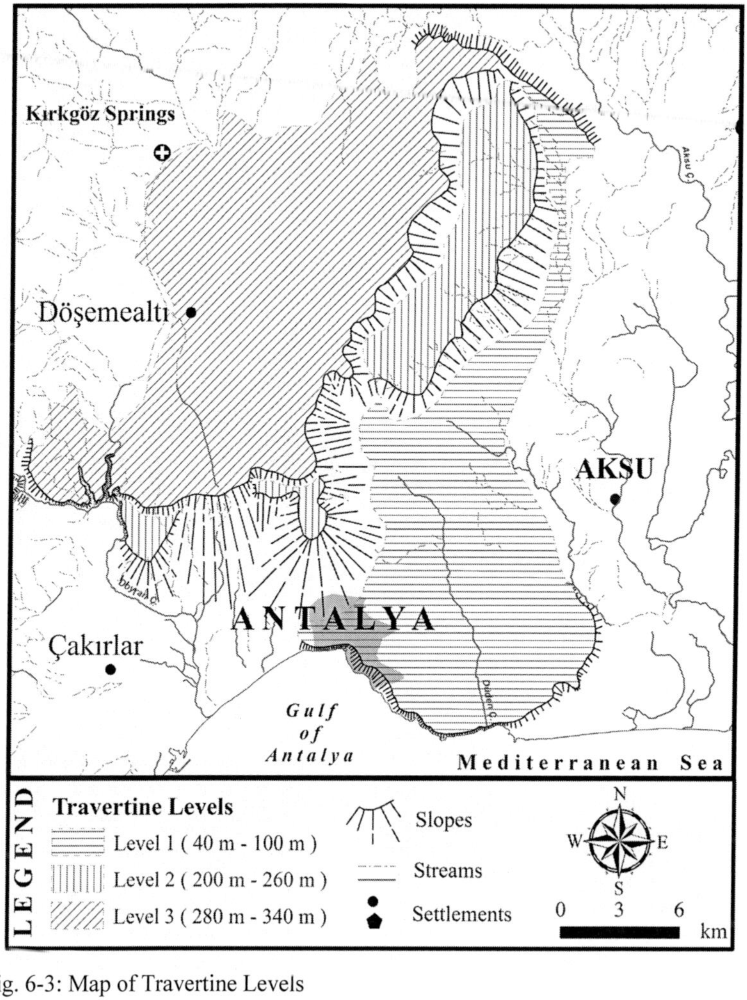

Fig. 6-3: Map of Travertine Levels

However, massive travertines are aragonite crystals and they very rarely contain plant fossils. Pores are formed when plant roots and stems in the environment decay through carbonate precipitation. Spongy travertines are the result of environments where water is deeper, massive warming is fast. Evaporation and carbon dioxide loss is high here. They look very porous and spongy. Because this structure causes high oxidation, they are brown in color.

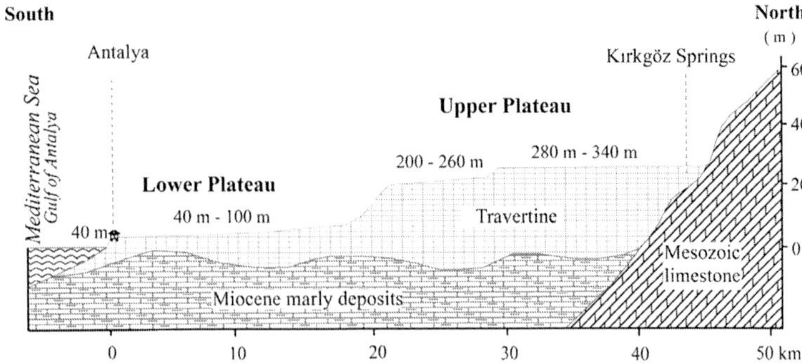

Fig. 6-4: Travertine terrace levels around Antalya

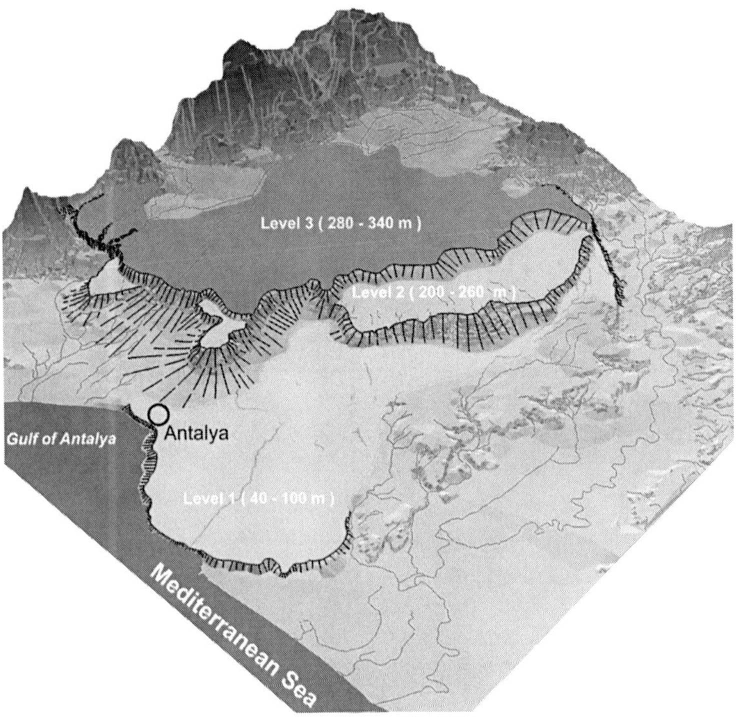

Fig. 6-5: Digital Elevation Model of travertine terrace levels.

Fig. 6-6: The contact line between upper part of the travertine plateau and the Mesozoic limestone.

There are lots of springs discharging from Mesozoic limestone and travertines in the Antalya travertine plateau. The most important of these are Kırkgöz Springs discharging from Mesozoic aged karstic limestone on the east of Mountain Katran 30 km north of Antalya (Fig. 6-2, 6-3). Important springs discharging from travertines are Düdenbaşı Spring, Kemerağzı Spring, Mağara Spring, Arapsuyu Spring and Duraliler–Okul Spring. Hurma Spring in the southwest of the area discharges from Antalya. Kırkgöz Spring Group, which discharges in a zone 1 kilometer NE – SW elongated at the contact of the Antalya Travertine Plateau and Mesozoic limestones, forms a lake in discharge point and some part of the water drained from this lake is allowed to flow to Ponor of Bıyıklı.

The Formation of Antalya Travertine Deposits

Geologically, Antalya travertine deposits lie on the Miocene sedimentary formation composed of alternating marl, sand stone, and soft limestone. This position indicates that the travertine deposits were formed after the Miocene. Indeed, during the Pliocene the uplifting of the Taurus Mountains continued upwards epirogenic movements, and the fluvial activities on the Taurus Mountains were accelerated under humid-mild climatic conditions. The deltaic and thick fluvial deposits spreading on the edges of the mountains and lowlands indicate the existence of severe fluvial erosion and deposition activities.

As to the formation of travertine deposits, the rich calcium bicarbonate water originating in the karstic lands of Taurus Mountains spreads through all of the lowland of the Antalya Gulf. The evaporation of water has contributed to the formation of travertine deposits. The existence of plant remnants within the travertine and the nodules covering the gravels and sands indicates the evaporation conditions and the formation of travertine deposits. Namely calcium carbonates encircling plant remnants as branches shoot, and nodules covering the gravel show the accumulation of calcium carbonates that have taken place on the shallow water.

Fig. 6-7: Travertine steps and the new settlements of Antalya

Terrace levels which are found at an elevation of 40-100 meters, 200-260 meters, and 280-340 meters imply three accumulation phases and/or erosional and sedimentational periods. The sequences in the travertine deposits alternate with sand, gravel, plant branches, and pure travertine layers show the existence of paleo-environmental conditions. The travertine and/or carbonates encircling the plant remnants and nodules indicate a shallow water condition. The travertine layers being thick and porous explain the stagnant condition in which the carbonates have accumulated in the shallow water. The sandy and gravelly layers show that torrential conditions in the northern part of Antalya Gulf have occurred.

On the other hand, thick fluvial deposits on the edge and depressions of the Taurus Mountains explain the intense erosional activities which occurred during the Pliocene period.

Fig. 6-8: Kırkgöz Springs at the northern edge of travertine deposits and southern limit of Taurus Mountains.

Karstic Features on/in Travertine Deposits

Travertine deposits contain some rich karstic landforms. This situation indicates the rekarstification process. In other words the travertine deposits have been subjected to karstification processes after they had already formed. One of the main karstic features is the formation of pan shaped dolines. Their diameter ranges from a few meters to more than 100 meters. Reddish Mediterranean soil has developed on which agricultural activities are being carried out. The steep walls of the dolines were dissected by lapiés, small holes, and/or cavities.

The Fluvial Process and Ground River-Water Systems

Karstification processes leading to the formation of karstic springs and ground river system are evident in both the Taurus Mountains and travertine deposits. The ground river system developed firstly at the end of the Mesozoic era, and continued up to the present time due to both the surface river systems and ground river systems shifting from the surface towards the deeper part as a result of an uplifting of the Taurus Mountains. Some of the springs having abundant water discharge are the main sources of the ground rivers. In other words, some karstic springs are fed by the ground river (Fig. 6-10).

Fig. 6-9: Düden waterfall formed at the knickpoint of Düden stream.

In the northern part of the travertine deposit fitting along the fault or flexure line between the Antalya travertine deposits and Taurus Mountains there are numerous karstic springs. They are termed "Kırkgöz springs" (forty springs) and they take place on the northern edge of the travertine deposits. The water of some lakes is drained by sinkholes and flows as a ground river before discharging on the lower terrace level. The small rivers fed by the karstic springs flow into the Mediterranean Sea in the vicinity of Antalya.

The river system can be divided into two as surface drainage and ground rivers. The surface drainage system is fed by the karstic springs which occur in the southern edge of the Taurus Mountains. The sinkholes produce the ground rivers. Surface rivers flowing within the Güvemuçurumu canyon valley cut their own valley through backward erosion. The waterfalls named Varsak, Kurşunlu, and Düden, which have been formed as the result of backward erosion, are found on the second terrace level. At the same time, these waterfalls fit the knick points along the stream profile. The enlargement of waterfalls can be related to the water turbulence as well. However, the following three physical effects can lead to travertine deposition at waterfall sites: aeration, jet-flow, and low-pressure effects (Zhang *et al.*, 2001). The elevation difference between the travertine springs and the main stream appears to be a result of Pleistocene glaciations during which karstic base-level lowering was either stopped or slowed down while down cutting in the main stream continued (Atalay 1987, 1996; Bayari, 2002).

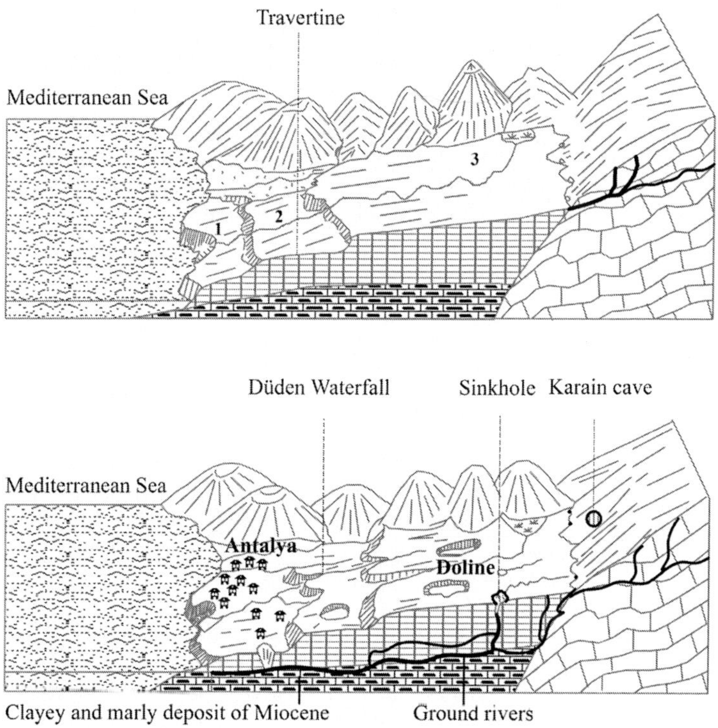

Fig. 6-10: Formation and evolution of Antalya Travertine

The ground rivers were formed through the shifting of surface rivers as a result of the intense karstification process and the lowering of base levels according to the Mediterranean Sea.

The karst window appearing in the vicinity of the village of Varsak shows the existence of the shifting processes of surface water. The abandoned river channels also explain the shifting of the surface water flow toward the deeper part of the travertine deposits.

The surface and ground rivers join on the first terrace level and flow into Mediterranean Sea forming waterfalls on the travertine cliff. The cliffs extending towards the northern part of the Mediterranean Sea have been formed as a result of wave erosion.

Conclusions

We can reach the following conclusions from this study:

1. Antalya travertine deposits are a product of the evaporation of rich calcium bicarbonate water originating from the karstic uplands of the Taurus Mountains.
2. Travertine deposits were mainly formed during the Pliocene and partly interglacial periods of Pleistocene.
3. The karstic springs existing on the travertine deposits lying between the southern part of the Taurus Mountains and the northern part of the travertine deposits supply both irrigation and drinking water for Antalya and its vicinity.
4. The water derived from the Taurus Mountains contributes both surface and ground river systems on the travertine deposits. Ground river systems have been developed as the result of intense karstification and a lowering of the Mediterranean Sea level. In other words, intense karstification processes and a lowering of the sea level are responsible for the ground river system and/or shifting surface draining to a deeper part.
5. Travertine deposits which have been subjected to karstification led to the formation of dolines on which citrus and Mediterranean vegetables are being cultivated.
6. The waterfalls have been formed as a result of the backward erosion of rivers and sea wave action on the coast.
7. The travertine deposits have also been deeply dissected or cut as deep as 40-50 meters by streams.

References

Alagöz, C. A. (1943) A research on Karstic landforms in Turkey. Turkish Geographical Society Publications, no. 1, Ankara.

Atalay, I. (1987) Introduction to Geomorphology of Turkey. Ege University Publication, Izmir.

—. (1996) Karstification and karstic landforms in Turkey. Karen Landforms (Eds. Fornos and Gines). Universitat de les Illes Balears, pp. 325-334.

—. (2003) Effects of the neotectonic movements on the karstification in Anatolia, Turkey. Acta Carsologica, 32: 195-204.

Aydar, C. and J.F. Dumont (1979) Observations on Landsat images on Antalya Travertines: Discussion the relationships between Neotectonics and hydrology. Journal of MRE, No. 92. p. 131-135.

Başal, A. and M. Ekmekçi (2000) Natural attenuation characteristics of the Antalya Travertine Plateau soil zone as a protective cover for the aquifer. Journal of Earthscience, 22 (2000), 95-107

Bayari, C.S. (2002) A rare landform: Yerköprü travertine bridges in the Taurids Karst Range, Turkey. Earth Surface Processes and Landforms Volume 27, Issue 6 , Pages 577-590

Chafetz, H. and R.L. Folk. (1984) Travertines: depositional morphology and the bacterially constructed constituents. Journal of Sedimentary Petrology, 54(1), 289-316.

Darkot, B. and S. Erinç, (1948) Antalya Travertine Terraces on the west of Aksu. University of Istanbul, Journal of Geography, vol 2, pp. 54-61.

Denizman, C. (1989) Kırkgöz Springs and Hydrologic research of Antalya Travertine plateau. University of Hacettepe, Inst. of Sciences. Master Thesis, 145 p. Ankara (unpublished, Turkish).

Ford, T.D. and H.M. Pedley (1996) A review of tufa and travertines deposits of the world. Earth Sciences Reviews, 41, 117-175.

Guo, L. and R. Riding (1998) Hot-spring travertine facies and sequences, Late Pleistocene, Rapolano Terme, Italy. Sedimentology, 45, 163-180.

Güldalı, N. (1970) Karstmorphologische Studien im Gebiet des polijesstem von Kestel (Westlicher Taurus, Türkei): Tübinger Geogr. Studien 40.

—. (1981) Lapie and doline types and their development around Korkuteli and Bucak. Journal of Geomorphology, no. 4, p. 81-98. Ankara.

Inan, N. (1985) The features and genesis of Antalya travertines. Geological Engineering. Ankara.

Kılıç, R. and S. Yavuz (1994) Relationship between geotechnical properties of the Antalya (Turkey) travertine. Bull. of IAEG, 50, pp. 43-50, Paris.

Koşun, E., A. Sarigül, B. Varol (2005) Lithofasies features of Antalya tufes. Journal of MTA, no. 130, pp. 57-70.

Lombardo, M., G. Calderoni, L. D'Alessandro and E. Miccadei (2003) The Travertine Deposits of the Upper Pescara Valley (Central Abruzzi, Italy): A Clue for the Reconstruction of the Late Quaternary

Palaeoenvironmental Evolution of the Area. Global Change and Protected Areas, vol. 9, Springer Netherlands

Marks, J.C., R. Parnell, C. Carter, E. Dinger and A. Haden (2006) Interactions between geomorphology and ecosystem processes in travertine streams: Implications for decommissioning a dam on Fossil Creek, Arizona. Geomorphology, Vol. 77, Issues 3-4. p. 299-307.

Özgül, N. and E. Arpat (1978) Structural units of the Taurus orogenic belt and their continuation in neighboring regions, selection of papers on the Eastern Mediterranean region. 28rd Congress of Ciesm in Athens, November 1972, Bulletin of the Geological Society of Greece, 10,1, 156-164,

Özkul, M., B. Varol, and M.C. Alçiçek (2002) Petrographic properties and depositional environments of Denizli Travertines. Journal of MTA, no. 125, p. 13-29.

Özüş, S. (1992) Geochemistry and hydrology of Antalya Travertines. University of Çukurova, Institute of Sciences. PhD Thesis, 179 p. Adana (unpublished, Turkish)

Pedley, H.M. (1990) Classification and environmental models of cool freshwater tufas. Sedimentary Geology, no. 68, pp.143-154.

Planhol, X.D. (1956) Contribution â l'etude geomorphologique de Taurus Occidental et de plaines bordieres. Revue de Geographie Alpine, no. 44, 609-685.

Poisson, A. (1977) Recherches Geologiques dans les Taurides Occidentales (Turquie). Orsay No. d'Ordre 1902, These presentee a 1'Universite de Paris-Sud (Centre d'Orsay) pour obtenir le grade de Docteur en Sciences.

Robertson, A.H.F. and N.H. Woodcock (1982) Wrench and Thrust tectonics along a Mesozoic-Cenozoic continental margin: Antalya complex, SW Turkey, Journal of the geological society, Vol. 139, part 2, pp. 147-163

Zhang, D.D., Y. Zhang, A. Zhu and X. Cheng (2001) Physical Mechanisms of River Waterfall Tufa (Travertine) Formation. Journal of Sedimentary Research; January 2001, volume 71, no. 1, p. 205-216.

CHAPTER SEVEN

TRENDS OF THE CLIMATOLOGICAL GROWING SEASON IN TURKEY

ECMEL ERLAT

Introduction

Climatological, phenological and satellite studies revealed that the number of frost events has decreased and the growing season has lengthened in many regions of the world during the last decades. In general, the extension of the growing season has been associated with recent global warming and with large-scale atmospheric oscillations such as the North Atlantic Oscillation (NAO) and Artic Oscillation (AO).). Observational analyses of air temperature data at global scales with respect to climate change confirm 0.76 °C rise since 1850. Warming since the 1990s has been particularly rapid with eleven of the twelve warmest years on record occurring since 1850 (IPCC, 2007).

The growing season may be determined from phenological observations, the normalized difference vegetation index (NDVI) from satellite data and climatologically from surface air temperatures. Differently from the phenological growing season which defined period of actual growth of plants, the climatological growing season could be defined as the entire period in which growth can theoretically take place (Carter, 1998). As pheonological and NDVI studies, many other studies have reported on the lengthening and shifts in timing of the climatological growing season in the twentieth century as being associated with increasing temperatures. For North America, Skaggs and Baker (1985) found a general increase in the growing season's length in Minnesota between 1899 and 1982, where the last freeze events showed a general trend towards earlier and last freeze events towards later occurrence. Cooter and Leduc (1995) found that there was a significant trend towards an earlier initiation of the growing season over a large portion of the northeastern USA for the period 1961-1990. Easterling *et al.* (2000) indicated that for the period 1910-1998, there was a slight decrease in the number of days below freezing over the entire USA, although there was regional variation in the trends. Robeson (2002) reported that the length of the growing season in Illinois increased by nearly one week over the last 100 years. Spring freezes occurred nearly one week earlier, whereas the date of first fall appeared unchanged over the last century. Feng and Hu (2004) showed that there was a significant decrease of annual frost days in the western USA associated with lengthening growing degree-days and unchanged annual frost days in the eastern USA matched with the small changes in growing degree-day. Schwartz et al. (2006) found that over the 1955-2002 period the onset of spring started 1.5 days earlier per decade across most temperate Northern Hemisphere land regions.

For Europe, Watkins (1991) found that the annual frost? Season had declined by about 2 days per decade in central England, due to the earlier advent of spring rather than a later end to autumn. As a result, the freeze season was an average of 28 days shorter in the 1980s compared to the 1850s. Heino et al. (1999) concluded that some decrease of frost days had taken place since the 1930s due to the very strong increase in minimum winter temperatures in northern and central Europe. Chmielewski and Rötzer (2002) showed that significant changes had occurred in the mean air temperatures from February to April and in the average beginning of the growing season in Europe since 1989. As a result, the average beginning of the growing season advanced by 8 days for the period 1969-1998. Scheifinger *et al.* (2003) found that most trend values of phenological and last frost events were negative for the period 1951 to 1997 in Central Europe. In particular, frost events based on the last occurrence dates of daily minimum temperatures below a certain threshold had been moving faster towards earlier occurrence dates, than phenological phases during the previous decades. Menzel *et al.* (2003) compared the changes of the climatological and phenological growing season in Germany for the period 1951-2000. They found that the growing season was lengthened by 0.11 to 0.49 days per year, depending on the criterion analyzed. Mean trends of phenological changes corresponded well to the climatological results in their seasonal occurrence.

Biological responses to climate change have also been reported in many papers for North America and Europe. Schwartz and Reiter (2000) found, with some regional differences, that spring season advanced 5-6 days earlier across North America according to the phenological events for the period 1959-1993. Their results were paralleled by progressing toward earlier dates for last frost events and trends towards longer growing seasons. Cayan *et al.* (2001) indicated that phenological indicators such as the series of lilac and honeysuckle first bloom dates and spring pulse dates tended to become earlier from 1957 to 1994 in the western USA. Wolfe *et al.* (2005) found that the first flowering date for the same clone of lilac had been monitored since the 1960s and showed an advance of 4 days for 72 sites across the northeast United States.

For Europe, it was found that phenological phases of plants in spring advanced by 6 days and autumn events delayed by 4.8 days over 30 years, depending on changes in air temperature. In contrast, spring was delayed in some of the Balkan stations as a result of regional patterns of climatic change (Menzel and Fabian 1999). Menzel (2000) revealed that spring events, such as leaf unfolding, advanced on average by 6.3 days and autumn events such as leaf coloring, delayed on average by 4.5 days. For the period 1959-1996, the average annual growing season lengthened on average by 10.8 days in Europe. Penuelas *et al.* (2002) reported that during the period 1952-2000, leaf unfolding advanced on average by 16 days, leaf fall delayed on average by 13 days and the flowering of plants advanced on average by 6 days in northeastern Spain. Phenological changes were correlated with warming in that region.

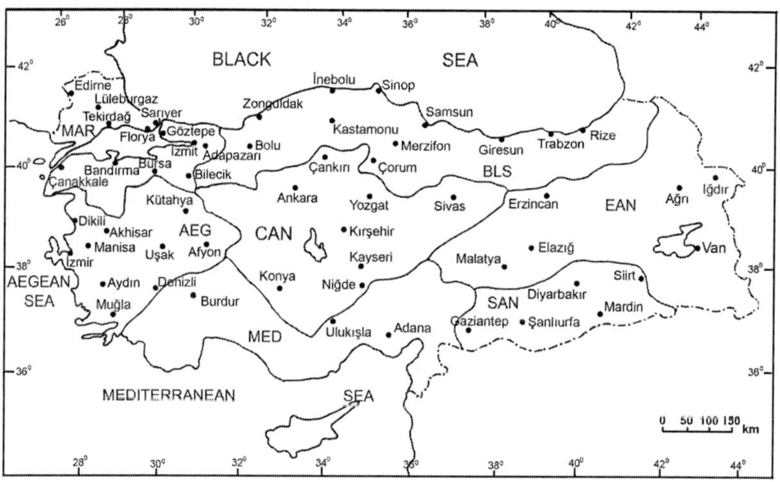

Fig. 7-1: Spatial distribution of the geographical regions of Turkey and location of 55 stations used in the study. BLS: Black Sea; MAR: Marmara; AEG: Aegean; MED: Mediterranean; SAN: South-Eastern Anatolia; CAN: Central Anatolia; EAN: Eastern Anatolia.

A number of studies were performed for variations and trends in surface air temperatures in Turkey. Türkeş *et al.* (2002) found slightly increasing trends for annual winter and spring mean temperatures, particularly over the southern regions of Turkey, and slightly decreasing trends for summer and in particular, autumn mean temperatures over the continental inner and northern regions for the period 1929 to 1999. Winter mean temperature series in Turkey are mostly random against any significant secular trend. Türkeş *et al.* (2002) also indicated that the daytime maximum temperatures show weak warming and cooling in comparison with significant warming of the nighttime minimum temperatures in many regions of Turkey, and in most seasons. Atalay (1976, 1983) used mean daily temperature above 8ºC to determine the vegetation period in Turkey. He found that vegetation period generally continued throughout the year in the Mediterranean climate. In central Anatolia, vegetation period was found to be between 180-160 days. The lowest number is over northeastern portions of the Anatolian Peninsula, which typically have 140 days.

Kadıoğlu and Aslan (2000) showed that there had been decreasing trends in the length of the growing season due to the earlier beginning of the autumn freeze season over the coastal areas of southern Turkey and eastern Anatolia. On the other hand, there was no apparent trend for the central Anatolian and the Black Sea regions of Turkey. Kadıoğlu and Şaylan (2001) showed that the accumulated growing-degree days above 5ºC have indicated a spatially coherent and statistically significant decreasing trend over the coastal areas of Turkey for summer and autumn for the period 1930s to 1990. However, growing-degree days were mostly random against any significant secular trend for the central and southeast Anatolia regions of Turkey for all months and seasons. Erlat (2005) found that annual frost day numbers had tended to decrease weakly at 38 out of a total of 55 stations for the period 1939 to 2003. Trends were statistically significant at 11 stations, 7 of which were at the 0.01 significance level. On the other hand, weak increasing trends in the annual frost day numbers were found at 17 stations over the inner parts of the Aegean region and northeastern part of central Anatolia. Results from examinations of the time-series plots showed that almost all years within the period 1960-1971 were characterized by less frost events. The decline in frost events had been intensifying particularly since 1976.

Consequently, the scope of the paper is determined as follows:

(a) To reveal the spatial distribution of the climatological growing season in Turkey;

(b) To determine the nature and magnitude of the serial dependence and long-term trends in the annual growing period by using non-parametric tests; and

(c) To detect the linear trend rates (in days per year) in the growing season series.

Study Area, Materials and Methods

In this study, the daily mean and minimum surface temperatures used were recorded at the 55 stations of the Turkish State Meteorological Service (TSMS). Stations were re-examined with respect to their length of record, geographical distribution over Turkey and missing values. The stations mostly having about a 67 year record length were selected for the present study. The spatial distribution of the 55 stations over the geographical regions is shown in Figure 7-1.

The climatological growing 'season' can be viewed as the entire period in which growth can theoretically take place, and should be distinguished from the growing 'period', which is the period of actual growth (Carter, 1998). GSL variations are a useful climatic indicator and have several important climatological applications (Robeson, 2002). In the literature, there exists a wide variety of growing season definitions. In most mid-latitude regions, temperatures are often used as the sole parameter to determine growing period length. Growing-season statistics based on a single-threshold air temperature can sometimes be misleading (Brinkman, 1979). Because of this, in this paper, we have used two different threshold temperatures (both *T*min and *T*mean) to determine the climatological growing season. First, we have selected the number of days between the date of the last spring and first autumn frosts to determine the climatological growing period. It means that for each station and for each year, the frost-free periods were identified (GSL *T*min < 0 °C). Secondly, the growing season is defined as the period between daily temperatures of >5°C for >5 days and daily temperatures of <5°C for >5 days (GSL *T*mean > 5 °C). By using these definitions, two different climatological growing season lengths were determined for each year.

The non-parametric *Mann–Kendall (M-K) rank correlation test* (WMO, 1966) was used to detect any possible trend in growing season length, and to test whether such trends are statistically significant. Before applying the test, original observations of x_i are replaced by their corresponding ranks k_i, such that each term is assigned a number ranging from 1 to N, reflecting its magnitude relative to magnitudes of all other terms. Then the P statistic is computed. P statistic is given by,

$$P = \sum_{i=1}^{N-1} n_i \tag{1}$$

M-K rank correlation statistic τ is derived from N and P by the following equation:

$$\tau = \frac{4P}{N(N-1)} - 1 \tag{2}$$

Distribution function of τ is the Gaussian normal for all N larger than about 10, with an expected value of zero and variance (τ_{var}) equal to

$$\tau_{\text{var}} = \frac{(4N+10)}{9N(n-1)} \tag{3}$$

and the significance test $(\tau)_t$ is then written as,

$$(\tau)_t = 0 \mp t_g \sqrt{\tau_{\text{var}}} \tag{4}$$

where, t_g is the desired probability point of the normal distribution with a two-sided test, which is equal to 1.960 and 2.58 for the 5 and 1 % levels of significance, respectively. Using **a** two-sided test of the normal distribution, null hypothesis of absence of any trend in the series is rejected for the large values of $|(\tau)_t|$ for the desired level of significance.

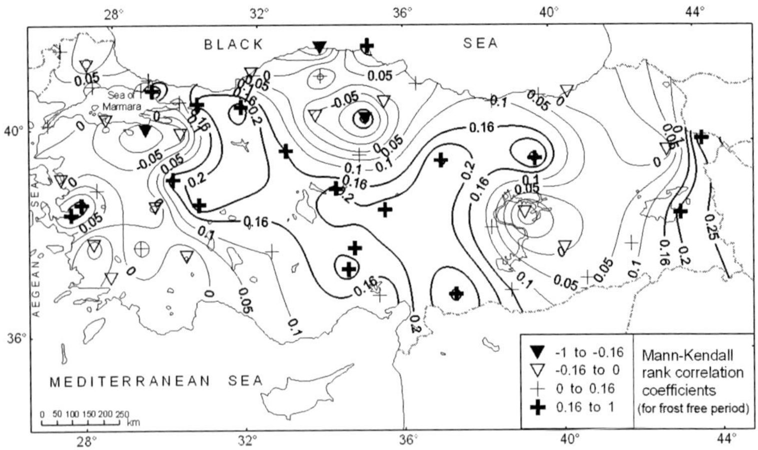

Fig. 7-2: Spatial distribution patterns of GSL for the threshold Tmin < 0ºC (a), Tmean > 5ºC (b) trends of 55 stations in Turkey according to the M-K test statistic

The least-squares linear regression equations were also calculated to detect linear trends in the growing period per year. In estimating linear regression lines, the simple least-squares approach was used, with time as the independent variable, and temperature values as the dependent variable. The statistical significance of each estimated X (β) coefficient was tested using the Student's *t* test for significance with (n-2) degrees of freedom (Türkeş *et al.*, 2002). In using the two-tailed test of the Student's *t* distribution, the null hypothesis for the absence of any trend is rejected for large values of $|t|$.

The Spatial Distribution of the Climatology of the Growing Season

The spatial distribution of the climatological growing season varies significantly from region to region within small distances, due to the change in continentality, distance to the sea, altitude and topography around Turkey.

Based on the frost free period (GSL *T*min < 0 °C), an average growing season length of more than 320 days is seen along the Mediterranean, Aegean and Black Sea coasts. Stations such as Anamur, Antalya and Alanya, which are located along the Mediterranean coast, have had a "never ending" growing season in many years. The average annual growing season length varies between 250-300 days around the coastal parts of the Marmara region, the interior part of the Aegean region and the west side of south-eastern Anatolia. Growing season length shows a sharp decrease towards the interior and eastern parts of the country, where the altitude is mostly over 1000 m. In central Anatolia, the average number of frost- free days is found to be between 170-200 days. The lowest number is over the eastern and northeastern portions of the Anatolian Peninsula, which typically have 150 days of frost-free conditions per year.

Average growing season length (GSL *T*mean >5 °C) varies between 184 days (Yozgat) and 346 days (Adana). As is expected, the distribution pattern of the average growing season length is very similar to the frost-free period. The Mediterranean coast of Turkey has the longest growing season with 350 days, followed by the Aegean and eastern Black Sea coasts with 320 days or more. The growing season shortens to180 days in the central and eastern parts of Turkey (Table 7-1).

The spatial distribution pattern of the inter-annual variability of growing season length was investigated by a coefficient of variation (CV). According to the threshold GSL *T*min < 0°C, distribution of CV over Turkey ranges from 8.1% at the Florya station in the Marmara region to

15.7% at the Kayseri station in the central Anatolia region (Table 7-1). Year-to-year variability of growing season length increases from the coastal and southern parts of the country to the eastern and interior regions of the country. Lowest year-to-year variability is evident at the stations of the southern and southwestern portions of the Anatolian Peninsula.

Trends and Changes in Growing Season

Results of this study show that the climatological growing season tended to lengthen significantly in Turkey over the 1939 to 2005 period. According to the Mann-Kendall test, GSL for the threshold *T*min < 0°C are characterized with mostly positive (lengthened) and insignificant negative (decreasing) trends over much of Turkey (Table 7-2 and Fig. 7-2).

Growing season lengths have significantly increased at 18 stations, 12 of which are at the 0.01 significance level (Table 7-2). On the other hand, results show significantly decreasing trends at 3 stations; only the Çorum stations are at the 0.01 significance level. Negative trends were observed mainly over the BLS and south of the MAR regions.

Similarly, there has been a general increase in GSL for the threshold *T*mean > 5°C. According to the Mann-Kendall test, statistically significant increasing trends in the GSL are found at 9 stations, 4 of which are at the 0.01 significance level over different parts of Turkey. Growing season length has tended to decrease insignificantly at four stations (Fig. 7-2).

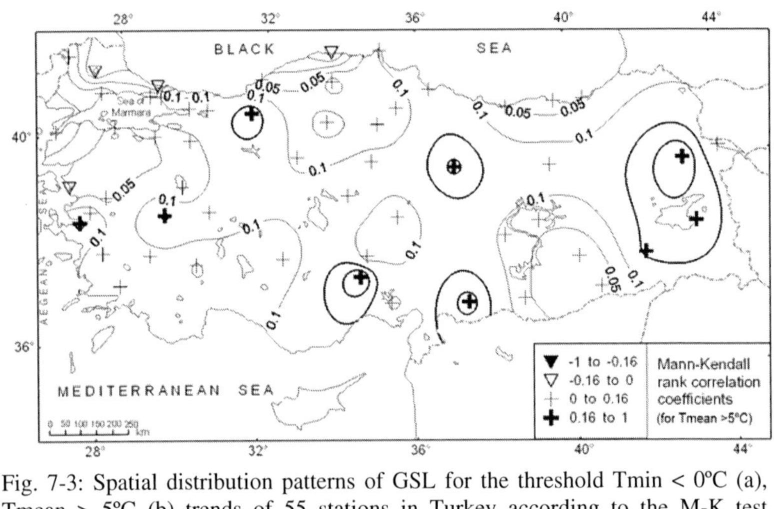

Fig. 7-3: Spatial distribution patterns of GSL for the threshold Tmin < 0ºC (a), Tmean > 5ºC (b) trends of 55 stations in Turkey according to the M-K test statistic τ.

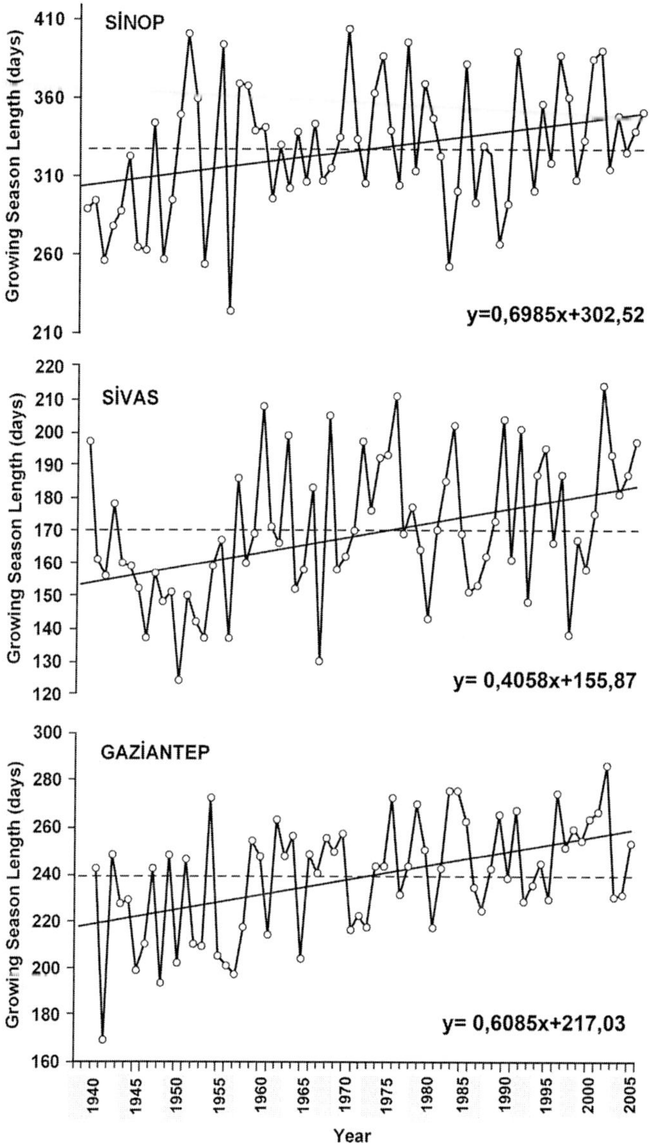

Fig. 7-4: Year-to-year variability in growing season length (GSL 1) (–•–), long term average (– –) and the least square linear fit to trend (**—**), for the selected 3 stations of Turkey. Trends and Changes in Growing Season

Trend Rates in Growing Season

The nature and magnitude of the least-squares linear regression lines coincide perfectly with the results of the M–K test at almost all stations. For growing season length, according to the non-freezing period (GSL *T*min < 0 °C), the series for Turkey has revealed a significant linear increasing trend at 21 stations, 14 of which are at the 0.01 significance level. Increasing trends are apparent in the CAN, MAR, AEG, and EAN regions of Turkey (Fig. 7-3). Cooling trends are significant at three stations (Çorum, Bursa and Elazığ).

The highest lengthened change, which is significant at the 0.01 level, is seen at Sinop station (BLS), with a trend rate of 0.70 days per year. Other highest significant linear rates per year for geographical regions are as follows: 0.59 days at Göztepe (MAR); 0.51 days at Izmir (AEG); 0.36 days at Ulukışla (MED); 0.61 days at Gaziantep (SAN) and 0.44 days at Kayseri (CAN). A trend rate of 0.38 days per year for the Iğdır and Erzincan stations (EAN) is significant at the 0.01 level.

Growing season length for the threshold *T*mean >5 °C is characterized by a linear increasing trend at 51 stations from 1939 to 2005. An increasing trend is significant at 16 stations, which are distributed over all regions of Turkey. Significant increasing trend rates per year vary between 0.63 (the strongest, Ulukışla) to 0.21 day per year. On the other hand, the GSL series of 4 stations is characterized by an insignificant decreasing trend.

Conclusions

Our study confirms that the increase in minimum and mean temperature, due to global warming and rapid urbanization in Turkey during recent decades, has lengthened the climatological growing season. The conclusions of the study are summarized as follows:

1. There is remarkable spatial variability in the climatological growing season length in Turkey. The spatial distribution of climatological growing season varies because of a change in distance to the sea, altitude and topography around Turkey. For growing season length according to the criteria of frost-free period, these values range from 330 days to 150 days per year.

2. Results showed an increasing trend in the GSL series for the two different thresholds over most regions of Turkey. For example, according to the Mann-Kendall test, GSL for the threshold *T*min < 0°C is characterized by positive trends at 37 stations. Similarly, growing season

length for the threshold *T*min < 0°C is demonstrated by a linear increasing trend at 40 stations. On the other hand, both trend tests have shown statistically insignificant negative trends at 15 and 18 stations which respectively take place over the south of the MAR and northwest of the MED regions.

3. Lengthening of the growing season depends on the definition used. For example, according to the Mann-Kendall trend results, the percentage of climate stations which showed significant positive trends of growing season length varies according to thresholds, ranging between 33 % for GSL *T*min < 0°C and 16 % for GSL *T*mean ≥5°C. Observed tendencies suggest that extended GSL is a more prominent feature for the threshold of a frost free period (*T*min < 0°C), due to an observed stronger increase in the daily minimum rather than mean temperatures. This was in perfect agreement with the findings of Türkeş *et al.* (2002), who found that the minimum temperature warming rates of spring and summer are generally stronger than those that exist in the spring and summer maximum and mean temperatures.

4. This present situation of increasing trends in GSL at many stations is considerably different from the results of a previous study that revealed decreasing trends in the growing season length (frost-free period) over the coastal areas of southern Turkey and eastern Anatolia, and there was no apparent trend for the central Anatolia and Black Sea regions of Turkey (Kadıoğlu and Aslan, 2000).

5. Our results are generally in agreement with the results of the studies achieved for the European countries (e.g. Chmielewski and Rötzer 2002; Menzel *et al.* 2003) and other regions of the world (e.g., Cooter and Leduc 1995; Robeson 2002; Folland *et al.*, 2001).

6. Our results have suggested that the growing season length has become longer in many regions of Turkey in this century. However, global mean surface temperatures are projected to increase by 1.8 to 4.0°C over the period 1990 to 2100. The length of the growing season is predicted to increase in response to increasing global temperatures (IPCC, 2007). Increasing GSL, together with a warmer growing season, is expected to cause some changes in crop types, water management and energy consumption around Turkey.

References

Atalay I. 1976. Distribution of vegetation periods of Turkey. Atatürk Üniv. Araştırma Dergisi 7: 247-279, Erzurum, Turkey. (In Turkish)

—. 1983. Vegetation Geography of Turkey. Ege Üniversitesi Basımevi, Izmir. (In Turkish)

Brinkmann W. A. R. 1979. Growing season length as an indicator of climate variations? Climatic Change Volume 2: 127-138.

Carter T. R. 1998. Changes in the thermal growing season in Nordic countries during the past century and prospects for the future. Agricultural and Food Science in Finland Volume 7: 161-179.

Cayan D. R., Kammerdiener S. A., Dettinger M. D., Caprio J. M., Peterson D. H. 2001. Changes in the onset of spring in the Western United States. Bulletin of the American Meteorological Society 82: 399-415.

Chmielewski F.M. and T. Rötzer 2002. Annual and spatial variability of the beginning of the growing season in Europe in relation to air temperature changes. Climate Research Volume19: 257–264.

Cooter E. J. and S. K. Leduc. 1995. Recent frost date trends in the north-eastern USA. International Journal of Climatology 15: 65-75.

Easterling D.R., Evans J.L., Groisman P.Y.A., Karl T.R., Kunkel K.E., Ambenje P. 2000. Observed variability and trends in extreme climate events: a brief review. Bulletin of the American Meteorological Society 81: 417–425.

Erlat E. 2006. Türkiye'de Don Olaylı Günlerde Gözlenen Eğilim ve Değişiklikler. Türk Coğrafya Kurumu Ulusal Coğrafya Kongresi–2005, İstanbul, Bildiri Kitabı s.201-208 (In Turkish).

Feng S. and Q. Hu. 2004. Changes in agro-meteorological indicators in the contiguous United States: 1951-2000. Theoretical and Applied Climatology 78: 247-264.

Heino R., Brazdil R., Forland E., Tuomenvirta H., Alexandersson H., Beniston M., Pfister C., Rebetez H., Rosenhagen G., Rösner S., Wibig J. 1999. Progress in the study of climatic extremes in Northern and Central Europe. Climatic Change 42: 151-181.

IPCC. 2007. Climate Change 2007: The Physical Science Basis, Fourth Assessment Report, Working Group I of the IPCC, Paris, February 2007.

Kadıoğlu M. and Z. Aslan. 2000. Recent trends of growing season length in Turkey" 2nd International Symposium on New Technologies for Environmental Monitoring and Agro-Applications Proceedings, 18-20 October 2000, Tekirdağ, pp. 295-303.

Kadıoğlu M. and L. Şaylan. 2001. Trends of growing degree-days in Turkey. Water, Air and Soil Pollution 126: 83-96.

Menzel A. and P. Fabian. 1999. Growing season extended in Europe. Nature 397: 659.

Menzel A. 2000. Trends in phenological phases in Europe between 1951 and 1996. International Journal of Biometeorology 44: 76-81.

Menzel A., Jakobi G., Ahas R., Scheifinger H., Estrella N. 2003. Variations of the climatological growing season (1951-2000) in Germany compared with other countries. International Journal of Climatology 23: 793-812.

Penuelas J., Filella I., Comas P. 2002. Changed plant and animal life cycles from 1952 to 2000 in Mediterranean region. Global Change Biology 8: 531-544.

Robeson S. M. 2002. Increasing growing-season length in Illinois during the 20 th century. Climatic Change 52: 219-238

Scheifinger H., Menzel A., Koch E., Peter C. 2003. Trends of spring time frost events and phenological dates in Central Europe. Theoretical and Applied Climatology 74: 41-51.

Schwartz M.D. and B.E. Reiter. 2000. Changes in North American spring. International Journal of Climatology 20: 929-932.

Schwartz M.D., Ahas R., Aasa A. 2006. Onset of spring starting earlier across the Northern hemisphere. Global Change of Biology 12: 343–351

Skaggs R.H. and D.G. Baker. 1985. Fluctuations in the length of the growing season in Minnesota. Climatic Change 7: 403-414.

Türkeş M, Sümer UM, Demir İ. 2002. Re-evaluation of trends and changes in mean, maximum and minimum temperatures of Turkey for the period 1929-1999. International Journal of Climatology 22: 947-977.

Watkins C. 1991. The annual period of freezing temperatures in Central England. International Journal of Climatology 11: 889-896.

WMO. 1966. Climatic Change. World Meteorological Organization (WMO), Technical Note No. 79, Geneva, 79 pp.

Wolfe D.W., Schwartz M.D., Lakso A., Otsuki Y., Pool R., Shaulis N. 2005. Climate change and shifts in spring phenology of three horticultural woody perennials in northeastern USA. International Journal of Biometeorology 49: 303–309.

CHAPTER EIGHT

TYPES OF MICRO MUD VOLCANOES IN TRANSYLVANIA, ROMANIA

ANDREA ADEL GAL, MELANIA LAKATOS

Introduction

Mud volcanoes are one of the world's most dynamic phenomena. They are distributed worldwide in various tectonic settings. Mud volcanoes can be defined as geological structures with a variety of morphological expressions, constructed mainly of mud that originates deep in the sedimentary succession. The occurrence of mud volcanoes is related to recent tectonic activity (particularly compressional); sedimentary or tectonic loading; hydrocarbon generation and the existence of thick, fine-grained, soft, plastic sediments deep in the sedimentary succession (Dimitrov, 2002; Kholodov 2002a, b).

The main driving force of mud volcanism is believed to be the abnormally high pore-fluid pressure and secondly the buoyancy pressure (Dimitrov, 2002). The abnormally high pore fluid pressure can be primarily a result of very high sedimentation rates and the fast burial of these sediments. As a result of the fast burial, sediments retain a big amount of water and organic material and they remain undercompacted. Under the weight of the overburden layers they become overpressured compared with the surrounding sediments. This state is accentuated by the decomposition of organic material and gas generation.

According to Dimitrov (2002) "the high water and gas content decreases the bulk density, shear modules and viscosity, making such sediments semi-liquid and able to flow. Such a sequence of low density, under compacted, over pressured mud or shale overlain by thick denser material, is mechanically unstable and frequently finds expression in mud or shale diapirs or in mud volcanoes." Pore fluid pressure can be intensified by compressional tectonic forces (folding, thrusting, over thrusting) as well (Kholodov 2002a; Dimitrov, 2002).

Hovland and Judd (1988) suggested a model for mud diapirism related to the formation of mud volcanoes. They assumed that the presence of diapirism facilitates the formation of mud volcanoes, as the migration of salt provides pathways for hydrocarbon migration, which is trapped inside the mud above the salt. As the gas rises and expands, it increases the pore pressure and contributes to the upward mobilization of the mud. If the salt is still mobile it can result in tensional faults that serve as pathways for the over pressured mud.

The Transylvanian Basin is rich in gaseous hydrocarbon (methane) and salt so it is not unexpected that it hosts many small size mud volcanoes spread all over the depression.

In comparison with the Azerbaijani mud volcanoes which are several hundreds of meters high and more then 1 km large, some of them having very large craters (300-350 m in diameter), the mud volcanoes of Transylvania (and usually in Romania) are miniature features having a maximum of 6 meters in height and 20 meters diameter. Despite these differences in size, there are some similarities in structure, shape and evolution.

The study and description of these small size mud volcanoes started in the 19th century but no detailed investigation had ever been done. The mud volcanoes of the Târnave Basin (southern part of Transylvanian Basin) had only been studied by Bányai in 1932. In his work, the preliminary morphological classification of these small size mud volcanoes and mud volcanic manifestation was performed. On the basis of the observations of the mud volcanoes from the Târnave Basin, Bányai distinguished four types of mud volcanoes (Fig. 8- 1, 8-2, 8-3 and 8-4). Whereas he used a very specific nomenclature, here we will use terms parallel with those used and defined by Kopf (2002): "bottomless mud pit" (similar to Kopf's mud pool), "maybug" (mud pot), "stack" (gryphon) and "white mud mounds" (high salinity mud mounds).

The genesis of the "bottomless mud pits" – mud pools – is related to the preexistence of a negative form where mud, water and gas infiltrate owning to the pressure of the upward moving gas (Fig. 8-1).

The "maybug" shaped mud pot is the result of the deposition of mud on a slope similarly to the travertine deposition. The discharged mud is deposited in an elliptic form, heightening its cone in the shape of a maybug (Fig. 8-2).

The "stack"-type gryphons resulted from the slow accumulation of liquid mud on a horizontal surface. The quantity of the discharged mud is very low, so it is unloaded at the vent, heightening very slowly the cone in a shape of a stack (Fig. 8-3).

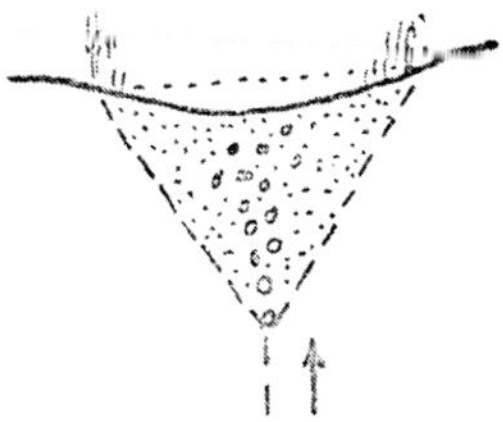

Fig. 8-1: "Bottomless mud pits" (mud pools)

Fig. 8-2: "Maybug" (mud pot)

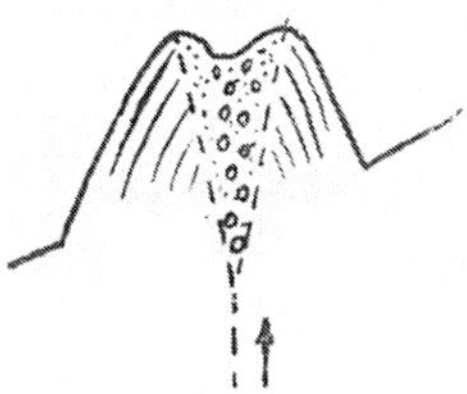

Fig. 8-3: "Stack" (gryphon)

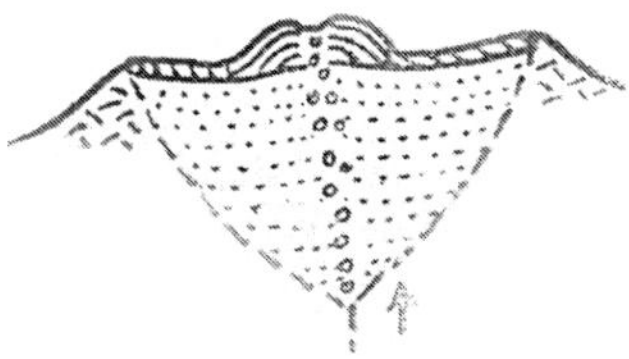

Fig. 8-4: "White mud mounds" (high salinity mud mound)

"White mud mounds" got their name because of their light color that is probably due to the presence of the salt (Fig. 8-4). These mud mounds arose through small eruptions launched by the methane as its way upward was blocked by crust which resulted from the drying of the mud. They have a crater filled with mud that mostly dries out and forms a crust where weak fluid discharge can be observed.

Study Area

The mud volcanoes in the present study are located in the Târnave Basin, in the southeastern part of the Transylvania Basin, in the central part of Romania (Fig. 8-5). These mud volcanoes are spread all over in several villages (Cobăteşti; Filiaş; Porumbenii Mici; Băile Szejke; Sângeorgiu de Pădure; Corund; Crăciunel; and Sânpaul) in the neighborhood of Odorheiu-Secuiesc.

The Transylvania Basin represents a post-Cenomanian sedimentary basin, locally comprising more than 5 kilometers of sediments, having a basement involving crystalline units, ophiolits and volcanits (Krézsek, 2006).

The sedimentary fill was complicated by the migration of salt and lateral tensions related to the uplift of the Carpathians. These processes led to the formation of domes that served as traps for the accumulation of methane resulting from the quickly deposited organic matter (Vancea, 1960; Paraschiv, 1975).

Objectives

The main purpose of this study is to monitor the different types of mud volcanoes, checking the classification proposed by Bányai, and examining the external and interior shape and structure with help of manual, shallow drillings.

Regarding the structure of these mud volcanoes, we proposed to identify mud chambers (in case they exist and are close to the surface), to define the characteristics of the upper part of the feeder channel. Finally, we would like to compare the sketches made by Bányai (1932) with the new sketches based on the information from the drillings.

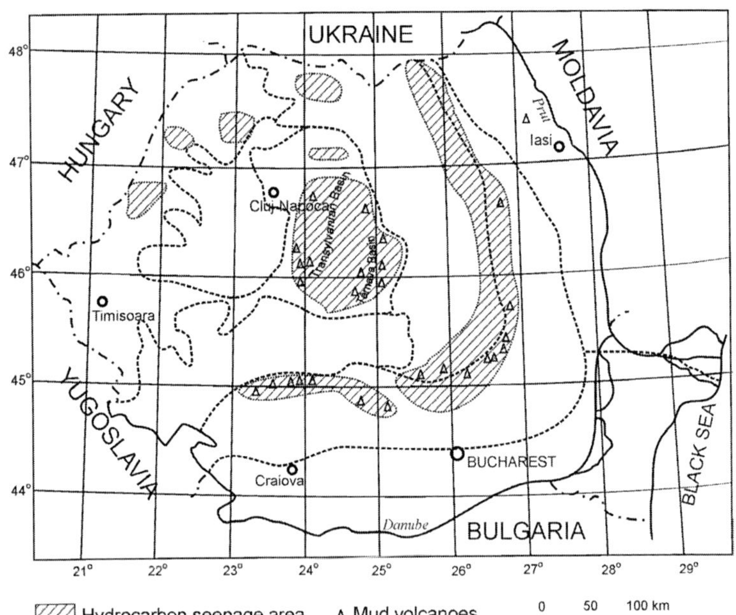

Fig. 8-5: Distribution of Mud Volcanoes and Hydrocarbon Seeps in Romania (after Paraschiv, 1984).

Results

After monitoring all of the mud volcanoes studied by Bányai (those that are still to be found) completed with the newly found ones, we observed that in spite of the various morphologies the mud volcanoes have some features in common. Based on the information from observations and drillings, we concluded that the classification of Bányai should be refined.

Taking into account the shape of the mud volcano, the subsurface morphology (as it could be concluded from the drilling information), the characteristics of the feeding channel and the presence or lack of near-surface mud chambers, we propose to separate three morphological types of mud volcanoes in the Târnave Basin:

1. Mud pools;
2. Gryphons;
3. Mud lumps (small mud volcanoes).

The terms were borrowed from Kopf's denomination of the various mud volcanic manifestations.

Mud pools. Mud pools are the most widespread in the Târnave Basin as well as in the Transylvanian Basin. This cate- gory includes all the negative forms that show very weak fluid mud and methane discharge.

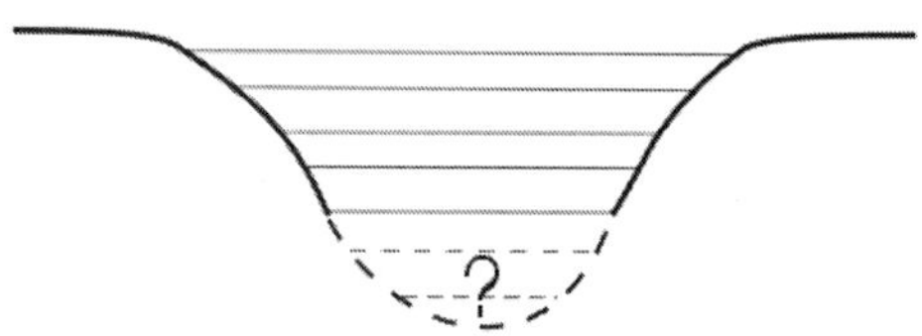

Fig. 8-6: Mud pools

The very liquid material (mud) that ascends through the feeding channel contributes to the liquefaction of the surrounding rocks and it results in a mud pool, surrounded by a swampy area. The vent of the feeding channel is, in some cases, hardly observable; its place varies as much as the feeding channel, looking for the easiest way to get to the surface.

The muddy water bubbles gently, the gas can be lit and it burns with a blue flame, though the origin of these gases hasn't been determined yet (if it is thermogenetic methane from depth or is it just the result of the superficial decomposition of the organic material).

The size of these features can vary between 1 meter to 15 meters, but the depth remained undefined because of the deficiency of the drilling set (the maximal length of the drill was 8.5 meters), though at about 8 meters depth we found the same homogenous, fluid mud, full of organic matter.

Mud pools are almost always present where there are other manifestations of mud volcanism: Cobăteşti, Filiaş, Porumbenii Mici, Sângeorgu de Pădure, etc.

Gryphons. The term gryphon has been defined by Kopf (2002) as a small cone or mud crater off the crest or at the flank of the mud volcano. In this study, the term gryphon is used for those simple positive forms that resulted from the deposition of muddy material on the surface through a small vent without taking into account if the surface is plane or aslope, or if they appeared in subaerial or aquatic conditions. The bottom of these microforms is represented by the surface where they occurred. The upper part of the feeding channel is represented by very narrow openings or fissures.

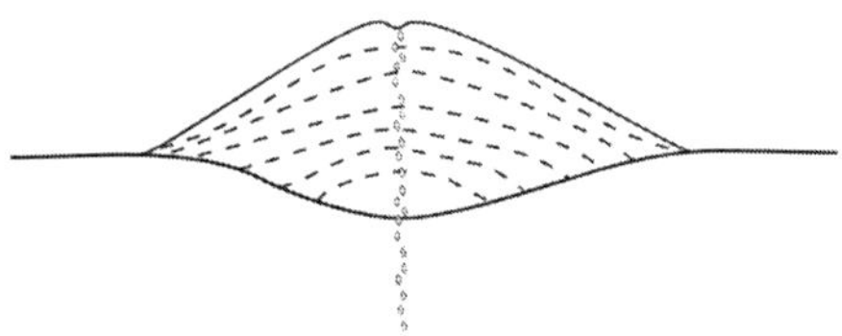

Fig. 8-7: Gryphon

The cone is heightened slowly by the deposition of mud at the vent. Depending on the viscosity of the mud, the cone becomes higher and steeper or smaller and flatter. The bubbling of the gas (probably methane) is evident only if the cone has a small crater in which the expelled mud accumulates. In this category can be included the gryphons from the following villages: Cobăteşti; Filiaş; Porumbenii Mici; B-le Szejke; Corund, etc.

Mud lumps. Mud lumps represent small mud volcanoes, according to Kopf (2002), though they are the biggest from the Târnave Basin. These have a more complicated structure than the first two types. Some of the mud lumps from the Târnave Basin were initiated by small explosions.

As a result of the drillings some mud chamber-like formations have been identified. The mud from the chamber is under pressure so that, when drilled, the mud and water (and probably gas too) escaped through the drilled vent.

It is still unknown whether the liquefaction of the material takes place at depth or in the near-surface mud chamber. The material in the "crater" is overlain by a solid or semisolid crust, while at the sides it is surrounded by a gleic clayey layer. This type has the fewest representatives in the

Târnave Basin: two (2) in Filiaş, two (2) in Porumbeni Mici and only one (1) in Cobăteşti.

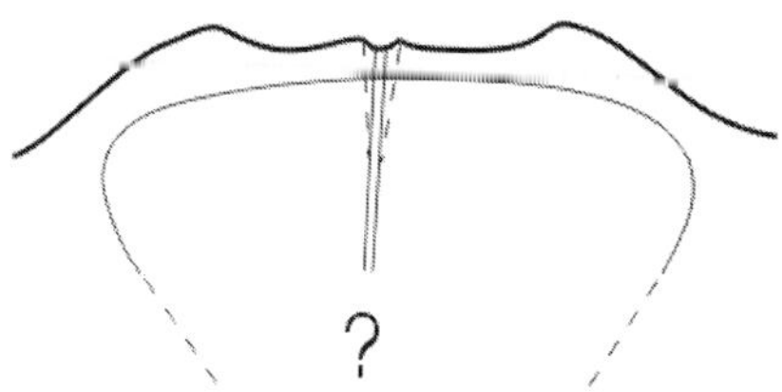

Fig. 8-8: Mud lump

Conclusion and Discussion

The mud volcanoes of the Târnave Basin are usually localized either on the gas domes or on anticlines. Due to the fact that the structure of the Transylvanian Depression is determined by the salt diapirism, we believe that the driving force behind the mud volcano formation is related to tensions produced by the migration of salt and consequently the migration of methane on the tensional faults produced by diapirism.

In the Târnave Basin, as well as in the whole Transylvanian Basin, the most characteristic are the small size mud pools, usually integrated in swampy areas. There are some gryphons and mud lumps that reach 3-4 m in height, and more then 20 meters in diameter.

As a result of drillings, we have come across some mud chamber-like formations, close to the surface. After the first drilling was performed, a big quantity of liquid mud escaped to the surface, so we think that this material was accumulated in a reservoir/mud chamber.

Using the drilling information we modified the structural sketches made by Bányai (1932) and refined his classification. We propose that in the Târnave Basin there are only three types of mud volcanoes (mud pools, gryphons and mud lumps) instead of four, as was defined by Bányai (1932). Further examinations need to be done to finalize this classification.

The bottom of the gryphon's edifice is a function of the preexistent topographic surface. Unfortunately, the bottom of the mud lumps's edifices and mud pools couldn't be defined, but it surely goes deeper than 8 m.

The gleyic character of the material from the flanks of the mud lumps is related to intermittent activity: in active periods the reduction of the mineral compounds takes place while in dormant periods oxidation is characteristic.

We couldn't draw conclusions from the results of the grain size analyses, as all of the analyzed samples from different types of mud volcanoes were included in the clay size limits.

A future plan is to find the relationship between the shape of the mud volcanoes and the material it consists of and to define the reason why within a small area (a few meters, for example, 4-5 meters) there can appear positive and negative forms as well. The study of these small size mud volcanoes can be very important in our understanding the dynamics and characteristics of other, more complex mud volcanoes.

References

Bányai J. 1932. "The Mud Volcanoes of Udvarhely County (in Hungarian)," Erdélyi Múzeum. Volume 17, Number 1, pp.1-17.

Dimitrov, L. I. 2002. "Mud Volcanoes—The Most Important Pathway for Degassing Deeply Buried Sediments." Earth Science Reviews, Volume 59 (1), pp. 49-76.

Hovland, M. and A. G. Judd. 1988. Seabed Pockmarks and Seepages. Impact on Geology, Biology and the Marine Environment. London: Graham and Trotman. 293 pp.

Ioanovici, V. and D. Rădulescu. 1968. "Geologic Map of Rumania," Foaia Odorhei, 1: 200.000. Bucureşti: Inst. Geol.

Kholodov, V. N. 2002(a). "Mud Volcanoes, Their Distribution Regularities and Genesis: (Communication 1. Mud Volcanic Provinces and Morphology of Mud Volcanoes)." Lithology and Mineral Resources, Volume 37, Number 3, pp. 197-209.

—. 2002(b). "Mud Volcanoes: Distribution Regularities and Genesis (Communication 2. Geological – Gochemical Pecularities and Formation model). Lithology and Mineral Resources, Volume 37, Number 4, pp. 293-310.

Kopf, A. J. 2002. Significance of mud volcanism. Rev. of Geophys, 40, 2

Krézsek Cs. And A. W. Bally. 2006. The Transylvanian Basin (Romania) and its Relation to the Carpathian Fold and Thrust Belt: Insights in Gravitational Salt Tectonics." Marine and Petroleum Geology, Volume 23, Issue 4, pp. 405-442

Paraschiv, D. 1975. "Geology of Hydrocarbon in Rumania" (in Romanian). Ed. Stud. Tehn. Econ., Bucureşti

Vancea, A. 1960. "The Neogen of the Transylvanian Basin" (in Romanian). Ed. Acad. RPR, Bucureşti.

CHAPTER NINE

THREATENED CYPRESS FOREST ECOSYSTEMS IN THE MEDITERRANEAN REGION

CUMHUR GÜNGÖROĞLU AND RUMI SABUNCU

Introduction

By observing Turkey's ecological diversity and high endemism ratio of that diversity, we can see that Turkey has an unusual biological structure. This has occured because Turkey is situated where three main phytogeographical regions meet:

1. Euro-Siberian: around the Black Sea coast and surrounding mountains;
2. Irano-Turanian: in Central, East and Southeast Anatolia;
3. Mediterranean: the Mediterranean coast and mountains;

The Mediterranean Region is mainly situated in the Taurus Mountain Belt (except inland in the Teke Peninsula and Lakes Region). The region is generally divided into three sub regions (Atalay, 2002):

1. Thermo-Mediterranean zone, where typical Mediterranean climate occurs;
2. Meso and oro-Mediterranean zones (mountainous region), where mountain forests are dominant;
3. An alpine zone, where steppes are dominant;

A typical Mediterranean climate is characterized by mild and rainy winters, and warm to hot dry summers, with high solar radiation and high rates of evaporation. Average annual precipitation ranges between 600 and 1000 meters and the average annual temperature is between 18-20°C.

The Mediterranean Region is one of the most negatively affected regions due to human activity (Naveh & Dun, 1973). This region has a high biological diversity and it is one of the 10 hot spots in Turkey (Medail & Quézel, 1997).

The vegetation of the Turkish Mediterranean Region is characterized mainly by xerophilous plants, due to characteristic climate conditions. In thermo-Mediterranean zones, from sea level up to 600-800 meters, garrigue (*phrygana*) and maquis are dominant. In addition, evergreen forests, olive (*Olea europaea*), terebinth (*Pistacia terebinthus*) and carob (*Ceratonia siliqua*) plants are widespread (Boydak *et al.*, 2006). The primary forest tree species of this zone is red pine (*Pinus brutia* Ten.).

Characteristic trees of the mountainous Mediterranean region (oro-Mediterranean) are Taurus cedar (*Cedrus libani)*, Cilicican fir (*Abies cilicica*) and black pine (*Pinus nigra*). The Cedar and Fir generally grow on the south-west portions of high elevation slopes, which are effected by moist winds; however, black pine grows also in inland regions. The main distribution elevation of Taurus cedar is between 1300-2000 meters. *Abies cilicica, Juniperus foetidissima, Juniperus excelsa, Acer monspessulanum, Acer hyrcanum, Fraxinus ornus* and *Quercus libani* are the main species which grow in the natural distribution area of Taurus cedar (Kayacık & Yaltırık, 1971). In higher elevation zones, scattered Juniper woodlands grow. In alpine zones, (over 2000 meters) mountain pastures and steppe vegetation occur.

The Mediterranean Region has various geomorphological structures. Therefore, different plant formations are seen in deep valleys and canyons. The Mediterranean cypress, (*Cupressus sempervirens* L.) which is the primary species of this formation, is naturally distributed in thermo-Mediterranean and meso-Mediteranean zones such as in the Aydıncık-Babadıl Valley, Antalya-Köprülü Canyon, Kemer-Kuzdere Valley, Dilek Peninsula Canyon and Fethiye-Saklıkent districts. Cypress is one of the rare tree species which has been blessed because of its fruit, pith, leaves, bark, fiber, wood, odor, color, shadow, fire analogous crown structure and monumental appearance (Sabuncu, 2004).

Study Area

The study area, Köprülü Canyon National Park, is located in the western part of the Taurus Mountains, between the elevations of 110 meters (Beşkonak Valley) and 2505 meters (summit of the Bozburun Mountain range), on a very heterogeneous geological and geomorphological structure. It lies between 31°03' 31''-31°14' 00'' longitude east and 37° 87' 36"-37° 25' 11" latitude north. It encompasses 37000 hectares in north-eastern Antalya province (Çetinkaya, 2002). *Cupressus sempervirens* grows in deep valleys in Köprülü Canyon National Park, where it is well protected against cold winds and is able to survive in the Köprüçay-conglomerate formation, which has a cleft system with extreme cliffy conglomerate hillsides or vertical conglomerate rock faces (Ayaşlıgil, 1987). The Park has four different plant climate zones. Cypress stands are distributed here in the vegetation zone for *Quercion calliprini* (Fig. 9-2) between altitude 289 meters and 1216 meters (Fig. 9-3).

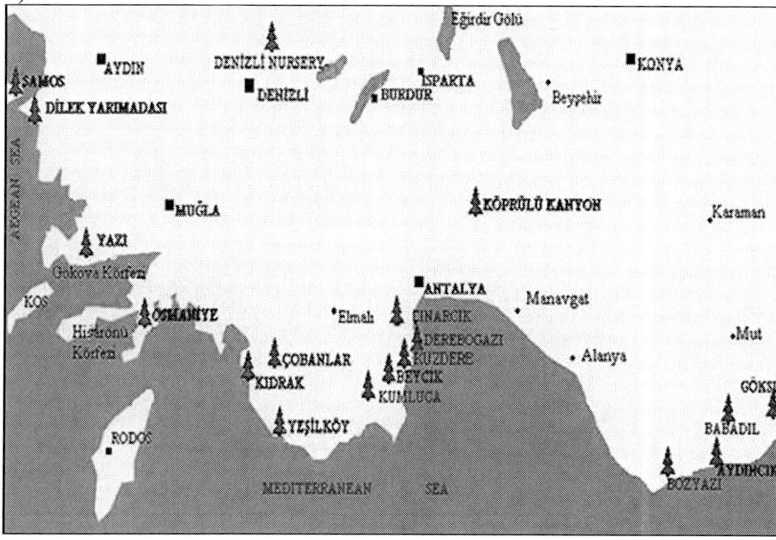

Fig. 9-1: Locations of *Cupressus sempervirens* var. *horizontalis* stands

Methods

Remote sensing and geographical information systems were used. For this study we used 1-m pan-sharpened satellite images merged from 1-m panchromatic and 4-m multi-spectral IKONOS satellite images. The IKONOS data were acquired on June 13, 2003 and the following methods were used to classify cypress stands:

- Normalized Difference Vegetation Index (NDVI). With this method, vegetation, bare rock, water areas, etc., were classified.
- Unsupervised image classification techniques were used to classify forest vegetation areas such as coniferous, deciduous and Mediterranean shrub land.
- Supervised image classification techniques were used to classify coniferous forest areas, and to define cypress stands. In addition to these,

ground control points, forest management and vegetation maps were used for this classification.

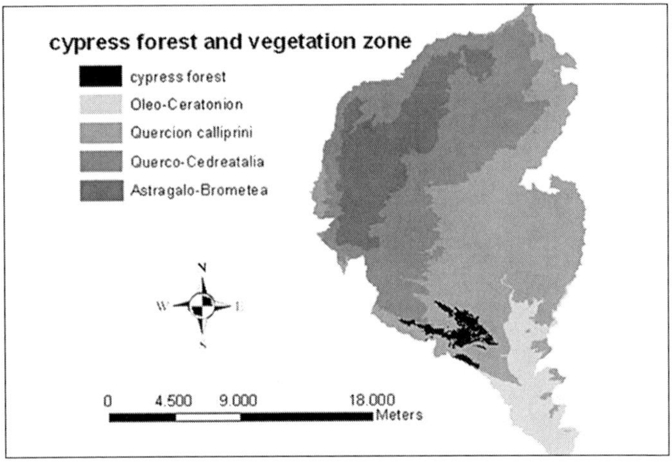

Fig. 9-2: Locations of *Cupressus sempervirens* var. *horizontalis* stands in the National Park Köprülü Canyon, with review of the vegetation zones

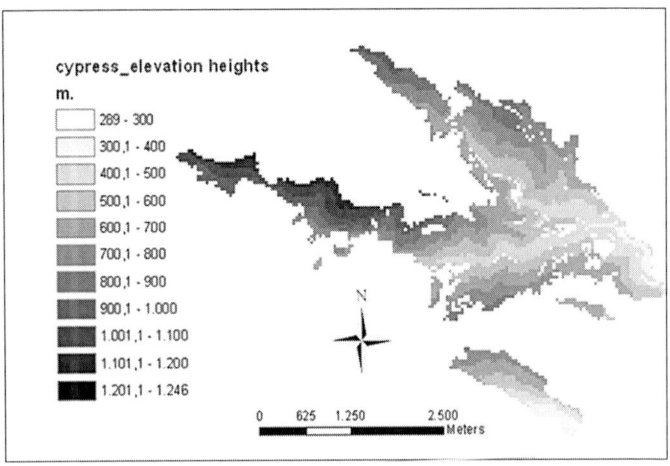

Fig. 9-3: Altitudinal distribution of cypress stands in the Köprülü Canyon National Park (digital elevation model)

GIS techniques were used to build a digital elevation model (DEM) with an Arc GIS spatial analyst. Elevation from sea level, aspect, and slope degrees were derived from DEM to understand site specific ecological conditions for the cypress forest of the Köprülü Kanyon National Park.

Findings and Results

The distribution of cypress in Turkey extends from the Dilek peninsula to the Göksu River, where typical Mediterranean climate prevails. The largest eastern distribution in the Mediterranean region of cypress stands (altitude 150 meters) are around the Sıhlar and Sahincik settlements along the banks of the Göksu River near Silifke. To the west, there are scattered small natural Cypress stands (altitude 70 meters), along the highway between Gülnar and Aydıncık. Cypress constitutes its widest (436 hectares, altitude 400-1100 meters) pure stand is found Antalya, Köprülü Kanyon National Park. Other Cypress areas around Antalya are Kemer-Derebogazi (altitude 200-250 meters), Cınarcık, Kuzdere, Beycik high plateau (altitude 500-950 meters) and Ulupınar Valley. Further west in Kıdrak Bay around Fethiye-Ölüdeniz are small groups of Cypress, which can be found at an altitude of 180 meters. It also appears naturally in the Esen Valley, along the road between Kayadibi and Bagliagac and Dalyan-Kargıcak cove (about 4 hectares), either in groups or mingled with Turkish red pine (*Pinus brutia* Ten.). It's widest (15-20 hectares) distribution on Resadiye Peninsula is seen around Değirmenbükü village. It also appears in Yazı (altitude 50 meters) and the Sındı and Hızırsah sites of the same peninsula (Sabuncu, 2004). Natural stands are also found in the Canyon Valley of Dilek Peninsula National Park and in the İcmeler region of Marmaris. The area covered by cypress forest in Turkey is around 1375 hectares (Neyişçi and Sabuncu, 2007).

Statistical Analysis for Genetic Study

The following linear model was used for the ANOVA:

$$Y_{ijkm}=\mu+R_i+P_j+F(P)_{k(j)}+RP_{ij}+RF(P)_{ik(j)}+E_{m(ijk)}$$

where:

Y_{ijkm}= *i. in replication j. belonging to population k. family m. belonging to seedling value;*

μ = *overall mean*;

R_i=*i. replication fixed effect* (i =1,...,3);

P_j =*j. population randomly effect* ~ Normal independent variable (NBD) $(0, \sigma^2_p)$, j=1,...,p;

$F(P)_{k(j)}$= *j. belonging to population k. family randomly effect*, ~NBD (0, $\sigma^2_{f(p)}$), k=1,...,f;

RP_{ij}=*i. replication j. population interaction randomly effect*, ~NBD (0, σ^2_{rp});

$RF(P)_{ik(j)}$=*i. in replication j. belonging population k. family interaction randomly effect*, ~NBD (0, $\sigma^2_{rf(p)}$);

$E_{m(ijk)}$ =*i. in block j. belonging k.family m. randomly effect*, ~NBD (0, σ^2_e).

NBD: Normal independent variable

According to the mixed model, Form of F Test for dependent variables is given in Table 9-2.

Table 9-1: Information about *Cupressus sempervirens* var. *horizontalis* populations and family numbers sampled for the genetic study

Populations	Location	Fam. Num.	Lat.(N)			Long.(E)			Altitude (m.)
AYDINCIK*	Merkez	4	36	08	55	33	20	05	30
BABADIL*	Gülnar	13	36	12	55	33	26	05	50
BEYCİK**	Kemer	7	36	29	55	30	25	05	900
BOZYAZI*	Merkez	2	36	04	55	32	55	05	20
ÇINARCIK**	Antalya	10	36	44	55	30	26	05	750
ÇOBANLAR***	Fethiye	3	36	34	55	29	20	05	70
DEREBOĞAZI**	Kemer	21	36	36	07	30	29	05	200
DİLEK	Kuşadası	4	37	39	55	27	10	06	150
GÖKSU*	Silifke	19	36	23	55	33	48	05	150
KIDRAK***	Fethiye	6	36	30	55	29	07	05	200
KÖPRÜLÜ KANYON**	Manavgat	68	37	11	55	31	08	05	700
KUMLUCA**	Merkez	4	36	26	55	30	25	05	350
KUZDERE**	Kemer	6	36	34	55	30	32	05	50
OSMANİYE****	Marmaris	9	36	45	55	28	11	05	400
SAMOS****	Greece	20	37	45	55	26	59	15	430
YAZI****	Datça	4	37	02	55	27	39	05	50
YEŞİLKÖY***	Kaş	4	36	16	55	29	22	05	60

*Eastern Mediterranean region populations, ** Middle Mediterranean region populations, *** Western Mediterranean region stands ****Aegean region populations

The GIS based image classification techniques showed five cypress stand types depending on different pure stand ratio of cypress within an area of 738, 93 hectares (Table 9-3). Previously, there was a 436 hectares pure stand of cypress in the National Park's Köprülü Canyon. (Neyişçi and Sabuncu, 2007). It has been found in this study that 738, 93 hectares, 352,

28 hectares. The pure cypress stands are distributed mostly between altitudes of 400 meters-800 meters. (Fig. 9-3, 9-4) and are located only on the valley plain and hillside of Kocadere and its tributary streams Değirmen Dere and Bağlıovası Dere, where it is well protected against cold winds. Stream Kocadere flow master stream Köprüçay which flow to the Mediterranean Sea.

Table 9-2: Form of F Test according to the applied experimental design of cypress seedling traits and expected mean squares

Source.	d.f.	Expected Mean Squares	F Test
Rep	r-1	$\sigma^2_e+10\sigma^2_{rf(p)}+63\sigma^2_{rp}+Q(rep)$	$KO_r/(0.5*KO_{rf}+0.5*KO_{rp}))$
Populasyon	(p-1)	$\sigma^2_e+10\sigma^2_{rf(p)}30\sigma^2_{f(p)}+107\sigma^2_{rp}+322\sigma^2_p$	$KO_p/(KO_{rp}+KO_f)-(KO_{rf})$
Rep*pop	(r-1)(p-1)	$\sigma^2_e+10\sigma^2_{rf(p)}+107\sigma^2_{rp}$	KO_{rp}/KO_{rf}
Aile(pop)	p(f-1)	$\sigma^2_e+10\sigma^2_{rf(p)}+30\sigma^2_{f(p)}$	KO_f/KO_{rf}
Rep*Aile(pop)	(r-1)p(f-1)	$\sigma^2_e+10\sigma^2_{rf(p)}$	KO_{rf}/KO_e
Hata	bpf(n-1)	σ^2_e	

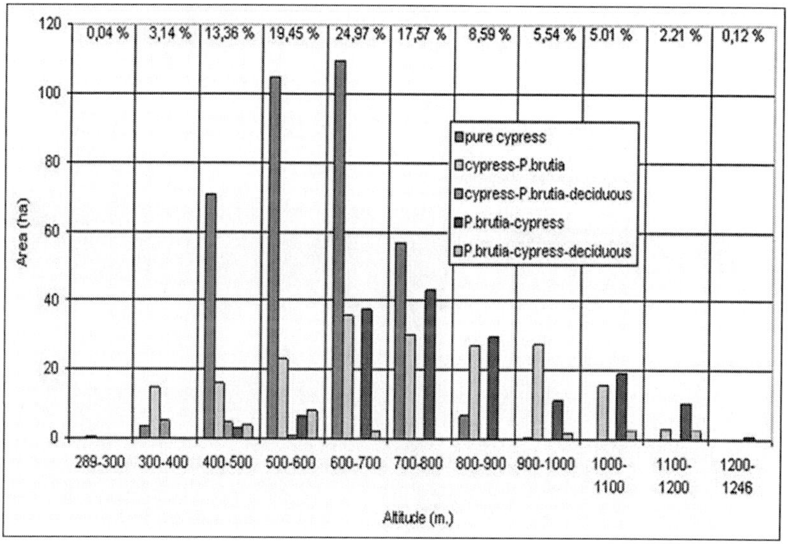

Fig. 9-4: Altitudes of cypress stand type and their total area in comparison with distribution rate

Table 9-3: *Cupressus sempervirens* L. stands type and their total area in comparison with distribution rate

Stands type	Total Area (ha.)	Ratio %	Pure Stand Ratio %
C. sempervirens pure	352,28	47,7	100
C. *sempervirens -Pinus brutia*	192,75	26,1	70-100
C. sempervirens -Pinus brutia-deciduous	10,7	1,5	20-70
Pinus brutia- C. sempervirens	161,7	21,8	10-20
Pinus brutia- C. sempervirens – deciduous	21,5	2,9	10-20
Total	738,93	100	

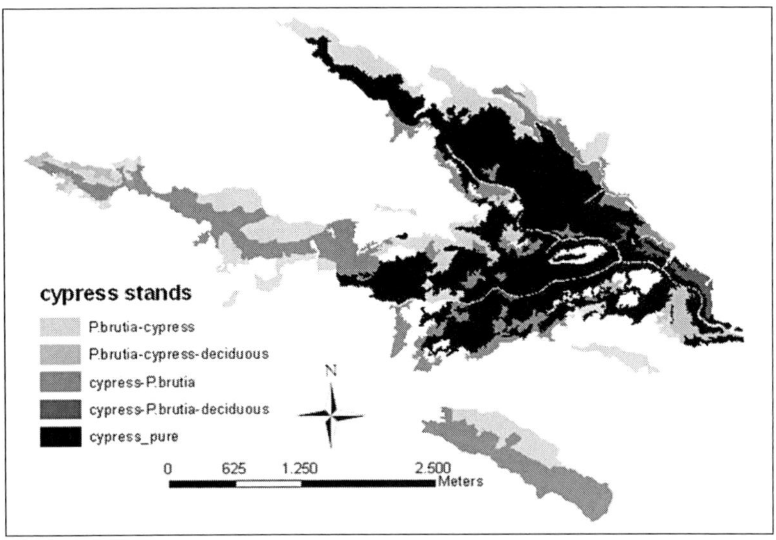

Fig. 9-5: Distribution of cypress stand types in the Köprülü Canyon National Park

By combining cypress stands with DEM data, it is seen that the site of these stands in Köprülü Canyon National Park has a marginal condition for growth. *Cupressus sempervirens* var. *horizontalis* grows in the National Park mostly on southerly exposed (Fig. 9-5 and 9-6) and cxtreme cliffy hillsides, (Fig. 9-7 and 9-8) which are important to protect especially against forest fires, illegal cutting and over grazing.

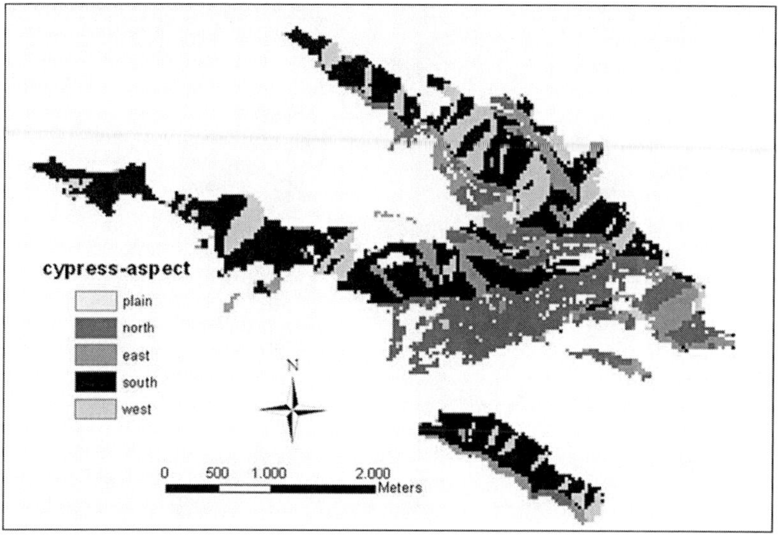

Fig. 9-6: Distribution of cypress stands in the Köprülü Kanyon National Park with aspect from digital elevation model

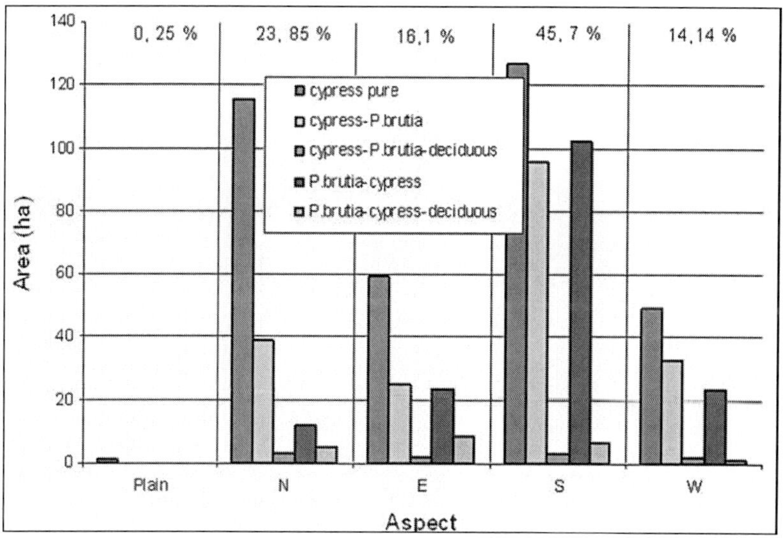

Fig. 9-7: Ground aspect of cypress stand types and their total area in comparison with distribution rate

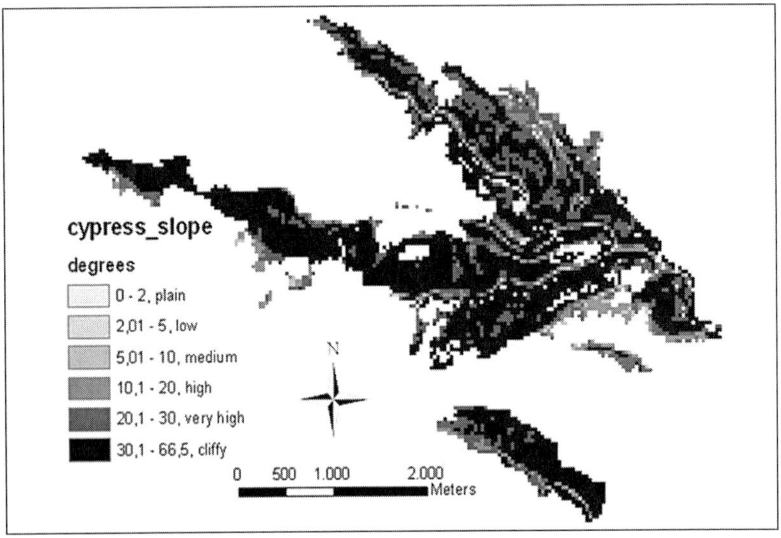

Fig. 9-8: Distribution of cypress stands in the Köprülü Canyon National Park with slope degrees from digital elevation model

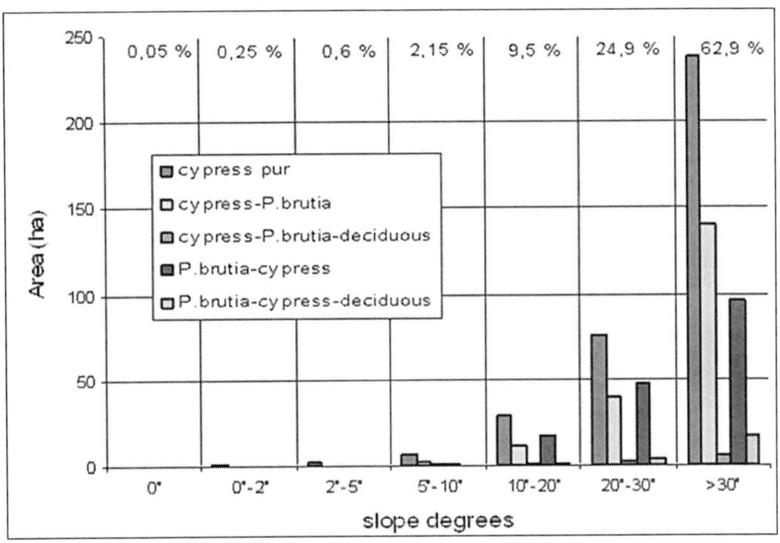

Fig. 9-9: Slope degrees of cypress stand types and their total area in comparison with distribution rate

The characteristic species of *Quercion calliprini* plant climate zones, which are distributed in the cypress stands, are: *Arbutus andrachne, Phillyrea latifolia, Quercus coccifera, Fontanesia philliraeoides* subsp. *philliraeoides, Laurus nobilis, Myrtus communis* subsp. *communis, Jasminum fruticans, Smilax aspera.* Other plant species from *Oleo-Ceratonion* plant climate zones such as *Olea europaea ssp. sylvestris*, *Cistus creticus, Euphorbia characias* subsp. *wulfenii* and *Micromeria myrtifolia* are distributed on the south cliffs and other thermically advantaged sites. The plant species from *Querco-Cedretalia libani* plant climate zones such as *Quercus infectoria* subsp. *boissieri, Fraxinus ornus* subsp. *cilicica*, *Salvia tomentosa* and *Clinopodium vulgare* subsp. *arundanum* are distributed on the north hillsides and so, need humidity. Therefore, cypress stands are located in a transition zone between *Oleo-Ceratonion* and *Querco-Cedretalia libani* plant climate zones and have different vegetation species depending on the aspect and altitude (Ayaşlıgil 1987). Furthermore, cypress forests of the National Park have the following endemic plants (Ayaşlıgil 1987) listed with Turkish IUCN Red List Categories (Ekim. *et al.,* 2000) (Table 9-4).

Table 9-4: The list of endemic plants of the cypress forest in Köprülü Canyon National Park

Endemic Plants	IUCN Red List Categories
Eremopa attalica	VU
Verbascum chrysochaete	LR (nt)
Pholomis leucophracta	LR (nt)
Galium floribundum ssp. airoides	LR (nt)
Verbascum nutadum var. nutadum	LR (nt)
Fraxinus ornus spp. cilicica	LR (lc)
Alkanna areolata var. areolata	LR (lc)

Analysis Results for the Genetic Studies

An analyses of variance result of seedling height and diameter is given in Table 5. The results of variance analyses are: There are significant differences among populations and families in the height and diameters of 1 year old seedlings; the variation between populations was also strongly related to separation; the distance between each population is quite far; the eastern Mediterranean populations of Bozyazi and Aydincik were the best performers for seedling height and diameter traits and highly significant differences (Pr<0.001) were found between and within the populations for height and collar diameter at age one (Table 9-6).

Table 9-5: Analyses of variance result of seedling height and diameter

Source of Var.	df	Mean square	Height F value	Pr>F	Mean square	Diameter F value	Pr>F
Replication	2	0.76	0.03	0.9745	0.371	0.78	0.4626
Population	16	1241.99	6.42	<.0001	12.817	8.61	<.0001
Rep*pop	32	26.91	0.82	0.7426	0.533	1.36	0.0985
Family(pop)	186	199.30	6.10	<.0001	1.348	3.43	<.0001
Rep*Family(pop)	372	32.69	2.20	<.0001	0.393	1.58	<.0001
Error	5481	14.88			0.249		

Table 9-6: Variance Component ± standard error, component of variance as a % of total variance, heritability (h^2_i, h^2_f) and population repeatability value.

	Height	V.C %	Diameter	V.C %
Population	4.995± 2.197	18	0.059±0.025	17
Pop*Rep	0±0	0	0.0014±0.0013	0
Family (pop)	5.598± 0.697	21	0.032±0.0048	9
Rep*Aile (pop)	1.735 ±0.229	6	0.015±0.0029	4
Error	14.88±0.2843	55	0.249±0.00475	70
h^2_i ±SE	0.25±0.024		0.11±0.015	
h^2_f ±SE	0.89±0.013		0.77±0.028	
R_p ±SE	0.73±0.092		0.84±0.060	

In this study, the genetic differentiation observed among cypress stands in the natural population in Turkey (G_{st}=31%) was above average for wind-pollinated conifers (G_{st}=6.8 % Hamrick and Godt 1989), endemic species and regionally distributed species (G_{st}=14 % Hamrick et al. 1992). This indicates that gene flow between populations is weaker than the other conifers. Moreover, differentiation between populations is increasing due to regional distribution and populations being represented by few trees.

Geographic range is also the best predictor of genetic variation among populations. The genetic structure of cypress trees is probably highly influenced by the effects of genetic drift in small populations and population divergence due to limited gene flow among isolated populations. Pine species which are distributed as scattered isolated populations have more genetic diversity among populations (e.g., *P. torreyana* (G_{st}= 1.00, Ledig and Conkle 1983); *P. halepensis* (G_{st}= 0.30, Scheller et al. 1985), and *P. muricata* (G_{st}= 0.20, Millar 1983) (Bastian and

Alia 2000). The highest values of observed and expected heterozygosity and the total genetic diversity among *C. Sempervirens* var. *horizontalis* recorded for the Turkish populations suggests that these populations might be relicts of the center of origin of this variety (Korol *et al.,* 1997).

Discussion

The conservation of forests and forest species in the Mediterranean basin is a complex issue, given the wide range of ecological situations, diverse forest uses and the pressure exercised by different cultural groups. Management of the various dynamic stages of forest systems demands consideration not only of forest communities and species *per se,* but also of associated ones on the edges of forests and in clearings. Although the fairly widespread distribution of most forest species would indicate that any threat to them is limited, the marginality of certain populations (based on their geographic isolation) and/or their ecological marginality (species linked to rare habitats) must be taken into account (Rameau & Olivier, 1991). Moreover, a fair number of trees represent key species whose preservation is vital for the sustainability of biogeochemical cycles.

In recent decades, socio-economic process has caused some changes in the land use of the Mediterranean upland in Turkey. The tourism industry has been developed in the Mediterranean region of Turkey in the last 20 years. Migration is continuing at a rapid pace, from rural to urbanised or industrialised areas and to regions with important tourism development. In Turkey, tourism is also responsible for large-scale migration to the Mediterranean and Aegean regions. This has lead to rapid and often unregulated urbanisation, which adversely affects the cultural and natural heritage of rural areas.

In this study of the variation of genetic characteristics, it was observed that within and among populations this variation is significant. It is attributed to the additive effect of genes. The population diversity and heritability are both high for seedling height, and the best nursery growth was observed for the Bozyazı and Aydıncık populations. The total variation observed among populations of the characteristics studied, was attributable to genetic components. The variation observed among populations was 18% and 17% for height and diameter respectively, however, within a population the variation was 21% and 9 % respectively. Individual heritability (h_i^2) values estimated for height and diameter at collar were 0.25 and 0.11 respectively. In addition, family heritability (h_f^2) values obtained were 0.89 and 0.77 for the height and diameter of the collar characteristic. Family heritability values were higher than the individual heritability values for both characteristics studied. An additional study analysed the site-specific growth conditions and vegetation units of the cypress forest of the National Park Köprülü Canyon. The study found that where the cypress forest is more prolific, it is situated on a very cliffy and southerly exposed hillside, which is especially important for protection against forest fires, illegal cutting and over grazing.

This paper studies threatened forest ecosystems in Mediterranean regions for the cypress case. *Cupressus sempervirens* natural stands are located in the Mediterranean and Aegean regions of Turkey, while some populations around Antalya (Çınarcık, Dereboğazı, Kalkan-Kapotaj) and Fethiye (Kıdrak, Kayadibi, Bağlıağaç) have been discovered more recently. According to the analyses of the ecological site condition and vegetation units of the natural distribution area, cypress can grow on sites ranging from a semiarid bioclimatic zone with warm winters, to a wet zone with cold and fairly harsh winters. This great plasticity of the species with regard to natural environmental factors, together with its relative resistance to fire, renders cypress one of the most valuable forest species for a large part of the Mediterranean region (Brofas at al.2006, Neyişçi and Sabuncu, 2007). Due to its qualities, the cypress tree has many uses. Besides its extensive ornamental use, it is used as a windbreaker and boundary hedge; whereas its use for soil protection and wood production in Turkey is relatively limited compared to other coniferous species such as Pinus brutia. The limited use of cypress for reforestation purposes can be attributed to the lack of knowledge about its real potential.

The pure cypress forest is located in Köprülü Canyon National Park, which is the largest cypress forest in Turkey. The complete area of the cypress forest is located inside the National Park. The formation of an actual management plan of the protected area has not yet been prepared for the National Park. Sensitive growth characteristics and the genetic importance of the cypress forest should be considered when a management plan is prepared.

Conclusion

According to Tsoumis (1985), Zacharis (1977) and Zohary (1973), cypress had a wider distribution and density in the past, but over-grazing, fires and intense exploitation of its wood for construction purposes, led to the degradation and restriction of the natural stands. In general, the cypress stands are degraded, sparse, and in many cases consisting of small-sized trees. These features are particularly evident along the supra-and mountain-Mediterranean vegetation belt, and also in the shallow stony soils of the lower altitudinal levels (thermo-meso- Mediterranean belts). Besides degradation, the extent of their distribution has decreased, leaving the southwest Mediterranean population in imminent danger of extinction. Quite possibly, this may have also happened in other parts of the Mediterranean and Aegean regions in the past, where today the cypress is absent and unable to re-establish itself, due to the seeds' inability to disperse over a long distance. For this reason, natural cypress populations must be immediately conserved by implementing both ex-situ and in-situ programs.

The problem with *in situ* conservation of the genetic resources of Mediterranean forest species through existing natural reserves lies in the fact that such reserves are very seldom created with this aim in mind. The designation of conservation zones and their size is still governed more often by local considerations or by subjective, aesthetic criteria than by a strict, multilevel analysis of the biodiversity of a given area. Although an archipelago of small reserves may contain more species, this approach is not ideal for the sustainable conservation of forest species populations. They will be conserved solely within areas large enough for their lifecycle to unfold in the best possible way, without suffering the effects found in border areas. Only extensive reserves can guarantee the development of normal disturbance cycles and assure maintenance of the large vertebrate populations needed to guarantee the sustainability of a whole range of landscapes and the distribution of plant species. Therefore, it seems best to set up systems incorporating large unbroken reserves surrounded by buffer zones, making up a mosaic with differently managed areas (Frankel, et al. 1995).

References

Atalay, I. 2002. Ecoregions of Turkey). Orman Bakanlığı yay. no: 163.

Ayaşlıgil, Y. 1987. Der Köprülü Kanyon Nationalpark-seine Vegetation und ihre Beeinflussung durch den Menschen. Landschaftsökologie Weihenstephan, Heft 5, Freising, 307 s.

Bastian, C. and Alia, R., 2000. What Might Be Useful Measures of Genetic Variability for Adaptive Traits Within Population of Scots Pine. Invest. Agr. : Sist. Recur. For. : Fuera de Serie no 1.

Boydak, M., Dirik, H. and Çalıkoğlu, M. 2006. Kızılçamın *(Pinus brutia* ten.) biyolojisi ve silvikültürü. Ormancılığı geliştirme ve orman yangınları ile mücadele hizmetlerini destekleme vakfı yayını (OGEM-VAK), Ankara, 364 pp.

Brofas, G., Karetsos, G., Dimopoulos, P. and Tsagari, C. 2006. Land Degradation and Development. The natural environment of Cupressus sempervirens in Greece as a basis for its use in the Mediterranean region 17: 645-659.
Çetinkaya, G. 2002. The progress report regarding "Research for the establishment of Köprülü Kanyon National Park as a biosphere reserve". Saint-Petersburg State University of Architecture and Civil Engineering, http://www.unesco.org/mab/bursaries/mysrept/2001/Cetinkaya/Gulay1.PDF
Dufour-Dror, J. M. 2002. A Quantitative Classification of Mediterranean Mosaic-Like Landscapes. Journal of Mediterranean Ecology 3:3-12.
Ekim, T., Koyuncu, M., Vural, M., Duman, H., Aytaç, Z., and Adıgüzel, N. 2000. "Türkiye Bitkileri Kırmızı Kitabı". Türkiye Tabiatını Koruma Derneği Yayını, Ankara, S. 178
Frankel, O.H., Brown, A.H.D. and Burdon, J.J. 1995. The conservation of plant biodiversity. Cambridge, UK, Cambridge University Press. 299 pp.
Hamrick, J. L., Godt, M. J. W., 1989. Allozyme Diversity in Plant Species. In: A. D. H. Brown, M. T. Clegg, A. L. Kahler, and B. S. Wier (eds.) Plant Population Genetics, Breeding and Genetic Resources. Sinauer Associates, Inc, Sunderland, MA, pp. 43-63.
Hamrick, J. L., Godt, M. J. W. and Sherman-Broyles, S. L., 1992. Factors influencing genetic diversity in woody plant species. New Forests, 6: pp.95-124
Kayacık, H. and Yaltırık, F. 1971. General aspects of Turkish forest. In: Plant life of South-West Asia by P. H., Davis, P.C. Harper and I. C. Hedge. pp. 283-291, Published by the Botanical Society of Edinburg.
Korol, L., Kara, N., Işık, K and Schiller, G., 1997. Genetic Differentiation Among and Within Natural and Planted *Cupressus sempervirens* L. Eastern Mediterranean Populations. Silvae Genetica 46 (2-3), p: 151-155.
Ledig F. T. and Conkle, M. T. 1983. Gene diversity and genetic structure in a narrow endemic, Torrey pine (*Pinus torreyana* Parry ex Carr). *Evolution* 37: 79-85
Medail, F and Quézel, P. 1997. Hot- spots analysis for conservation of plant biodiversity in the mediterranean basin. Ann. Misouri. Bot. Gard. 84: 112-127
Millar, C. I. 1983. A steep cline in Pinus muricata. *Evolution* 37 [2]:311-319
Naveh, Z. and Dun,J. 1973. The human degradation of Mediterranean Region. In: T.T. Kozlowski and C. E. Ahlgren (eds.) Fire and

ecosystems, pp. 401-434, Akademic Pres, New York, San Francisco, London.

Naveh, Z., and P.Kutiel 1986. Changes in the Mediterranean vegetation of Israel in response to human habitation and land use. Pages 259-296 In: G. Woodwell (eds.) The earth in transition: patterns and processes of biotic impoverishment. Cambridge University Press, Cambridge.

Neyişçi, T., and Sabuncu, R. 2007. Le Cyprès dans les pays Meda. MedCypre; Un Project Interreg pour renforcer la cohénsion des peuples du Bassin Méditerranean.Palermo, Italie 7-9 Février 2007.

Quézel, P. 1985. Definition of the Mediterranean region and the origin of its flora. In C. Gomez-Campo (eds.) Plant conservation in the Mediterranean area. Geobotany 7, W. Junk, Dordrecht, the Netherlands, pp. 9-24.

Pons, A. & Quézel, P. 1985. The history of the flora and vegetation and past and present human disturbance in the Mediterranean region. In C. Gomez-Campo, (eds.) Plant conservation in the Mediterranean area. Geobotany 7, p. 25-43. Dordrecht, the Netherlands, W. Junk.

Rameau, J.C. & Olivier, L. 1991. La biodiversité forestière et sa préservation. Intérêt patrimonial de la flore, de la végétation et des paysages forestiers. Rev. For. Fr., 43, num. sp. 1991: 19-27.

Sabuncu, R. 2004. Dallı servide (*Cupressus sempervirens* var. *horizontalis*), fidan büyüme karakterleri açısından genetik çeşitlilik. Orman Bakanlığı Yayınları, Teknik Bülten No: 22

Scheller, G., Conkle M. T. et Griswald L. 1985. Local differentiation among mediterraneen populations of Aleppo pine in their isozymes. Silvae Genetica 35: 11-19

Shoshani, M. 2000. Satellite remote sensing of natural Mediterranean vegetation: a review within an ecological context. Progress in Physical Geography 24:153-178.

Thirgood, J.V. 1981. Man and the Mediterranean forest. New York, Academic Press. 194 pp.

Tsoumis, G. 1985. The depletion of forests in the Mediterranean region. A historical review from ancient times to the present. Scientific Annals, Department of Forestry and Natural Environment, Aristotle University, Thessaloniki, Greece, Vol. 28(11), 265-301 (In Greek and in English).

Zacharis, A. 1977. Forests of Crete. Apo tin Archaiotita os Simera (From Antiquity to Today). Ministry of Agriculture - General Forestry Division, p. 146.

Zohary, M. 1973. Geobotanical Foundations of the Middle East, volumes 1, 2. Gustav Fischer Verlag, Stuttgart, Germany.

CHAPTER TEN

LAND DEGREDATION AND SOIL CONSERVATION ON THE MOLDAVIAN PLATEAU, ROMANIA

ION IONITA

Introduction

The Moldavian Plateau, located in eastern Romania and extending about 27,000 square kilometers, is one of the most severely eroded agricultural areas in the country. Clayey-sandy Miocene-Pliocene layers with a gentle gradient of 7-8 m km-1 NNW-SSW have outcropped from sedimentary substratum (Jeanrenaud, 1971). The climate is temperate continental with a mean annual temperature of 8.0 – 9.8 0C. Average annual precipitation varies from about 460 millimeters at the lower elevation in the southern part to 670 millimeters in the central and north-western area with elevations up to 587 meters. Natural vegetation cover was drastically changed by man particularly during the last two centuries. Mollisols and argiluvisols (forest soils) are among the most common soils and have been used for crop production. The main land use stratification is cropland (58%), pastures and meadows (16%) and forest (13%).

Under these circumstances, at present soil erosion, gullying, landslides and sedimentation have been recognized as major environmental threats. The total erosion is averaging 15 - 30 t ha-1 yr-1. By 1960, the traditional agricultural system on slopes consisted of up-and-down hill farming. Most of the land was excessively split into small plots, each under one hectare in size. Since 1960, awareness of soil erosion and the adoption of conservation practices has increased. By the end of 1989 as much 0.9 million hectares, representing 71 percent of the agricultural land with erosion potential were adequately protected under different conservation practices. Romania's new land property Law Number 18/1991 has resulted in the revival of the old traditional up-and-down hill farming.

Methods

In order to evaluate the severity of erosion and sedimentation there was a need to get reliable information. Current methods of erosion and deposition assessment may be divided into three approaches, namely long-term monitoring of experimental plots, repeated field survey of gully erosion features, and field measurements to identify spatial patterns of sedimentation.

The effect of soil cover on runoff and soil losses was determined using research plots over the period 1970–1999 at the Central Research Station for Soil Erosion Control Perieni-Barlad.

A substantial field database that covers almost 20 years by monitoring representative gullies located in the southern part of the Moldavian Plateau near the city of Barlad was obtained. Three main areas of monitoring gullies were explored: aerial photographs from flights of the 1960s and 1970s, classical leveling work, and repeated survey through a particular close stakes grid after 1980.

An Eijkelkamp auger kit was used to collect data on filled cross-sections of discontinuous gullies. By period leveling over previous check-iron plates deployed on the gully basin floor it was possible to monitor sediment deposition. The Caesium-137 technique was used to get information on dating specific levels of sediments and documenting erosion and sedimentation rates along the discontinuous gullies.

Since autumn 1996, fifteen (15) reservoirs have been gradually selected to estimate sedimentation rates. By means of the same Eijkelkamp hand auger equipment a lot of holes were firstly drilled in each site to determine the original reservoir floor. Then, at least one pit was dug in each reservoir. After manual sampling of the sediments, the Gamma spectrometry was used to get the Caesium-137 depth profile and to assess reservoir sedimentation rates.

Results and Discussion

Land degradation: The Perieni runoff plots were set up on the left valley-side of Tarina catchment with 12% slope and slightly eroded mollisoil. Generally, data collected here over a 30-year period on soil and water losses indicate the following (Ionita, 2000b, Ionita et al., 2006):

-Mean annual precipitation is 504.3 mm, and precipitation which caused runoff and erosion falls as rain during the crop-growing months of May through October;

-About 26 percent (133.5 millimeter) of the annual precipitation induced runoff and or erosion for continuous fallow and 18.5 percent (93.5 millimeter) for maize;

-Runoff ranges from 36.5 mm under continuous fallow with the peak of 12.0 mm during July and 17.7 mm under maize with the peak of 6.5 mm during June;

-Soil loss is averaging 33.1 t ha-^{1}yr-^{1} for continuous fallow with the peak of 12.8 t ha-1 during July and 7.7 t ha-^{1}yr-^{1} for maize with the peak of 3.7 t ha-^{1} during June.

One mention must be made, namely that on strong eroded wooden soils, the value of the soil loss is doubling. According to Motoc et al. (1998) and Ionita (2000b) data collected under continuous fallow plot and processed by using a 3-year moving average revealed that over the period 1970-1999 there were three soil erosion peaks, namely in 1975, 1988 and 1999.

Figure 10-1 illustrates the spatial distribution of the gullies within the Moldavian Plateau as identified by Radoane et al. (1995). There are obviously two areas depicted by a higher gully density, namely, the northern “cloud” developed on clayey layers, mostly consisting of small discontinuous gullies, and the southern “cloud” around the city of Barlad developed on loamy-sandy layers where the valley-bottom, continuous gullies are prevailing.

Results on discontinuous gullies have indicated that during a variable period of 6-18 years, the mean gully head advance was 0.92 m yr^{-1} and ranged from 0.42 to 1.83 m yr^{-1}. The mean areal gully growth was 17.0 m^{2} yr^{-1} and varied between 3.2 and 34.3 m^{2} yr^{-1}. Both values indicate a slow erosion rate (Ionita 1998, 2000a, 2003, 2006, Ionita *et al.*, 2006). Moreover, the average annual regime of gullying is a pulsatory one that is best described by great fluctuations. Of the total gully growth, 60% resulted from only five years (1980, 1981, 1988, 1991 and 1996).

Conventional measurements on sedimentation by using check-iron plates along the floor of discontinuous gullies over the period 1987-1997 indicate a higher rate of aggradation in the upper half of the gully floor. This finding supports the development of a short, steeper reach within the gully floor as a critical location for gully renewing.

Similar values were obtained by using the Caesium-137 depth profile. Furthermore, it was possible to estimate the age of gullies as being between 23-48 years and to determine that discontinuous gullies are delivering most of the sediment needed for their own aggradation. The major result of this pulsating evolution reflects a dynamic balance between two simultaneous processes, erosion and sedimentation, inside a single system.

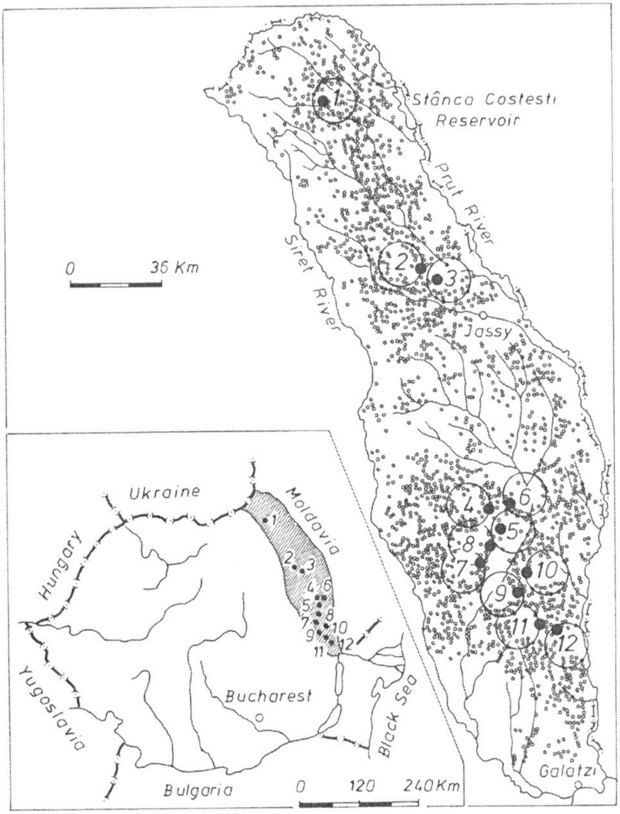

Fig. 10-1: Gully Distribution on the Moldavian Plateau, Romania (Radoane *et al*, 1995)

As for continuous gullies (Fig. 10-2), linear gully head retreat, areal gully growth and eroded material rates were quantified for three periods (1961-1970, 1971-1980 and 1981-1990). Results indicate that gully erosion has decreased since 1960 (Ionita 1998, 2000a, 2006, Ionita et al, 2006).

Fig. 10-2: Valcioaia Gully, Falciu Hills, Romania, May 10, 2005

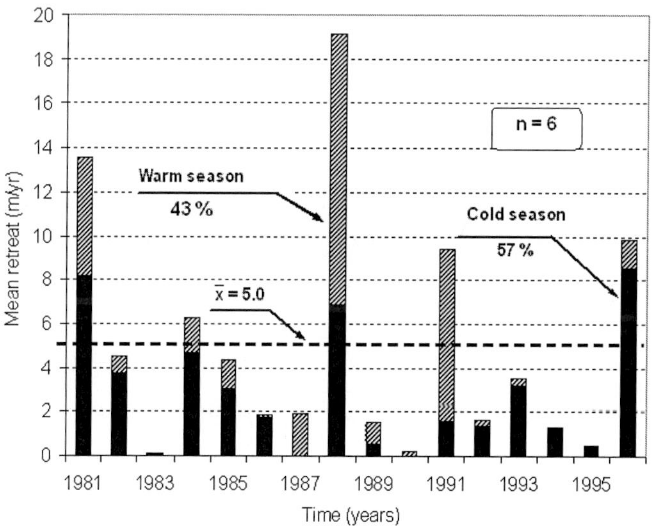

Fig. 10-3: Measured Mean Rate of Gully Head Retreat on the Moldavian Plateau, Romania, 1981-1996,

The mean gully head advance ranged from 19.8 m yr^{-1} during the 1960s and 12.6 m y^{-1} in the 1970s to 5.0 m yr^{-1} during the 1980s. This gullying decline is due to rainfall distribution, and the increased influence of soil erosion control. The mean gully head retreat of 12.5 m yr-1 over the 30-year period (1961-1990) was accompanied by a mean areal gully growth of 366.8 m2 yr^{-1} and a mean erosion rate of 4,168 tons yr^{-1}. The continuous gullies also showed a pulsatory development, as illustrated in Fig. 10-3. The critical period for gullying for the 1981-1996 period is during the four months from mid-March to mid-July in an area with mean annual precipitation around 500 millimeters. Another main finding of this 16-years stationary monitoring was that 57 % of the total gullying occurred during the cold season, with the remainder during the warm season. Of the total gully growth, 66% results from only four years (1981, 1988, 1991 and 1996) when precipitation was higher.

Field measurements performed in small catchments during flash streamflows allowed identifying two types of sediment delivery scenarios, synchronous and asynchronous (Ionita, 1998, 1999, 2000a). Even for very rare occurrences, the synchronous scenario, mostly associated with quick thawing, shows very high sediment concentration, exceeding 300.0 g l-1 at the basin outlet and low values, up to 40.0 g l-1 upstream of gullies in the upper basin. Gullying is the major sediment source. The asynchronous scenario commonly occurs and is characterized by higher water discharges and fluctuating sediment concentrations as described by Piest et al (1975).

Another major degradation process is associated with landslides that during the previous century recorded a peak over the period 1968-1973. The denudation rate by landslides was evaluated by Pujina (1997) as averaging 36.0 mm yr-1 within the Barlad Plateau over the period 1968-1992. The total erosion on the Moldavian Plateau of Eastern Romania averaged 15 - 30 t ha-1 yr-1 (Motoc, 1983).

Reservoir sedimentation was studied in catchments ranging in size from 329 to 50,978 hectares and most of them are poorly covered by forest (Ionita et al., 2000). Sediments in the reservoirs containing Caesium -137 are usually more recent than late 1960s. The Caesium-137 inputs derived from the Chernobyl accident are prevailing with respect to those derived from testing of nuclear weapons. By taking advantage of Caesium-137 as a tracer it was possible to date a particular alluvial level and to estimate accurately the rate of deposition from the depth of material overlying this specific level as shown in Fig. 10-4.

Results obtained indicate an obvious regional differentiation in the magnitude of sedimentation rates. Since April 1986 the mean value of the reservoir sedimentation rate for the entire Moldavian Plateau is 4.6 cm yr^{-}

[1]. The highest deposition rates are located in the southern part of the study area, the Tutova Rolling Hills, and averaging 7.9 cm yr^{-1}. The sedimentation rate for the northern area, the Moldavian Plain, is 2.4 cm yr^{-1} being by far the most reduced value. Probably, the best interpretation of the regional differentiations in reservoir sedimentation rates would be that they are mostly consistent with the main changes in the texture range of the underlying layers from clay in the north to sand in south.

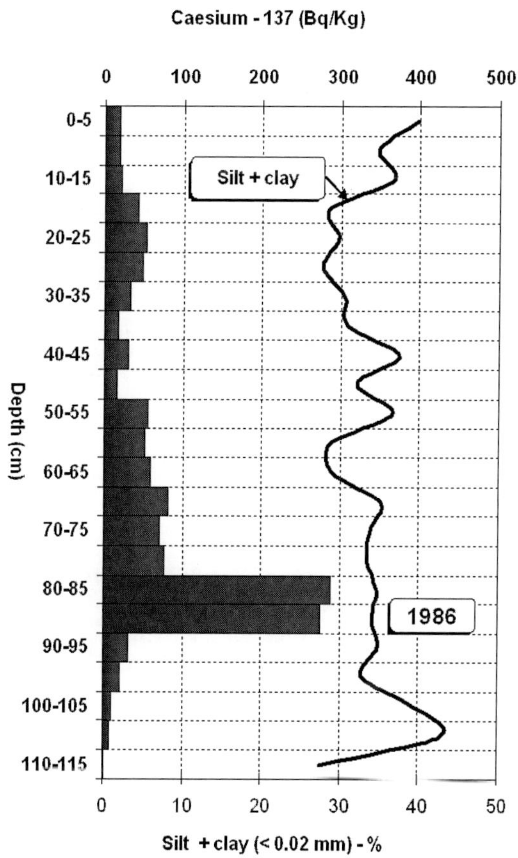

Fig. 10-4: Depth Distribution of 137Cs and the Silt + Clay Content in the Horgeşti Reservoir, Middle Racatau Basin, Tutova Rolling Hills, Romania, June 11th 1998

Soil Conservation

By 1960, the traditional system on sloping agricultural land in Romania consisted in up-and-down hill farming. Some 85 percent of this land was excessively split into small plots, each under one hectare in size. The tillage operations were made by animal traction, by hand and there was not any input of chemical fertiliser. There was no concern about the soil erosion threat and there was but a minimum awareness of conservation practices, excepting local areas.

Since 1960, the awareness of soil erosion and adopting conservation practices had been progressively increased. The area under those small plots was turned into co-operative farms. About 15 percent of the agricultural land, which belonged to proper farmers regarding ownership size, was changed into state farms. Practising conservation involves changes, all of which have costs. Nationwide, the first important objective was to perform ploughing on or nearly on the contour as one of the oldest and simplest conservation tillage. Then, based on the experience gained by the Central Research Station for Soil Erosion Control Perieni-Barlad, in each hilly county was set-up a representative farm under conservation practices. The total area of those particular units amounted to 65000 hectares.

By the end of 1989 as much as 0.9 million hectares, representing 71 percent of the agricultural land, with erosion potential in the Moldavian Plateau, were adequately protected. In some counties, such as Vaslui, Bacau, Galati, Iasi, the approach of best management practices resulted in appropriate erosion control at basin scale. The annual expenditure in conservation projects was state investment as non-reimbursable funding.

It should be mentioned that over the period 1960-1975 the main concern was focused on dam structures within drainage networks, especially on gullies. Then, over the period from 1975 to 1990, priority was given to implementing proper conservation practices on slopes, but mainly in huge agricultural units.

The new Landed Property Law Number 18/1991 includes two provisions which are not of a nature to create conditions for the extension of conservation measures. One of these stipulates that the land reallotment has to be usually done on the old sites. This means in most cases that the plots will be up-and-down the hill led out. The second refers to the successors' right up to the fourth degree! That enables an increasing rate of land splitting which is higher than that before the Second World War and allotments to about 2 million people who never had a connection with agriculture. Meanwhile, another law, Number 1/2000, was promulgated and which has in view forestland splitting on a large area for private ownership.

The major effect of the above-mentioned laws is revival of the old traditional agricultural system. So, the up-and-down hill farming method was "on the screen" (revived) again.. Ownership boundary lines interfere with conservation tillage and practices. It is not surprising that small plot layout is a significant obstacle to voluntary erosion control. Another problem for the last decade is that the state ceased funding for soil erosion control programs, and such an investment does not represent a priority for landowners.

The main changes in land management, especially arising from the up-and-down hill farming after 1991, produced evidence of a high and significant increase in erosion and sedimentation rates in comparison with the previous contour farming. Figure 10-5 provides an image of the strong impact which Law Number 18/1991 had on agricultural practices and soil erosion and deposition, which was not everywhere evenly distributed (Ionita et al, 2000).

During the 1982-1985 period, appropriate conservation practices were implemented in the upper Racatau basin, mostly used as cropland. After 1991, the area was gradually converted to an up-and-down hill farming system and the rate of soil erosion and sedimentation doubled. This finding is supported by the depth distribution of Caesium-137 that shows two peaks. The first pick clearly occurs below a depth of 70 cm (37.0 Bq kg^{-1}) and is connected to the Chernobyl accident of April 1986, while the second is related to the revival in 1993 of the up-and-down hill farming. The mean deposition rates rose from 5.0 cm yr^{-1} for the period 1986-1992 to 10.0 cm yr^{-1} over the period 1993-1996.

The 137Cs activity during the former period was associated with low 137Cs input (3-5 Bq kg^{-1}) indicating fairly low rates of soil erosion, whereas the latter period had severe rill-interrill erosion with the most significant 137Cs inputs along this profile (up to 59.7 Bq kg^{-1}). Therefore, this particular 137Cs depth distribution shows that 86% of the total inventory occurred in the top 40 centimeters, representing the period 1993-1996 of up-and-down hill farming.

It was concluded that the factors used to develop early soil and water conservation programs are not relevant to the present – day situation. Furthermore, it is obvious that activities promoting real soil erosion control on the Moldavian Plateau of Eastern Romania extended over the 30-year period from 1960 to 1990.

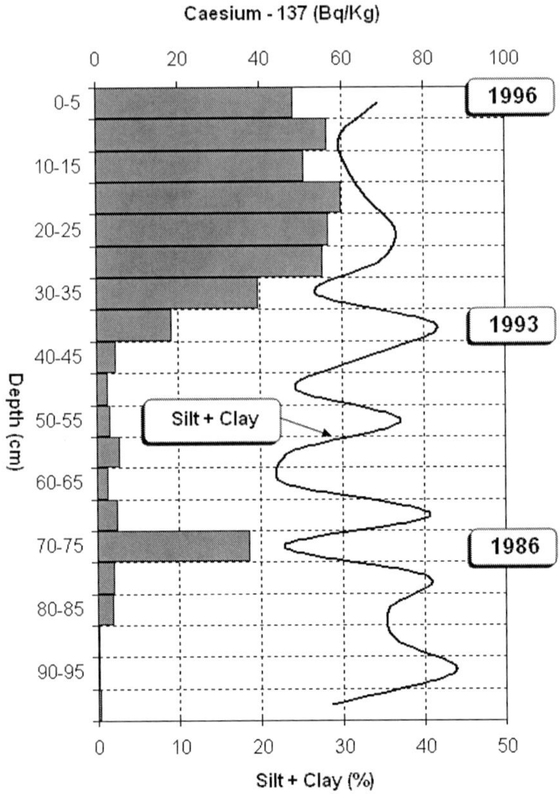

Fig. 10-5: Depth Distribution of 137Cs and the Silt + Clay Content in the Bibirești Reservoir, Upper Racatau Basin, Tutova Rolling Hills, Romania, May 13th 1997

Conclusions

On slightly eroded mollisoil soil, losses averaged 33.1 t ha-1yr-1 for continuous fallow and 7.7 t ha-1yr-1 for maize over 1970-1999 while on strong eroded wooden soils these values are double. The average annual regime of gullying is a pulsatory one that is best described by great fluctuations. Most of the discontinuous gullies exhibit both proper gullying, mainly if not exclusively, by gully head advance, and aggradation of the gully basin floor.

In regards to continuous gullies, the gullying decline over the period 1960-1990 was due to the rainfall distribution and the increased influence of soil erosion control. About 57% of the total gullying over 1981-1996 was associated with the cold season, especially in March due to the influence of the freeze-thaw cycles. Of the total gully growth, 66% resulted from only four years (1981, 1988, 1991 and 1996), when precipitation was higher.

The estimated mean reservoir sedimentation rates in three geomorphologic subunits varied between 2.4 and 7.9 cm yr-1 with an average rate of 4.6 cm yr-1 after April 1986.

Since 1960 the awareness of conservation practices has increased. By implementing the provisions of the Land Reform Law Number 18/1991 the old traditional up-and-down hill farming is "on the screen" or "in play" again, creating the probability of significantly higher levels of erosion.

References

Ionita, I. 1998. "Studiul geomorfologic al degradarilor de teren din bazinul mijlociu al Barladului". Teza de Doctorat, (Geomorphological study of the Land Degradation in the Middle Catchment of the Barlad River. Ph. D. thesis). Universitatea "Alexandru Ioan Cuza ", Iasi, 287pp.

—. 1999. "Vegetation, Land Use and Erosion Processes," in Sediment Delivery Scenarios for Small Watersheds. Symposium Proceedings. Bucharest: Institute of Geography. pp. 66-73.

—. 2000a. "Formarea si evolutia ravenelor din Podisul Barladului" (Formation and Development of Gullies within the Barlad Plateau). Iasi: Editura Corson. 169 pp.

—. 2000b. "Geomorfologie Aplicata. Procese de degradare a terenurilor deluroase" (Applied Geomorphology. Land Degradation Processes in Hilly Areas). Iasi: Editura Universitatii "Alexandru Ioan Cuza," 247 pp.

—. 2003. "Hydraulic Efficiency of the Discontinuous Gullies," in J. Poesen and C. Valentin, eds. "Gully Erosion and Global Change". Elsevier-Catena. Volume 50. Issues 2-4. pp. 369-379.

—. 2006. "Gully Development in the Moldavian Plateau of Romania" in Katharina Helming, J.L.Rubio and J. Boardman, ed. Soil Erosion Research in Europe. Elsevier-Catena, Special Issue, Volume. 68, Issues 2-3, pp. 133-140.

Ionita, I., R. Mărgineanu, and C. Hurjui. 2000. "Assessment of the Reservoir Sedimentation Rates From 137-Cs Measurements in the Moldavian Plateau." Acta Geologica Hispanica. Volume 35. Number 3-4. Special Issue "Assessment Of Soil Erosion and Sedimentation

Through the Use of 137-Cs and Related Techniques", Queralt, I., F. Zapata, and Agudo Garcia, eds. Barcelona. Spain. pp. 357-367.

Ionita I., M. Radoane and S. Mircea. 2006. "Romania." in John Boardman and Jean Poesen, eds. Soil Erosion in Europe. Amsterdam-London-New York-Oxford-Paris-Tokyo: John Wiley. pp. 155 – 166.

Jeanrenaud, P. 1971. "Geologia Moldovei centrale dintre Siret si Prut". Rezumatul tezei de doctorat. (Geology of Central Moldavia Between the Siret and Prut Rivers. Abstract of the Ph. D. thesis), Universitatea "Alexandru Ioan Cuza ", Iasi.

Motoc, M. 1983. "Ritmul mediu de degradare erozionala a solului in R. S. Romania". (Mean Rate of Soil Degradation by Erosion in Romania). Buletinul Informativ al Academiei de Stiinte Agricole si Silvice, nr.13., Bucuresti, pp. 47-65.

Motoc M., I. Ionita, and D. Nistor. 1998. "Erosion and Climatic Risk at the Wheat and Maize Crops in the Moldavian Plateau." Romanian Journal of Hydrology & Water Resources ". Volume 5. Number 1-2. National Institute for Meteorology and Hydrology. Bucharest. pp. 1-38.

Piest, R. F., M. Joe Bradford, and M.G. Wyatt. 1975. "Soil Erosion and Sediment Transport from Gullies". U.S. Department of Agriculture (for official use).

Pujina, D. 1997. "Cercetari asupra unor procese de alunecare de pe terenurile agricole din Podisul Barladului si contributii privind tehnica de amenajare a acestora". Teza de doctorat, Universitatea Tehnica "Gheorghe Asachi", Iasi.

Radoane, M., N. Radoane, and I. Ichim. 1995. "Gully Distribution and Development in Moldavia, Romania." Elsevier – Catena. Volume 24. pp. 127-146.

CHAPTER ELEVEN

IMPACT OF RAINFALL SPATIAL VARIABILITY ON RUNOFF USING SWAT MODEL

MANEL MOSBAHI, SIHEM BENABDALLAH AND RACHED M. BOUSSEMA

Introduction

Water resource management and hydrologic modelling studies are intrinsically related to the spatial processes of the hydrologic cycle. The ability of hydrological models to accurately predict water availability at a catchment outlet depends to a great extent on how well spatial input data describe the relevant characteristics. For this reason, proper implementation of these models requires decisions to be made about not only the size of the spatial grids such as DEM (Chaplot, 2005), land-use and soil map (Muttiah and al, 2001, Romanowicz and al, 2005) but also the rain gauge density.

Precipitation is the key input variable that drives flow and mass transport in hydrologic systems (Beven, 2001). Accurate spatial and time representation is essential to runoff prediction (Faurès et al, 1995). The sensitivity of watershed models to spatial rainfall variability is becoming a key issue, especially in semi-arid areas. Indeed, rainfall input is subject to uncertainty as a result of an imperfect knowledge of its spatial distribution.

In this study, our main objective is to assess the impact of spatial rainfall variability effect on runoff within a semi-arid catchment located in northern Tunisia using the Soil and Water Assessment Tool (SWAT). The simulations were realized using daily data collected over a 14-year period for eight rain gauges. The investigation is based on comparing the water discharge measured at the watershed outlet to those generated by the model from four different scenarios of gauge concentration.

Materials and Methods

Study area: The study site is the Sarrath River Basin, located in northwest Tunisia, presented in Fig. 11-1. The river originates in the semi-arid Atlas Mountains of eastern Algeria. This region is part of the large Tunisian Medjerda catchment; it encompasses an area of 1520 km² with an elevation range from 573 to 1350 meters and average slope steepness from 3 to 72 %.

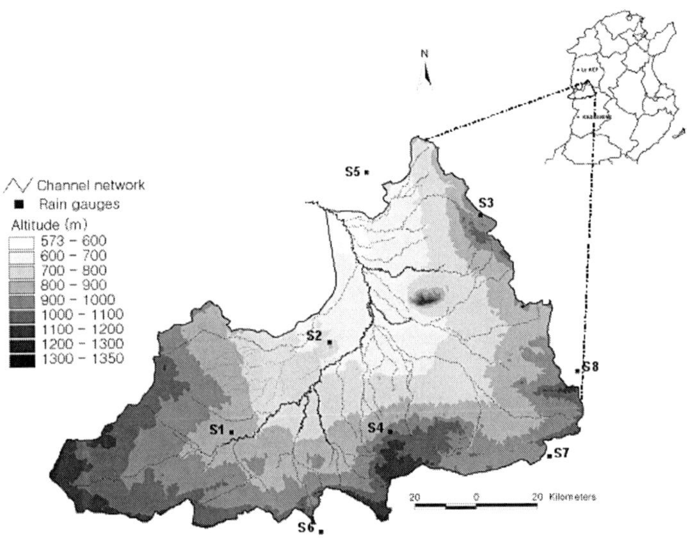

Fig.11-1: Catchment location and available rain gauges distribution

The area under study lies in the sub-humid to Mediterranean humid bio-climatic region. It is characterized with an extreme variability in annual and inter-annual rainfall coupled with hot summers and mild winters. The rainy season extends from September to May, with intense precipitation in autumn. In fact, for the 1990-2003 period, the average annual precipitation ranges from 343 to 563 mm with a variation coefficient ranging between 0.32 and 0.47. The annual temperature ranges from 4 to 28°C. July is the hottest month with maximum monthly temperatures valucs between 19 and 28°C. The coldest month is usually January with maximum monthly temperatures between 4 and 10°C. The Sarrath catchment is poorly covered with vegetation with the exception of some degraded forests and dense brushes on hilly areas. Majors crops grown in the basin are corn, representing 31 % of the total area, and tree cultivation, covering only almost 4 % of the whole catchment. The rest of the basin consists of pasture (9%), uncovered land (8%) and husbandry (4%).

The major soil types found in the Sarrath River Basin include calcareous brown soils, fluviosols and vertisols

Model description: The SWAT model is a physically-based time continuous simulating model that operates on a daily time step. It is a public domain model supported by the United States Department of Agriculture-Agricultural Research Service (USDA-ARS) (Arnold et al, 1998). Model design objectives were to predict the impact of land management practices on water, and sediment and agricultural chemical yields in large ungaged basins over long periods of time.

Major model components include: hydrology, weather, erosion, nutrients and pesticide fate. The hydrologic model is based on the water balance equation in the soil profile which is subdivided into multiple layers that support soil water processes including precipitation, infiltration, surface runoff, evapotranspiration, lateral flow and percolation. The soil percolation component of SWAT uses a storage routing technique to predict flow through each soil layer in the root zone. Thus, flow occurs when field capacity of soil layer is exceeded if the layer below is not saturated. The downward flow is governed by the saturated conductivity of the soil layer. Lateral subsurface flow in the soil profile is calculated simultaneously with percolation. Ground water flow contribution to total stream flow is simulated by routing shallow aquifer storage component to the stream (Arnold et al, 1993).

Surface runoff volume from daily rainfall is predicted with the modified SCS curve number method (USDA, 1972). The peak runoff rate is calculated according to the rational formula (USDA, 1986). Thus runoff is reduced by channel transmission losses that infiltrate to an underlying aquifer streambed.

Evaporation of soil water and transpiration by plants are evaluated as functions of potential evaporation and plant leaf area as described by Richie (1972). The Hargreaves equation (Hargreaves et al., 1985) was used in this study to determine the potential evapotranspiration based on daily temperatures.

The watershed parameterisation and the model input were derived using the SWAT ArcView Interface, which provides a graphical support to the desegregation scheme and allows the construction of the model input from digital maps.

The basic data sets required to develop the model input are: topography, land use, soil and climatic data. The Digital Elevation Model (DEM) was generated using contour lines taken from the 1/50000 scale national topographic maps. The cell resolution with an interval of 50 meters was used to generate the derived physical characteristics of the watershed. The soil and land use layers were obtained from the Soil and Water Conservation Agency (CES). They were produced from Landsat Thematic Mapper images and from soil maps.

Climatic data, including maximum and minimum temperatures over a 14 year-period were collected from two weather stations. Daily precipitation data were gathered for eight rain gauges over the same period.

The studied area is portioned into 24 sub-catchments and into 156 discrete computational units called Hydrologic Response Units (HRUs) which are a particular combination of topography, soil and land cover.

Analysis and discussion: Scenarios for rain gauges density

In this study, we analysed daily rainfall series for 8 rain stations located within and around the Sarrath catchment, denoted by the symbol "S" as it is shown in Figure 11-1. Other gauges with missing data were excluded from this study.

The average yearly rainfall amount over the study area was 459 millimetres, with values ranging from 224 millimetres in 1994 to 796 millimetres in 2003 (Table 11-3). The mean annual rainfall varies from 343 millimetres for S_3 to 563 millimetres for S_2 (Table 11-1). The central gauges contribute by the maximum rainfall rate.

Simulated runoff is computed using 4 different scenarios of rain gauges concentration. The first scenario consists of the use of a single rain gauge (S_2) located in the central part of the catchment and represents the highest yearly rainfall. In the second scenario, we applied three rain gauges (S_1, S_2, S_5) following the drainage of the river main channel.

In the third one, we tried to use five well-distributed rain gauges (S_1, S_2, S_3, S_4, S_5) within the watershed. The last scenario consists of adding three additional peripheral rain gauges (S_6, S_7, S_8) taking into account mountainous areas.

Table 11-1: Annual rainfall amount (mm) of the different rain gauges over the period 1990-2003

	S1	*S2*	*S3*	*S4*	*S5*	*S6*	*S7*	*S8*
1990	619	672	413	751	605	643	730	640
1991	428	466	215	453	398	302	332	376
1992	543	822	550	632	481	384	552	496
1993	232	332	192	309	176	156	319	279
1994	132	279	133	225	239	140	256	229
1995	348	608	348	420	371	341	363	343
1996	276	632	411	363	453	368	336	329
1997	338	640	411	443	472	344	431	413
1998	271	509	277	418	374	301	322	285
1999	364	582	367	420	373	357	370	380
2000	203	464	278	220	311	264	227	225
2001	185	360	204	369	269	256	263	251
2002	348	597	275	299	327	271	322	244
2003	717	915	726	706	696	697	726	545
Mean	357	563	343	431	396	345	396	360
CV*	0,47	0,32	0,46	0,38	0,35	0,45	0,41	0,35

* CV=S.D/MV ; S.D.: standard deviation; MV: mean value

The Thiessen method was used to determine buffered areas for each rain gauge within the catchment. The extension Arial Rain of ArcView was used to compute the weighted Thiessen factors as well as the polygonal borders for each rainfall station designed for each scenario. The results are shown in Table 11-2.

Table 11-2: Weighted Thiessen factors for each scenario

	Rain gauges	S_1	S_2	S_3	S_4	S_5	S_6	S_7	S_8
	Altitude (m)	830	856	980	970	633	950	930	950
Weighted Thiessen factor (%)	Scen 1	-	100	-	-	-	-	-	-
	Scen 2	18	45	-	-	37	-	-	-
	Scen3	33	14	17	28	8	-	-	-
	Scen4	30	14	12	14	8	7	7	8

Table 11-3: Annual rainfall amount for each scenario computed using weighted Thiessen factors

	Scen 1	*Scen 2*	*Scen 3*	*Scen 4*	*Mean*
1990	672	637	626	630	641
1991	466	434	402	388	423
1992	822	646	603	576	662
1993	332	256	257	251	274
1994	279	238	188	192	224
1995	608	474	407	396	471
1996	632	502	387	383	476
1997	640	524	434	427	506
1998	509	419	354	341	406
1999	582	465	411	404	466
2000	464	359	266	267	339
2001	360	295	271	260	297
2002	597	453	356	351	439
2003	915	799	742	729	796
Mean	563	464	407	400	459

As seen in Table 11-3, different rain gauge concentrations gave lower overall annual precipitation. On average, a difference of 30 percent is observed between Scen 1 and Scen 4.

Besides the annual discrepancy in annual precipitation, we should stress the daily variation observed in the recorded rainfall for the different stations. Figure 11-2 shows a 1:1 plot of the daily precipitation observed at S_1, S_2 and S_5. Daily precipitation is evidently variable across spatial locations. Simple regression analysis shows that the coefficient of correlation decreases as the location of the gauge is placed downstream (Fig 11-2). Furthermore, Figure 11-3 gives the 1:1 plot of daily precipitation with respect to S_3 and S_4. Even though, these gauges are located at similar altitude, they are not well correlated with S_1.

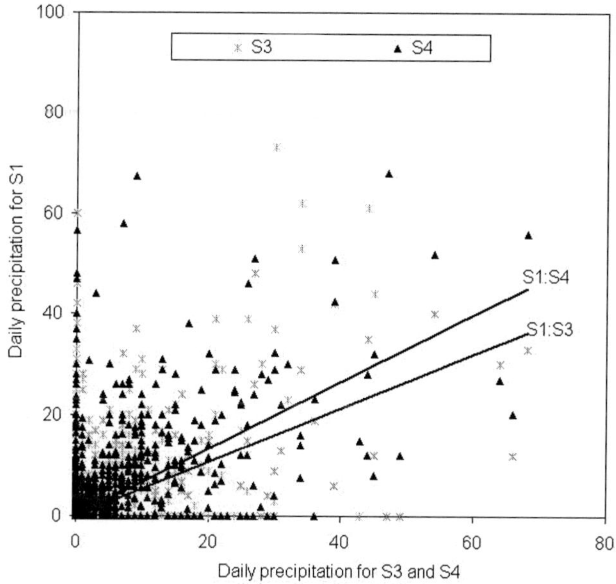

Fig. 11-2: Plot 1:1 of the daily precipitation observed at S_1, S_3 and S_4

Model prediction

For this study, SWAT was run at a daily time step for the 14 year period (1990-2003). Model parameters calibration was not attempted. Figure 11-2 shows the simulated daily water discharge at the outlet compared with observed daily water discharge obtained from Water Resources Authority. It is noted that simulated runoff systematically overestimated lower and medium peaks but underestimated high peaks (peaks for November 1990 and January 2003), with some rare exceptions.

We noted as well that simulation results produced hydrological responses correlated with rainfall data, with peaks that were not necessarily present in the measured water discharge. This is eventually due to the uncertainties within the observed data which are currently subject for discussion with the Water Resources Authority.

Impact of spatial rainfall variability on runoff prediction

Assuming that observed discharges are associated with some reading errors and the model results are more realistic, an attempt to assess the effect of rainfall spatial pattern and estimated model runoff is carried herewith.

In order to well represent the results, we focus on the hydrologic years 1990, 1995 and 2003, two of which are wet years and the other a dry one (see Table 11-1). Model runs were carried out for the four described scenarios of rain gauges distribution.

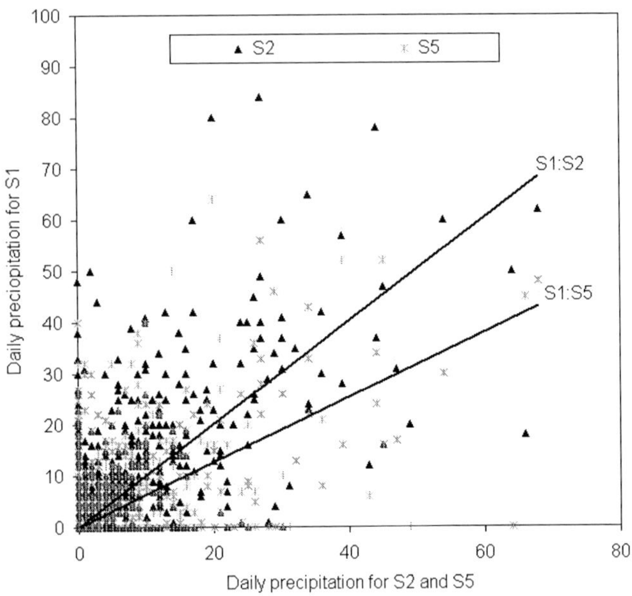

Fig. 11-3: Plot 1:1 of the daily precipitation observed at S_1, S_2 and S_5

For the year 1990, the main peak runoff occurred in the month of December, it was much higher for scenario 1 reaching 238 m^3/s. This value decreased as we added more rainfall stations in scenario 2 (with 3 gauges). However, we did not notice large difference as we added more gauges in scenarios 3 and 4. This is due to the spatial representation of the gauges in scenario 3.

The additional peripheral gauges outside the catchment boundary did not improve the runoff results substantially.

Model runs for the wet year 2003 confirmed the observation made for the previous simulation. In fact, the two main peaks for the months of October and December decreased as we added more gauges up to the third scenario. Further, the number of gauges improves the results to only a certain limit. Once the spatial rainfall pattern is ensured, the additional stations are not needed.

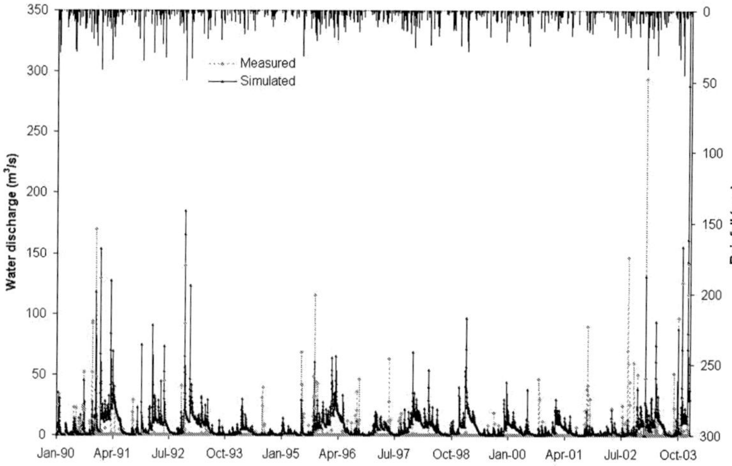

Fig. 11-4: Comparison of measured and simulated daily water discharge in the Sarrath catchment

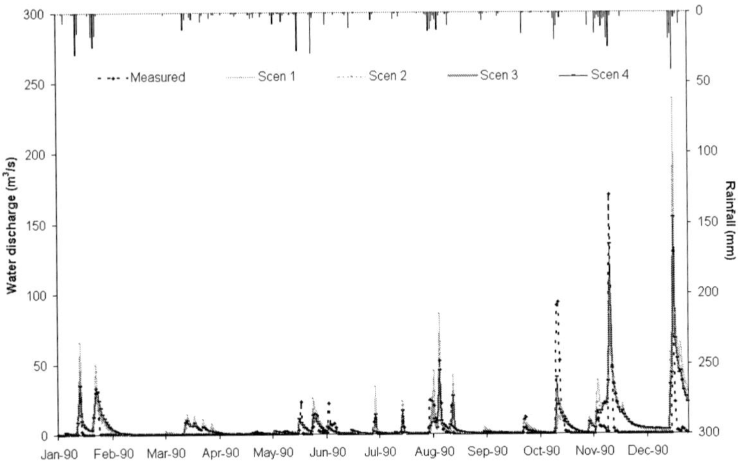

Fig. 11-5: Daily measured and simulated water discharge for 1990

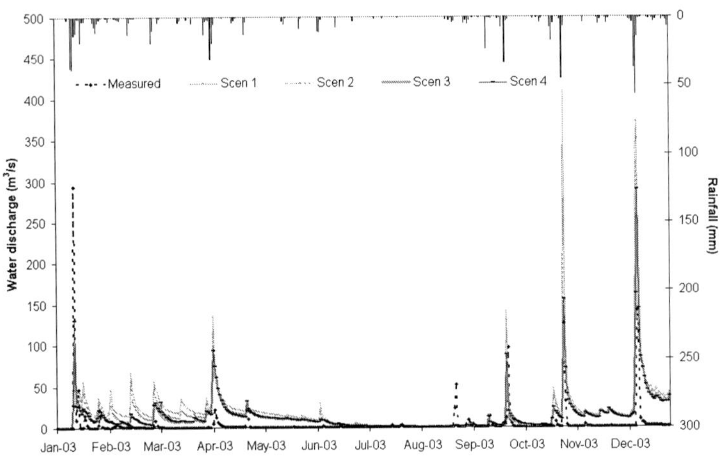

Fig. 11-6: Daily measured and simulated water discharge for 2003

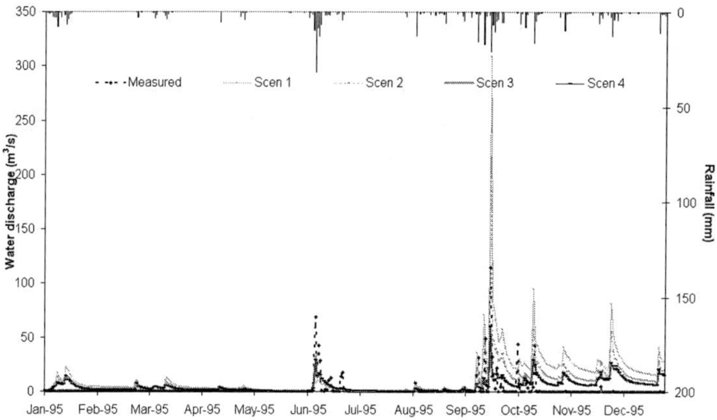

Fig. 11-7: Daily measured and simulated water discharge for 1995 using 8, 5, 3 and 1 rain gauge

For the dry year (1995), as model runs given in Figure 11-5 shows, there were pronounced differences between scenarios. It appears that the use of single gauge (S2) can induce inaccurate predictions of runoff compared to measured flow. Increasing the number of gauges (5 or 8 gauges) considering their location greatly reduced peaks.

Thus a better representation of rainfall spatial variability is essential to improve prediction and it is more acute issue for the semi-arid regions and especially in dry years.

Conclusion

Consideration of rainfall spatial variability is very important in studying runoff prediction with SWAT. In general, it is a crucial issue when dealing with semi-arid catchments. In fact, using a single rain gauge can lead to large uncertainties in runoff prediction.

In addition, model runs showed a general trend to overestimate the low level peaks at lower rain gauge density, leading sometimes to unrealistic water discharge peaks.

However, increasing the number of gauges without considering their spatial distribution in the catchment is not sufficient to improve results. It appears that by using five well-distributed gauges, we obtain the same results as the use of 8 gauges. Thus, the location of gauges is a crucial parameter to be considered prior to hydrologic modelling. Threshold values for rain gauge concentration need further research to ensure good modelling results in semi-arid regions.

Acknowledgements

The authors wish to thank the Water Authority for supplying the measured data and the engineer Faouzi Batti from the Water and Soil Conservation Authority of the Ministry of Agriculture for his interest and help in this work.

References

Arnold, J.G., Allen, P.M and Bernhardt, G. 1993. «A comprehensive surface-groundwater flow model» Journal of Hydrology, 142:47-69.

Arnold, J.G., Srinivasan, R., Muttiah, R.S., and Williams, J.R. 1998. «Large area hydrologic modeling and assessment. Part I: Model development» Journal of the American Water Resources Association 34 (1), 73–89.

Beven, K.J. 2001. «Rainfall-runoff modelling» The Primer, Wiley, Chichester.

Chaplot, V., Saleh, A., Jaynes and D.B. 2005. «Effect of the accuracy of spatial rainfall information on the modelling of water, sediment and NO3-N loads at the watershed level» Journal of Hydrology, 312:223-224.

Chaplot, V. 2005. «Impact of DEM mesh size and soil map scale on SWAT runoff, sediment and NO_3-N loads predictions» Journal of Hydrology, 312: 207-222.

Faurès, J.M., Goodrich, D.C., Woolhiser and D.A., Sorooshian, S. 1995. «Impact of small-scale spatial variability on runoff modelling » Journal of Hydrology, 173:309-326.

Hargreaves, G.H and Samani, Z.A. 1985. «Reference Crop Evapotranspiration from Temperature» Applied Engineering in Agriculture, 1:96-99.

Muttiah, R.S., Wurbs, R.A. 2001. «Scale-dependent soil and climate variability effects on watershed water balance of the SWAT model» Journal of Hydrology, 256:264-285.

Romanowicz, A.A., Vanclooster, M., Rounsevell, M and La Junesse, I. 2005. «Sensitivity of the SWAT model to the soil and land use data parameterisation: a case study in the Thyle catchment, Belgium» Ecological Modelling, 187: 27-39.

CHAPTER TWELVE

SOIL EROSION IDENTIFICATION USING GIS TECHNIQUES IN GOLBAŞI (ADIYAMAN) LAKES BASIN

MURAT KARABULUT, MEHMET GÜRBÜZ AND MUHTEREM KÜÇÜKÖNDER

Introduction

Soil erosion is a naturally occurring process on all types of land. The agents of soil erosion are water and wind, each contributing to a significant amount of soil loss each year worldwide (Garg and Harrison, 1992). Erosion is the movement of soil particles by the action of water, wind, and gravity. It is a process that is always occurring but it can happen faster if the land is misused. Erosion is a serious problem in many parts of the world as well as in Turkey (Yüksel *et al.*, 2007). On the whole, soil erosion is a very serious problem in arid and semiarid regions and the process is greatly accelerated by human activity, specifically when vegetation is destroyed or plant residues are buried through tillage. Farming, deforestation and grazing livestock are all examples of human activity that degrade the natural balance and accelerate the rate of soil loss. Considering the slow formation of soil and the importance of agricultural production to human survival, the severity of this problem can not be ignored.

When natural vegetation is removed from land and soil is disturbed, soil erosion can accelerate and lead to environmental problems such as stream degradation, wetlands damage, lost reservoir capacity, sediment deposition on private and public property, and increased flooding and water pollution. These effects adversely impact both aquatic life and the quality of life. Soil erosion related suspended sediment concentrations in surface water can shorten the useful life of many reservoirs and dams.

Suspended sediment concentration also has an effect on the biologic life in those valuable areas. Sediments also have a significant impact on the quality of drinking, recreational and industrial water, because it can serve as a carrier and storage agent of many kinds of pollutants such as phosphorus, nitrogen and other kinds of agricultural chemicals (Brown, 1984; Bhargava and Mariam, 1990; Han et al., 1994, Karabulut, 2004).

An examination and monitoring of soil erosion is very important for sustainable land management. Many studies have been conducted to determine erosion prone areas by traditional methods, which are time consuming, expensive and labor intensive. Because of the dynamic character of erosion and the requirement of large amount of spatial data, GIS techniques have become remarkably effective at the core of studying erosion. The utility of GIS in monitoring, mapping and analyzing erosion problems has increased, due to the fact that this method allows an examination of large amounts of data (Suri and Hofierka, 1994).

In order to assess land conditions and predict the potential impacts of erosion, comprehensive techniques and models need to be created. Creating a model is usually difficult. However, the use of GIS can simplify this task by integrating different types and levels of data, information, and human knowledge. Modeling is used to understand geographic processes and evaluate strategies for efficient operation. Minimum input data for modeling erosion includes hydrological data (rivers and its properties, drainage density), land use (agricultural, natural vegetation, wetlands, residential or other), topographic (slope steepness and length) and soil data (soil types, erodibility) and climatic (rainfall and storms) data. All these data can be used to model erosion problems in such areas by using GIS techniques (Garg and Harrison, 1992; Çullu and Dinç, 1994; Suri and Hofierka, 1994; Desmet and Govers, 1995; Lanza and Siccardi, 1995). Several models have been developed to predict soil erosion such as USLE, RUSLE, WEPP and AGNPS (Mellerowicz et al., 1994; Toy and Osterkamp, 1995; Savabi et al., 2005; Yüksel et al., 2007).

One of the negative consequences of the steep terrain of Turkey is soil erosion. The causes of soil erosion in Turkey are a function of the steep, mountainous topography and the unique cultural forces that have shaped the country throughout its history. According to the General Directorate of Reforestation and Erosion Control (GDREC, 2007), 86% of the Turkey's land faces significant erosion problems. Indeed, only 14 % of the land represents zero or weak erosion potential. Areas most susceptible to erosion on agricultural land include much of Turkey, such as the Interior, East and Southeastern Anatolia.

The aim of this study is to investigate erosion problems in the Gölbaşı Lake (Adıyaman) Basin by using a Universal Soil Loss Equation (USLE) model with the aid of GIS.

Study Area

The area selected for this study is located in a small part of Aksu River watershed, including the Gölbaşı lakes depressions and their surrounding areas. This area is approximately 100 kilometers away from the city of Kahramanmaraş and roughly 80 kilometers from the city of Adıyaman (Fig. 12-1). The elevation ranges from 830 m to 1480 meters. The mean annual precipitation is around 734 millimeters and average temperature is 14° C. Vegetation types consist of oak, juniper and terebinth. The dominant agricultural products are wheat, barley and Antep pistachio (*Pistacia vera*). The general soil types consist of brownish and reddish brownish soils. The land is partially cultivated, while the rest is covered with scattered vegetation and wetland.

Materials and Methods

The most widely used soil erosion mapping tool is a Universal Soil Loss Equation (USLE) that is an empirical equation derived from various investigations (Ndunda, 2005; Akyürek and Okalp, 2006). In general, USLE model estimates soil erosion by rain drop impact and surface runoff. In this study, USLE was used in an ArcGIS 9.1 environment to calculate the soil erosion index for Gölbaşı (Adıyaman), Turkey. A model was developed to execute the USLE formula using data from the study area. The USLE is defined as follows (Doğan and Güçer, 1976, Wijesekera and Samarakoon, 2001, Ndunda, 2005, Irvem et al., 2007):

$$A = R\ L\ S\ K\ C\ P$$

Where:
A = average annual soil loss in tons per acre,
R = rainfall and runoff erosivity factor,
L = slope length factor,
S = slope steepness factor,
K = soil erodibility factor,
C = cover and management factor,
P = supporting and conservation practices factor.

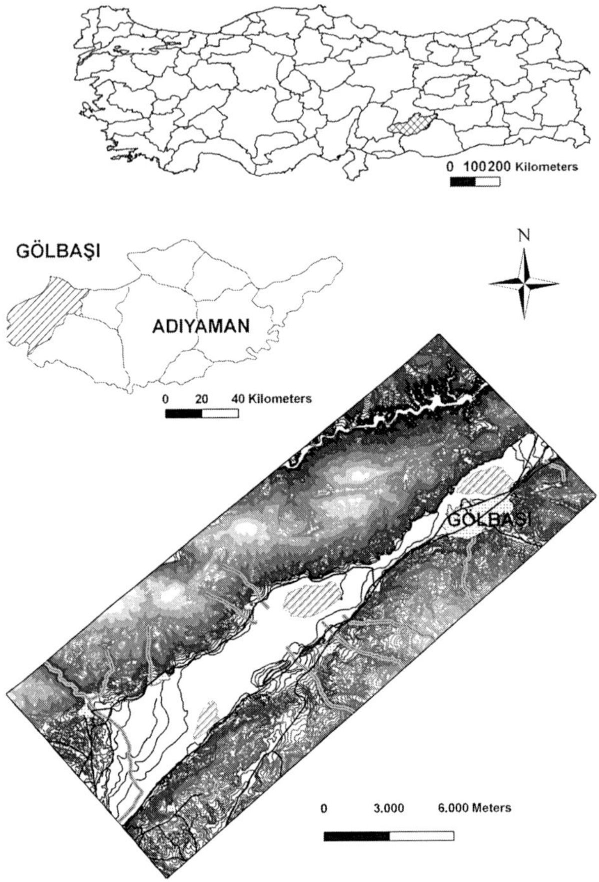

Fig. 12-1: Location of the study area.

Despite some limitation of this equation, it is still one of the best techniques which can be used to estimate soil losses as average annual erosion per unit area as a function of the major factors of sheet and reel erosion. In this study, each factor was used as a thematic layer in GIS model. Rainfall data, digital topographic map, soil and land use maps at 1:25,000 were used. GIS files were created for each factor and integrated into ArcGIS 9.1 Spatial Model (Fig. 12-2). Determinations of the various erosion factors are explained in following sections.

Soil Erodibility Factor (K). K factor defines the ability of soils to resist detachment and transport, based on the physical characteristics of each soil type. The K factor map was generated from a soil map at a scale 1:25.000 provided by the Ministry of Agriculture (1981). Soil slope 9% and slope length 22.1 meter were the main criteria for K factor determination (Table 12-1 and 12-2, Fig. 12-3) (Doğan and Güçer, 1976).

Table 12-1. K values and their erosivity levels (Doğan and Güçer, 1976)

K values	Erosivity levels
0-0,05	Very Low
0,05-0,10	Low
0,10-0,20	Moderate
0,20-0,40	High
0,40-0,60	Very High

Slope length and slope gradient factor (LS). Naturally, soil erosion tends to increase with the steeper slope of a field due to running greater amount of soil off by water erosion. Similarly, soil erosion by water also increases when the slope length increases due to the greater accumulation of runoff. The Digital Elevation Model (DEM) for this research was developed from vector contour lines using a topographic map scale 1/25.000 and the LS factor layer is then generated from DEM using the following formula (Figure 12-4) (Hudson, 2005).

*LS=1.6 *POW(([flowacc]*resolution)/22.1,*
*0.6)*POW(Sin([slope]*0.01745)/0.09, 1.3)*

Table 12-2: Soil groups with their K values used in this study (İrvem and Tülücü, 2004).

Soil Groups	K Values
Brown soil	0.20
Colluvial	0.18
Non calcic brown forest	0.15
Brown forest	0.20
Red brown Mediterranean	0.15
Hydromorphic alluvial	0.15
Red brown soil	0.20
Residential area / River bed	0.001

Cover and management factor (C). Soil loss from the land will increase if the land has no vegetative cover or very little vegetative cover and/or crop residues because vegetation intercepts rainfall, reducing their kinetic energy before hitting the ground. In this study the C factor was generated from digital soil maps at a scale of 1:25.000 provided from the Ministry of Agriculture and TM images. C factor classification procedures were completed based on previous works in Turkey (Table 12-3, Fig. 125) (Doğan and Güçer, 1976).

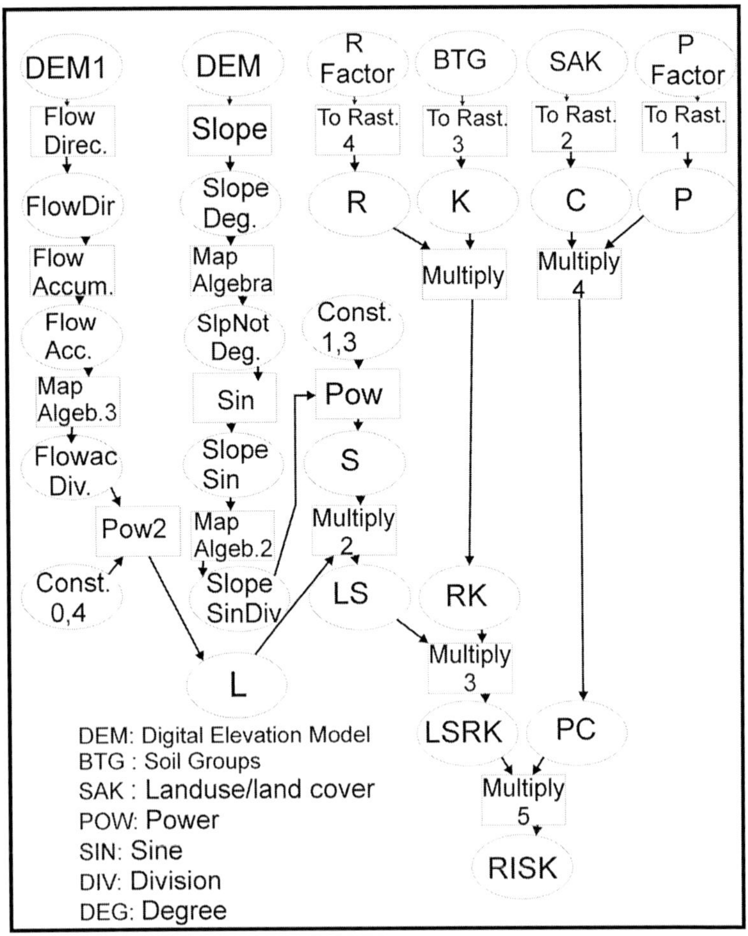

Fig.12-2: Created spatial analysis model for erosion risk assessment of Gölbaşı.

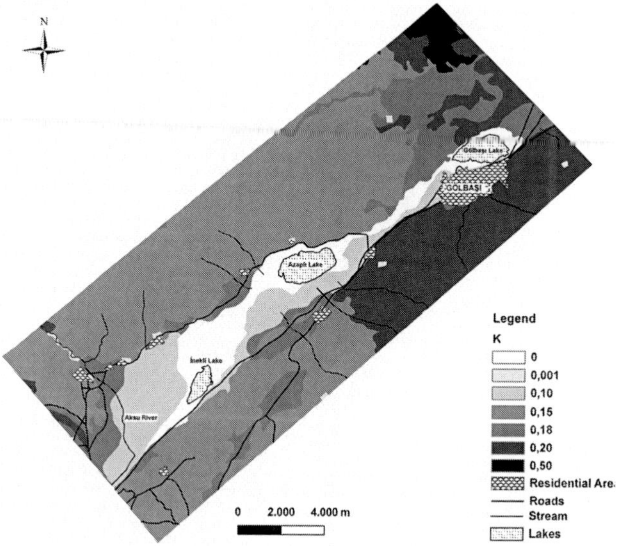

Fig. 12-3: K factor map of the study area.

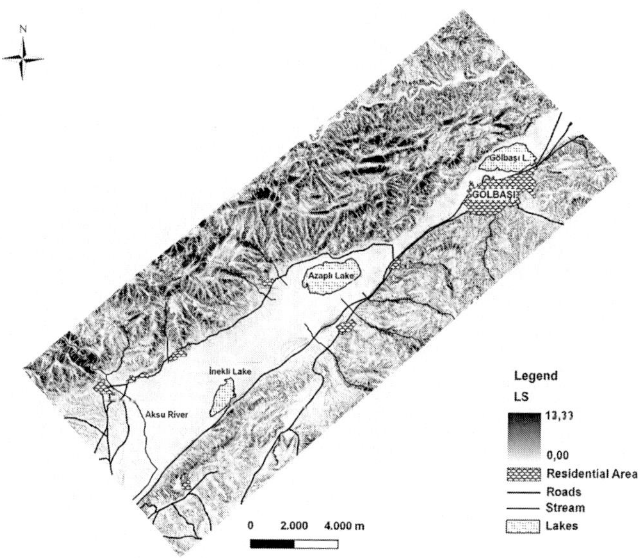

Fig. 12-4: LS factor map of the study area.

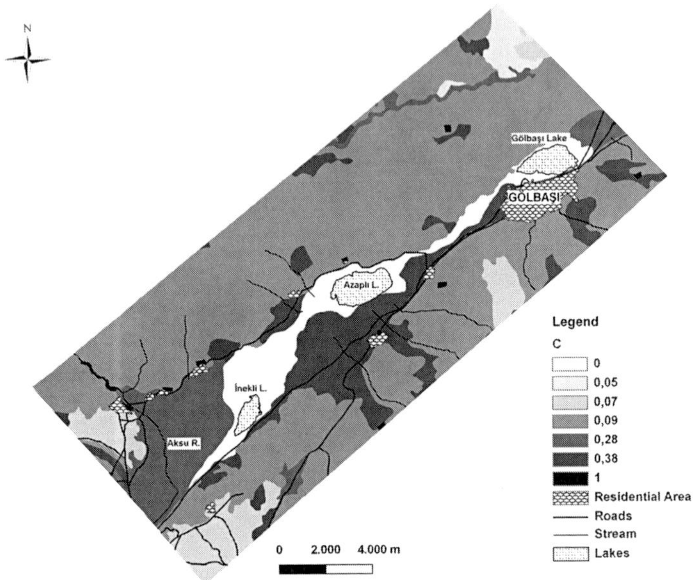

Fig. 12-5: C factor map of the study area.

Table 12-3: Land use and associated C values (İrvem and Tülücü, 2004).

Landuse	C values
Dry farming (fallow)	0.38
Dry farming (non-fallow)	0.07
Irrigated farming	0.28
Forest / Orchard	0.05
Scrub / Vineyard / Grassland	0.09
Residential area / River bed	1.00

Rainfall and runoff erosivity factor (R). Both rainfall and runoff factors must be considered in the determination of water erosion risk areas. The impact of raindrops on the soil surface can break down soil aggregates and disperse the aggregate material. Thus, R factor represents the erosive power of rain drops and run-off water. In this, the study R factor was determined by using works of Doğan and Denli (1999) (Fig. 12-6).

Support Practice Factor (P). Because no information in regard to the P-factor is available for this area, a value of 1 was assigned to the model presented in this study (Fig. 12-6).

Results and Discussions

In this study, soil erosion risk was modeled in the Gölbaşı Lake Basin by integrating the USLE with GIS. The soil erosion map resulting from the spatial overlay of USLE factors in the study area is presented in Fig. 12-7. Table 12-4 presents corresponding quantitative soil loss, in addition to spatial information. Figure 12-7 shows erosion risk in the study area, which is expressed in five classes, ranging from very low risk to very high risk.

The result of this study gives an erosion range of 0 to 25 tons per hectare per year (tons/ha/yr). Variation in the results is due apparently to variations in the values of each of the factors used. It is found that very severe erosion risk areas cover only around 1.38 % in the area. The study shows that only a 386.5 hectare area is under high and very high erosion risk. An area of 4.83% (1344.90 hectares) has a moderate risk of erosion. These areas require adaptation of suitable conservation measures to protect the present valuable soil. The rest of the land (11525.80 hectares 41.41%), represents a weak or very weak erosion risk (14574.50 hectares, or 52.36% of the total land), due to valuable vegetation cover and topographic conditions. Large percentages of these areas are located in regions with moderate to high erosion potential, where the inappropriate cultivation practices or crop rotation result in accelerated soil erosion. The higher erosion values are concentrated in steeper parts of slopes with accelerated movement of water, indicating medium or high erosion risk. The lower erosion values are located in less steep parts of the land, with slowed movement of water, indicating the areas with low erosion risk. Relatively flat areas, which are occupied mostly by wetland, are affected only by non significant erosion. A soil loss from these parts is not critical. Most probably, these low lying slopeless areas cause a slow movement of water, indicating the areas of medium to high sediment deposition. In this study, the above-mentioned areas are located in the bottom of the Gölbaşı depression, a depression which was shaped by tectonic activities.

Our erosion prediction model indicated that erosion response is much more sensitive to slope steepness and cover management factors than to other environmental variables. Therefore, in the area which was studied, priority must be given to protection of natural vegetation on steep slopes to reduce the erosivity effects on soil loss.

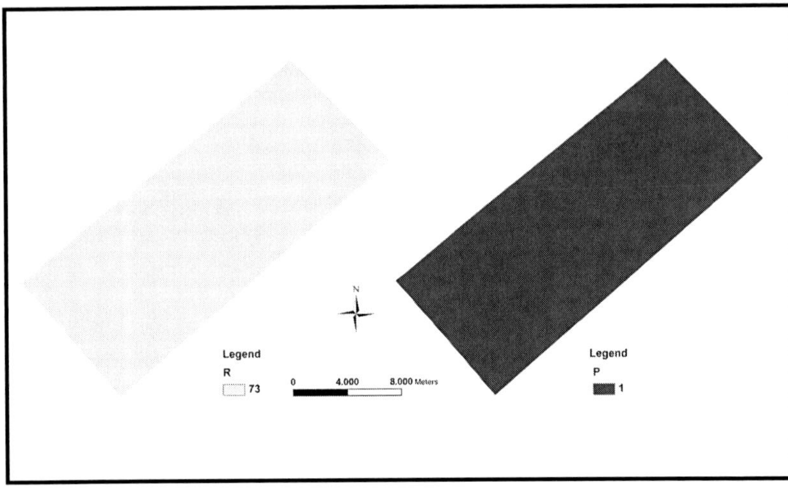

Fig.12-6. R and P factor maps of the study area

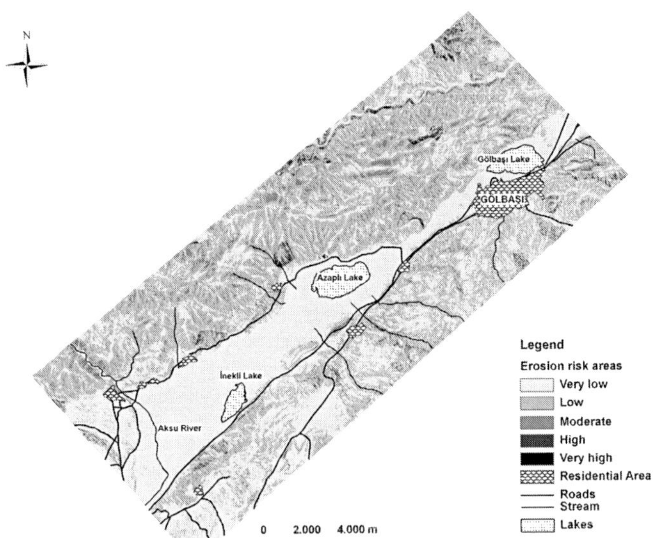

Fig. 12-7: The soil erosion map resulting from the spatial overlay of USLE factors in the study area.

Table 12-4. Area and percentage erosion hazard.

Erosion Risk	Area (ha)	Percent (%)
Very Low	14574.50	52.36
Low	11525.80	41.42
Moderate	1344.90	4.83
High	312.30	1.12
Very High	74.20	0.26
Total	27831.70	100.00

Conclusions

The USLE model which is used in this study is a useful technique for predicting soil erosion. This study showed that the combination of USLE and GIS presents a promising way for identification, mapping and interpretation of the spatial pattern of eroded soil. The results indicate that the use of GIS techniques can improve the mapping accuracy of USLE when applied to affected areas. Preliminary field investigations confirm those results. On the other hand, information gathered from field works can be used for accurately assessing model data and can provide reliable parameters for the USLE model.

Erosion risk maps provided here can be used in several ways. Firstly, these maps could help to identify natural and accelerated erosion prone areas. Variation of erosion density on the map can be compared with topographic conditions, ecologic status and cultural information to determine the nature of the problem. Secondly, the maps not only provide information on erosion risk areas, but also represent sediment deposition areas. Overall, these maps can produce valuable information about the condition of land to support governmental decision making.

In conclusion, it must be kept in mind that modeling can sometimes produce inappropriate results due to erosion factor values. However this study proved that it is possible to analyze multi-layers of data spatially and quantitatively within a watershed using GIS.

References

Akyürek, Z.and K. Okalp. 2006. A fuzzy-based tool for spatial reasoning: A Case study on soilerosion hazard prediction, 7th International Symposium on Spatial Accuracy Assessment in Natural Resources and Environmental Sciences. M. Caetano and M. Painho (eds.), 5-7 July, Portugal.

Bhargava, D. and D.W. Mariam. 1990. Spectral relationships to turbidity generated by different clay materials, *Photogrammetric Engineering and Remote Sensing* 56: 225-229.

Brown, L.R. 1984. The Global Loss of Topsoil", *Journal of Soil and Water Conservation* 39:162-165.

Çullu, M.A. and U. Dinç. 1994. Coğrafi Bilgi Sistemleri Yardımıyla Şimdiki ve Potansiyel Toprak Erozyon Alanlarının Belirlenmesi, *1. Ulusal CBS Sempozyumu Bildiri Kitabı*, 18-20 Ekim 1994, Trabzon.

Desmet, P.J.J. and G. Govers. 1995. GIS-Based Simulation of Erosion and Deposition Patterns in an Agricultural Landscape: A Comparison of Model Results with Soil Map Information, *CATENA* 25:389-401.

Doğan, O. and Ö. Denli. 1999. Türkiye'nin Yağış-Kuraklık-Erozyon İndisleri ve Kurak Dönemleri, KHGM Ankara Araştırma Enstitüsü Müd., Genel Yay. No. 215, Teknik Yay. No. 60, Ankara.

Doğan, O. and C. Güçer. 1976. Su Erozyonunun Nedenleri-Oluşumu ve Üniversal Denklem ile Toprak Kayıplarının Saptanması, KHGM Merkez TOPRAKSU Araştırma Enstitüsü Müd., Genel Yay. No. 41, Teknik Yay. No. 24, Ankara.

Garg, P.K. and A.R. Harrison. 1992. Land Degradation and Erosion Risk Analysis in S.E. Spain: A Geographic Information System Approach, *CATENA*, Vol.19, p.411-425.

GDREC, General Directorate of Reforestation and Erosion Control, http://www.agm.gov.tr, visited on May 3rd, 2007.

Han, L., D.C. Rundquist, L.L. Liu and R.N. Fraser. 1994. The Spectral Responses of Algal Chlorophyll in Water With Varying Levels of Suspended Sediment, *International Journal of Remote Sensing* 15:3707-3718.

Hudson, P.F. 2005. Soil Erosion Modeling Using the Revised Universal Soil Loss Equation (RUSLE) In a Drainage Basin in Eastern Mexico, Environmental GIS Lab Notes, (http://www.utexas.edu/depts/grg/hudson/grg360g/EGIS/labs_04/Lab9/lab9_soil_erosion_05.htm).

İrvem, A. F., Topaloğlu. and V. Uygur. 2007. Estimating Spatial Distribution of Soil Loss Over Seyhan River Basin in Turkey, *Journal of Hydrology* 335:3-4.

İrvem, A. and K. Tülücü. 2004. Coğrafi Bilgi Sistemi ile Toprak Kaybı ve Sediment Verimi Tahmin Modelinin (EST) Oluşturulması ve Seyhan-Körkün Alt Havzasına Uygulanması *Ç.Ü. Fen Bil. Enst. Dergisi* 1:1-7.

Karabulut, M. 2004. An Examination of Surface Water Using Close Range Remote Sensing Techniques, Proceedings on 3rd GIS Days in Turkey, October 6-9 (in Turkish).

Lanza, L. and F. Siccardi. 1995. The role of GIS as a tool for the assessment of flood hazard at the regional scale, In: *Geographical Information Systems in Assessing Natural Hazards*, A. Carrara and F. Guzzetti (eds.) Kluwer Academic, Netherlands, p. 199-217.

Mellerowicz, K.T., H.W. Rees, T.L. Chow and I. Ghanem. 1994. Soil conservation planning at watershed level using the Universal Soil Loss Equation with GIS and microcomputer technologies: a case study, *Journal of Soil and Water Conservation*49(2):194-199.

Ministry of Agriculture. 1981. Ceyhan Havzası Toprakları, Tarım Orman ve Köy İşleri Bakanlığı, Köy Hizmetleri Genel Müdürlüğü, Yay.No. 285, Havza No: 20, Raporlar Serisi: 69, Ankara.

Ndunda, P. 2005. Soil erosion estimation model for San Timoteo, www.esri.com/industries/university/model_builder/powerpoints_2005/ndunda_soil_erosion_san_timoteo.pdf

Savabi, M.R., D.C. Flanagan, B. Hebel and B.A. Engel. 1995. Application of WEPP and GIS-GRASS to a small watershed in Indiana, *Journal of Soil and Water Conservation* 50(5):477-483.

Suri M. and J. Hofierka. 1994. Soil Water Erosion Identification Using Satellite and DTM Data, Proceedings of EGIS/MARI 94 Conference in Paris, France, 1994, s. 937-944.

Toy, T.J. and W.R. Osterkamp. 1995. The Applicability of RUSLE to Geomorphic Studies, *Journal of Soil and Water Conservation* 50(5):498-503.

Wijesekera N.T.S. and L. Samarakoon. 2001. Extraction of Parameters and Modelling Soil Erosion Using GIS in a Grid Environment, *Proceedings of the Asian Conference on Remote Sensing*, Singapore, November.

Yüksel, A., A.E. Akay, M. Reis and R. Gündoğan. 2007. Using the WEPP Model to Predict Sediment Yield in a Sample Watershed in Kahramanmaraş Region, *International Congress on River Basin Management*, 22-24 March, Antalya.

CHAPTER THIRTEEN

THE GENERAL VEGETATION CHARACTERISTICS OF CEDAR FORESTS ON TAURUS MOUNTAINS

ALI KAVGACI

Introduction

The genus *Cedrus* includes four species. These are *Cedrus libani, C. deodora, C. brevifolia* and *C. atlantica. C. libani* only has distribution in Turkey. In addition to this, it also has distribution in Lebanon and Syria. But the forests of this species in these countries were destructively consumed in the past. Due to that, today there are only a few remnants of these forests in Lebanon (Seamaan and Haber, 2003; Yeşilkaya, 1994). *C. libani* mainly distributes on Taurus Mountains in Turkey. Beside this, there are small remnant forests of *C. libani* in the Middle Black Sea Region around Erbaa and Niksar and in the Çaykışla Valley in Emirdağ-Afyon (Günay, 1990). The cedar forests in Turkey that covers about an area of four hundred thousand hectares (Anon. 2006) are the densest and the most magnificent cedar forests in the world.

Cedrus libani is very important tree species, in terms of culture, science, economy and esthetic, (Boydak, 1996). According to the testaments written in Arabic, Phoenician, Greek and Roman languages, because of the unique property of its woods (Boydak and Asan, 1990), it was used in the construction of temples, palaces vessels, luzories buildings, furniture and coffin for mummies in the past (Kayacık, 1948 and 1980). Also it was mentioned as a symbol of greatness, glory, honour, force and solidity, mortality, magnificence of kingdom, appreciation, worldwide expansion of power and richness (Mayer and Sevim, 1959). All of these mentioned above indicate that *Cedrus libani* played a very important role in social and cultural life of several countries in the past.

Cedar generally constitutes dense forest between 1000 and 2000 m altitudes on Taurus Mountains. While in some regions, it descends to 500m. (Fethiye-Muğla), it also rises up to 2400 m elevation (Aydos Mountain-Adana). In those belt, cedar has either pure stands or mixed stands with other tree species, such as *Abies cilicica*, *Pinus nigra, Pinus brutia,* Oak species (*Quercus cerris, Quercus infectoria, Q. libani* etc.) *Juniperus excelsa* and *J. foetidissima*. Atalay (1990) mentioned that the cedar forests appear on the limestone having different facies and geologic eras, such as marl, conglomerate, sandstone, chlorite –sericite schists, clayey schists, quartzite, quartzite-schists, peridotite-serpantine complex and flysches of Eocene, Oligocene and Cretaceous. The karst structure of cedar sites is especially important for the ecology of cedar (Kantarcı, 1990; Boydak, 1996).

Although it is very complicated to climatically classify the Cedar forests, it is generally under the effect of Mediterranean climate type as macroclimate. The climate on the upper zones of Taurus Mountains, on which Cedar has the largest distribution, characteristically belongs to Mediterranean climate. But it also shows some private properties. The climate at this region is identified as Mediterranean mountain climate, which has strong winter, high light intensity, big temperature differences and less cloud. (Saatçioğlu, 1975).

In terms of synsistematic, the cedar forests on Taurus Mountains take place under two different alliances depending on to the class Quercetea pubescentis and the order Querco-Cedretalia (Akman et al., 1978). These alliances are Lonicero-Cedrion and Abieti-Cedrion. While Lonicero-Cedrion distributes on Western Taurus, Abieti-Cedrion has distribution on Middle and Eastern Taurus on which *Cedrus libani* makes mixed stands with *Abies cilicica* (Aksoy and Özalp, 1990).

Abieti-Cedrion (Quezel and Pamukçuoğlu, 1973)

This alliance has distribution on the Middle and Eastern Taurus Mountains. It appears on the hills and ridges on the upper of the *Pinus nigra* distribution zone, which nearly rises up to 1450 m. In addition to that, the soil type of these fields is usually loamy soil. *Abies cilicica*, a typical shade tolerant species, has stand structures, which are close to unevenged multi-cohort forests. At *Abies cilicica* distribution areas, the cedar was pushed to the steep slopes, relict rocky sites, high Mediterranean drier sites and hills. But inner Mediterranean parts, the competition ability of *Abies cilicica* decreases and it only takes place on the narrow riverbeds (Mayer and Aksoy, 1986). The characteristic species of Abieti-Cedrion are; *Abies cilicica, Quercus petraea* subsp. *pinnatiloba, Sorbus torminalis* var. *orientalis, Coronilla varia* subsp. *libatonica, Corydalis solida* subsp. *tauricola, Tanecetum cilicicum, Cephalorrhynchus tuberosus, Scilla cilicica, Acer hyrcanum* subsp. *tauricolum, Cyclamen cilicium* (Quezel and Pamukçuoğlu, 1973).

Thlaspi cataonicum-Cedrus libani forest

This is an endemic community. It was identified in Feke-Meran Valley and Bekir Mountain (Akman et al, 1979; Mayer and Aksoy, 1986; Akman, 1995). It ranges between 1600-1900m altitudes. In addition to that, it appears on the north, west, and east aspects with middle inclination on the limestone main rocks.

Campanula involucrata, Scilla bifolia, Veronica syriaca, Thlaspi cataonicum, Geum heterocarpum, Helleborus vesicarius, Rosa pulverulenta and *Quercus petraea* subsp. *pinnatiloba* are the distinguishing species of this community. *Brunnera orientali, Corydalis solida* subsp. *tauricola, Tanacetum cilicicum, Coronilla varia* subsp. *libanotica, Cephalorrhynchus tuberosus*, *Symphytum palaestinum, Abies cilicica* subsp. *cilicica, Cicerbita mulgedioide,* which are characteristics for Abieti-Cedrion alliance and *Bunium microcarpum, Pinus nigra* subsp. *pallasiana, Doronicum caucasicum, Anemone blanda, Galium peplidifolium, Turritis laxa, Geranium macrostylum, Viola parvula, Quercus libani, Quercus cerris, Sorbus umbellata, Juniperus drupacea, Vicia lunata, Cerastium fragillimum, Potentilla kotschyana, Cyclamen coum* subsp. *anatolicum* and *Juniperus excelsa*, which are characteristics for Querco-Cedretalia order, accompany with these species. The scrub layer of this community is formed by some deciduous tree species, such as *Quercus libani, Q. cerris, Sorbus umbellata* and *Carpinus orientalis*.

Quercus petraea subsp. pinnatiloba-Cedrus libani forest

This is another endemic community (Akman, 1973). It ranges from 1400m to 1800m altitudes on the Amanos Mountain, which is at the eastern part of Taurus Mountains. It distributes on rocky habitats, which contain limestone, quartzite, and greywacke main rocks. At these sites *Fagus orientalis* and *Abies cilicica* are more dominant than *Cedrus libani* according to stand competition. Because of this, *Cedrus libani* disappears on the climax sites and it only distributes on more extreme sites such as hills with shallow soil, steep ridges and steep rocky slopes. *Cedrus libani* forms mixed stands with *Quercus cerris, Abies cilicica* and *Pinus nigra* on these kind of habitats.

Sesleria anatolica-Cedrus libani forest (Quezel and Pamukçuoğlu, 1973)

Sesleria anatolica-Cedrus libani forest distributes on Middle Taurus Mountains. Yet it is not sufficiently known. However, the distribution of this community is restricted by *Abies cilicica,* which has high competition ability. Some development stages of this community were seen in Elmalı Valley and Bolkar Mountain (Quezel and Pamukçuoğlu, 1973). Mayer and Aksoy (1986) mentioned that this community has relation with *Lecokia cretica-Abies cilicica* forest. The cedar usually forms mixed stands with *Pinus nigra*, which were negatively affected by overgrazing in Ermenek and some other valleys, of which the main rock is limestone.

Thlaspio microstyli-Cedretum libani (Varol and Tatlı, 2001)

This association was identified in Çimen Mountain in Kahramanmaraş by Varol and Tatlı (2001). At this region, which has less rainy and cool Mediterranean climate, *Cedrus libani* constitutes mixed stands with *Abies cilicica* subsp. *cilicica*. This forest ranges between 1450-1650m. *Thlaspio microstyli-Cedretum libani* forest has the characteristic species of Abieti-Cedrion alliance. The distinguishing species of this forest are; *Cedrus libani*, *Abies cilicica* subsp. *cilicica*, *Thlaspi microstylum*, *Pilosella xmacrotricha*, *Rubus discolor*, *Potentilla, detommasii*, *Astragalus fraxinifolius* and *Onosma sieheanum*. In addition to these species, *Quercus petraea* subsp. *pinnatiloba*, *Brunnera orientalis*, which are characteristic for Abieti-Cedrion and also *Vicia cracca* subsp. *alpestris*, *Bunium pauciflorum* var. *junceum*, *Pinus nigra*, *Ranunculus reuterianus*, which are characteristic for Querco-Cedretalia join in to the floristic composition of *Thlaspio microstyli-Cedretum libani* association.

Cicerbito-Cedretum libani (Serin and Eyce, 1994)

Cicerbito-Cedretum libani was identified by Serin and Eyce (1994) in Hadim-Aladağ Region in Konya. Cedar forms pure or mixed stands with *Abies cilicica* at this region. While this community ranges between 1700-1800m altitudes, it appears on rough and rocky sites. In addition to these, it develops on red Mediterranean soil and brown forest soils without lime. The distinguishing species of this community are *Cedrus libani, Cicerbita variabilis, Orthurus heterocarpus, Galanthus elwesii, Umbilicus erectus*. In addition to these species, *Arabis caucasicaa* susbsp. *brevifolia, Veronica macrostachya* subsp. *sorgerae, Thlaspi perfoliatum Abies cilicica* subsp. *cilicica, Corydalis solida* subsp. *solida, Campanula involucrata, Cyclamen coum* var. *cilicium,* which are characteristics for Abieti-Cedrion join in to the floristic composition of *Cicerbito-Cedretum libani*.

Lathyro variabilis-Cedretum libani (Duman, 1995)

Cedrus libani takes places on the southern slopes of Engizek Mountain in a belt between 1000-2000 m altitudes and on the northern slopes of Gizi Mountain between 1400-1900m altitudes (Duman, 1995). The soil of this community is light asidic, without clay and has rich organic material. The distinguishing species of this community are *Astragalus commagenicus*, which is a local endemic species and *Lathyrus variabilis*. In addition to these *Cephalorrhynchus tuberosus, Asyneuma amplexicaule* and *Corydalis solida*, which are characteristics for Abieti-Cedrion join in the floristic composition of the community.

Alysso mouradicae-Cedretum libani (Serin, 1996)

This community was identified by Serin (1996) in Beyşehir province in Konya. It appears on limestone. The topography is rough at these sites. Cedar constitutes dense pure forests and these forests ranges from 1600 m to 1800 elavation. *Cedrus libani, Alyssum mouradicum* and *Odontutes aucheri* are the distinguishing species of this community. *Arabis caucasica* subsp. *brevifolia, Campanula involucreta, Veronica macrostachya,* subsp. *sorgarea, Corydalis solida* subsp. *solida, Thlaspi perfoliatum, Abies cilicica* subsp. *isaurica, Cyclamen cilicicum* var. *cilicicum*, which are characteristic for Abieti-Cedrion accompany with those species.

Abieto-Cedretum libani (Ocakverdi and Çetik, 1987)

Abieto-Cedretum libani was identified by Ocakverdi and Çetik (1987) in Seydişehir province in Konya. *Abies cilicica* straggly accompany with *Cedrus libani* at this region. This community ranges from 1550 m to the upper belts and appears on limestone. *Cedrus libani, Abies cilicica* subsp.

isaurica, Hieracium huber-morathii and *Crocus chrysanthus* are the distinguishing species of the community. *Arabis caucasica* subsp. *brevifolia* and *Thlaspi perfoliatum*, which are characteristic for Abieti-Cedrion accompany with these species. In addition to these species, some characteristic species of the Fagatelia sylvatica, such as *Corydalis solida* subsp. *solida, Moehringia trinervia, Festuca heterophylla* and *Galium rotundifolium* join in to the floristic composition of the community. This could be the result of microclimatic structure of the region (Ocakverdi and Çetik, 1987).

Lonicero-Cedrion (Quezel and Pamukçuoğlu, 1973)

Lonicero-Cedrion distributes on the Western Taurus. It takes place on limestone and dolomite main rocks. *Cedrus libani* mostly forms pure stands at this region and *Pinus brutia* and *P. nigra* straggly join in to the stand structure. The characteristic specieses of Abieti-Cedrion disappear at this region. *Cedrus libani, Acer hyrcanum* subsp. *sphaerocarpum, Lonicera nummularifolia* subsp. *glandulifera, Digitalis* cariensis are the characteristic species of the alliance (Quezel and Pamukçuoğlu, 1973; Çetik, 1976). Lonicero-Cedrion is peculiar to Western Taurus Mountains. *Bunium microcarpum, Anemone blenda, Cyclamen coum* var. *coum, Vicia stenophylla, Aemlanchier ovalis* and *Sorbus umbellatae*, which are the characteristic species of the Querco-Cedretalia order, are the dominant species of the alliance (Mayer and Aksoy, 1986). Mayer and Sevim, (1958) pointed that the cedar forests in the Western Taurus Mountains are the most magnificent ones of all and due to the disappearance of *Abies cilicica* and the limited distribution of *Pinus nigra*, *Cedrus libani* has a large ecologic amplitude in this region. At the high mountain zone of this region, *Cedrus libani* forms mixed stands with some deciduous tree species such as *Populus tremula, Acer hyrcanum* and *Acer platonoides*.

Alliaria officinalis –Cedrus libani (Akman et al. 1979)

Alliaria officinalis –Cedrus libani forest, which has wide distribution and shows local differences, ranges from 1600 to 2000m and mainly distributes on limestone main rock (Akman et al., 1979). The distribution of this community is restricted in Elmalı-Antalya Region. It usually has multi cohort stand structure. The physiognomic structure of this community is formed by old and tall cedar individuals, which have wide diameters. The distinguishing species of the community are *Cedrus libani, Alliaria officinalis, Rosa pulverulenta, Chamaecytisus pygmaeus,*

Saponaria glutinosa, Cicerbita variabilis, Scrophularia scapolii and *Cicerbite longirostris.*

This community has 3 different subcommunities. These are *Paeonia turcica* subcommunity, *Festuca varia* subcommunity and *Cerastium fragillimum-Galium peplidifolium* subcommunity. *Paeonia turcica, Acer platonoides, Populus tremula, Ulmus montana* and *Achillea grandiflora* are the distinguishing species of the *Paeonia turcica* subcommunity. This subcommunity mainly occurs on plain fields on the shadow aspects on the middle and high mountain zones. The main rock is limestone.

The distinguishing species of *Festuca varia* subcommunity are *Quercus libani, Daphne oleoides, Festuca varia, Ebenus boisseri* and *Pentaptera borquetii.* It appears on dry sites, which take place on the shadow aspects with sallow soil. The soil type of this subcommunity is rendzina or kserorendzina.

Cerastium fragillimum-Galium peplidifolium subcommunity doesn't include the distinguishing species of the other subcommunities. *Cerastium fragillimum, Galium peplidifolium, Alyssum contemptum Silene aegytiaca, Briza spicata*, and *Myosotis stricta* are the distinguishing species of this subcommunity. This subcommunity appears on the depth soils with low inclination and also which are acidic in upper soil zones. The number of therophytes in the herb layer of the community is high.

Oryzopsis holciformis-Cedrus libani (Akman et al. 1979)

This community distributes on the sunny slopes of the low mountain zone in Elmali region (Akman et al. 1979). The soil type of the community is rendzina. Cedar at this region either forms pure stands or mixed stands with *Juniperus excelsa* and *J. foetidissima.* In addition to these species, *Quercus coccifera* straggly join in to the stand structure at the lower distribution zone of the community. The distinguishing species of the community are *Oryzopsis holciformis, Delphinium fissum* subsp. *anatolicum, Arrhenatherum elatius, Campanula michauxioides, Lapsana communis* and *Acer sempervirens.*

Astragalus oxytropifolius-Cedrus libani

Astragalus oxytropifolius-Cedrus libani forest was determined by Bekat (1987) in Barla Mountain in Eğirdir. *Cedrus libani, Astragalus oxytropifolius* and *Viola heldreichiana* are the distinguishing species of this community. *Lonicero nummularia,* characteristic for Lonicero-Cedrion, accompanies with these species. In addition to these, *Berberis crataegina, Cotoneaster nummularifolia, Geranium tuberosum* subsp. *tuberosum, Ranunculus euterianus, Doronicum orientale, Juniperus excelsa, Cerastium brachypetalum* subsp. *roeseri, Galium debile, Sorbus umbellata* var. *cretica, Juniperus foetidissima, Millium vernale* and *Bunium microcarpum* subsp. *sicrocarpum*, which are characteristic for Querco-Cedretalia, join in the floristic composition of the community. *Astragalus oxytropifolius-Cedrus libani* forest, which ranges from 1450 to 1700 m altitudes, appears on limestone main rocks and brown forest soils.

Vicia cracca subsp. stenophylla-Cedrus libani

This community was also determined in Elmali region by Çetik (1976). It distributes on metamorphic limestone and ranges from 1450 to 1950 m altitudes. C*otoneaster nummularifolia, Berberis crataegina, Lonicera etrusca, L. nummularifolia, Anthemis carnea* and *Millium vernale* are the distinguishing species of this community. This community was resulted from the degradation of the cedar forests (Çetik, 1976). Because of that, Mayer and Aksoy (1986) pointed that more comprehensive studies on this community should be realized.

Conclusion

In Turkey, *Cedrus libani* mainly occurs on the ecologically and biologically rich habitats and extreme sites. However, as it was taken into account that, these forest was intensively under the effect of human being for centuries, the ecologic and biologic importance of cedar forest can be understood more clearly. On the other hand, the biologic diversity of cedar forest hasn't sufficiently been defined yet. Although some general knowledge were obtained, thanks to some local phytosociologic researches, it is possible to clearly determine the biologic and ecologic structure of the cedar forest with a wider phytosociologic work, which contains all distribution areas of cedar.

As it is widely acknowledged, the vegetation maps and tables resulted from the phytosociologic researches give comprehensive information on the biologic richness of a region. While a releve of a phytosociologic work define the Alfa diversity, all releves belonging to a community reflects the Beta diversity and the vegetation table reflects the Gama diversity (Ewald, 2003). From this point of view, not only forming a vegetation map for cedar forest, but also forming it for all forests in Turkey is important to clearly understand the biologic and ecologic values of them.

References

Anonim (2006) Orman varlığımız. Orman Genel Müdürlüğü yayını, 160 s.

Akman, Y. (1973) Apercu preliminaire sur les conditions phytosociologiques de la chaine de l'Amanus dans la Regio du Hatay (I-II-III). Com de la Fac-des Sci-d'Ank. Serie C, Tome 17.

—. (1995) Türkiye Orman Vejetasyonu, Ankara Üniversitesi Fen Fakültesi Yayını, 448 s. Ankara.

Akman, Y. Barbéro and Quézel (1978) Contribution à l'étude de la végétation forestiére d'Anatolie méditerranéenne. Phytocoenologia 5(1) 1-79.

Akman, Y. Barbéro, M. and Quézel, P. (1979) Contribution à l'étude de la végétation forestiéere d'Anatolie méditerranéenne. Phytocoenologia, 5(2) 189-276.

Aksoy, H. and Özalp, G. (1990) Türkiye'de Sedirin (*Cedrus libani* A. Rich.) orman toplumları. Uluslararası Sedir Simpozyumu, Ormancılık Araştırma Enstitüsü muhtelif yayınları no. 59, s. 93-102, Ankara

Atalay, İ., (1990) Regioning of seed transfer of cedar (Cedrus libani A. Rich) in Turkey. Uluslar arası Sedir Simpozyumu. Ormancılık Araştırma Enstitüsü Yayını, no 59, p.166-179.

Bekat, L. (1987) Barla Dağı (Eğirdir)'nın Vejetasyonu. Doğa Turk Botanik Dergisi. 11:3:270-305

Boydak, M. and Asan, Ü. (1990) Monumental forests and trees of *Cedrus libani* in Turkey. Uluslararası Sedir Simpozyumu, Ormancılık Araştırma Enstitüsü Muhtelif yayınları no. 59, s. 847-853, Ankara

Çetik, R. (1976) The phytosociological and ecological studies of the cedrus woodland vegetation of Çığlıkara and Bucak at Elmalı. Comm. Fac. Sc. Univ. Ankara C2:20:1-37

Duman, H. (1995) Engizek Dağı (Kahramanmaraş)vejetasyonu. Turk J Bot 19 (2):179-212. Akman, Y. 1995. Türkiye orman vejetasyonu. Ankara Üniversitesi, 450 s.

Ewald, J. (2003) A critique for phytosociology. Journal of Vegetation Science, 14: 291-296.

Günay, T. (1990) Afyon-Emirdağ Yukarı Çaykışla vadisinde stebe geçiş kuşağında yeni tespit edilen bir sedir (*Cedrus libani* A. Rich.) kalıntı meşceresi. Uluslararası Sedir Simpozyumu, Ormancılık Araştırma Enstitüsü Muhtelif yayınları no. 59, s. 53-63, Ankara

Kantarcı, M.D. (1990) Türkiye'de sedir ormanlarının yayılış alanlarında ekolojik ilişkiler. Uluslararası Sedir Simpozyumu, Ormancılık Araştırma Enstitüsü Muhtelif yayınları no. 59, s. 12-25, Ankara

Kayacık, H. (1948) Akdeniz mıntıkasında ve bilhassa İtalya ve Türkiye'de ağaçlandırmanın temel şartları. Orman Genel Müdürlüğü Yayını özel sayı, 79, 136 p.
—. (1980) Orman ve park ağaçlarının özel sistematiği, 1. cilt, Gymnospermae, İÜ. Orman Fakültesi yayını no. 2642/281, 388 s. İstanbul.
Mayer, H. and Aksoy, H. (1986) Walder Türkei. Gustav Fischer Verlag, 290 s. Stuttgart..
Mayer, H. ve Sevim, M. (1959) Lübnan Sediri, Lübnan'daki 5000 yıllık tahribatı, Anadolu'da bugünkü yayılış sahası ve bu ağaç türünün alplere tekrar getirilmesi hakkında düşünceler (Çeviren. Necmettin Çepel), İÜ. Orman Fakültesi Dergisi, Seri B, Cilt IX, Sayı II, s. 111-142.
Ocakverdi, H., Çetik, R. (1987) Seydişehir Maden Bölgesi (Konya) ve çevresinin vejetasyonu. Turk J Bot 11(1): 102-148.
Queézel, P. and Pamukçuoğlu, A. (1973) Contrubution a l'etude phytosociologique de quelques grouqements foreotiers du Taurus. Feddes Rep. Sond 84, Heft 3, Seite 185-229
Saatçioğlu, F. (1975) Akdeniz subtropik bölgede orman gençleştirmesi sorunları. İÜ. Orman Fakültesi Dergisi, Seri A, cilt XXIV, Sayı 2, s. 84-127.
Semaan, M. And Haber, R. (2003) In situ conservation of *Cedrus libani* in Lebanon. Acta Hort. (ISHS) 615: 415-417.
Serin, M. (1996) Dedegöl (Anamas) Dağı'nın doğu kısmı ile Kurucuova-Yeşildağ-Beyşehir-Konya) ve çevresinin vejetasyonu. SÜ. Fen Derg. 13:28-49.
Serin, M. and Eyce, B. (1994) Hadim (Konya) Aladağ (Orta Toroslar) ve çevresinin vejetasyonu. Tr. J. of Botany, 18:201-227.
Yeşilkaya, Y. (1994) Sedirin ekolojisi, Ormancılık Araştırma Enstitüsü Yayınları, El kitabı dizisi 6, s. 53-79.
Varol, Ö. and Tatlı, A. (2001) The vegetation of Çimen Mountains (Kahramanmaraş). Tur. J. of Botany, 25:335-358.

CHAPTER FOURTEEN

THREE EXAMPLES TO EXPLAIN THE FORMATION AND DEVELOPMENT OF THE WEST TAURUS POLJES

NURDAN KESER AND ALI ÖZEL

Introduction

The Sarıbelen, Bezirgan and Akyazı poljes, all of which are the subject of this study, exist on Teke Peninsula, on the descending area of the West Taurus Ak Mountains to the Mediterranean Sea. These poljes are near the villages of Kalkan, Kaş, the town of Antalya, and they are at 7, 20, and 31 kilometers north-west of Kalkan respectively (Fig. 14-1). The study is limited to the Darboğaz stream (Bezirgan) drained with the poljes, the Dargaz stream (Akyazı), the Kaputaş stream and the watery part of the Kısık stream (Sarıbelen). The total area of the study is 275 km^2. Office work supporting the study included constructing draft maps drawn from 1/25000-scale topographical maps and 1/100.000-scale geological maps. Based on the information gathered through field observations, the geomorphologic maps of the poljes have been prepared. In the first and second sections of the present paper, the geology of the field is analysed and the geomorphologic characteristics of the poljes, their formation and development are explained, while in the third chapter, the regional geomorphologic evolution is explained. In this study, the aim is to shed light on the evolution of the formation and development of the bigger and complex poljes by using the geologic- geomorphologic characteristics of smaller poljes.

Geologic Characteristics

Stratigraphic Characteristics: The poljes are on the western part of the Teke Peninsula, where autocthonous and allocthonous rocks exist. The

Bey Mountains autocthons composed of rocks of angular unconformity are represented by neritic limestones and detritic formations deposited during Upper Cretaceous-Miocene period. On the surface of the area exist the Bey Mountains composed of Jurassic-Upper Cretaceous neritic limestones belonging to the Bey Mountains autocthons, Burdigalian limestones with algaes (member of Gömüce), the Sinekçi formation composed of argillaceous limestone (member of Kıbrısdere) and claystone (member of Çayboğazı), the Kasaba formation composed of Upper Burdigalian-Lower Langhian conglomerate and grit, and the Felenk Mountain conglomerate composed of Langhien rough conglomerates. Also, the allocthonous rocks existing on the Bey Mountains autocthons as a tectonic cover are represented by Upper Lutetian-Lower Burdigalian grit and the Elmalı formation (Yeşilbarak nappe), the Mandırkaya formation composed of Liassic- Cretaceous neritic limestones and the Kozağaç formation (Dumanlıdağ nappe) composed of Upper Paleocene- Lutetian breccias (Şenel, 1997).

Most of these rocks on the surface of the area are carbonated formations with an amount of decomposition characteristics. The geomorphologic characteristics of the poljes depend mainly on the stratigraphic features of these rocks such as their lithology and the width and the thickness of their expansion areas, the thickness of their layers and values of their slope. When their location around the poljes are analysed, it can be seen that one side of the poljes is composed of limestone sequences, while the other is composed of claystone, siltstone, argillous and gritty limestone or clastic rocks (Fig.14-2).

Chemical and Petrographical Characteristics: The rock units in the study area have 83-97 % $CaCO_3$ and 1-8 % $MgCO_3$ (Keser, 2004b). Generally, the porosity of these rocks, which have a micritic quality (calcite clay with micro crystals), and higher solubility than sparit and dolomite (Folk, 1968), has suitable values for karstification (1-10%). The mountain units which have the relative highest $CaCO_3$ rates and porosity are the limestones of the Bey Mountains and Mandırkaya formations. During the field observations, it was seen that these rocks form intensive karst areas around the poljes, and the huge scale, multi-period dolines and uvalas on them are really outstanding. Among them, the Bey Mountains formation is the most important unit for the development of poljes. This unit forms the western side of the Sarıbelen, the southern side of the Akyazı, and the eastern and western sides of the Bezirgan poljes, and it starts at theKalkan-Finike shoreline included in our study zone and stretches to a wide area on the north; from Elmalı and Korkuteli to Isparta

(Şenel, 1997). Within this expansion, this unit shows varying rates of karstification, forms intensive karst environments in the Western and Middle Taurus, and is identical to various Jura-Cretaceous limestones with different names (Monod, 1977; Dumont, 1976; Uysal et all, 1980; Öztürk et all., 1987; Şenel, 1997). During the start and development of the formation of the relevant poljes, the existence of pure limestone sequences with a high potential for karstification, at least on one of their sides, has been really effective. Looking at the geologic map of Turkey, it is seen that almost all the poljes along the west and north of the West Taurus, including the Teke zone, generally exist in contact with autocthonous Jura-Cretaceous limestones and dissoluble or poor-soluble rocks with nappe quality. That the rocks which are generally lateral correlatives of limestones in the area have detritic feature due to external factors shows that, besides karstic decomposition, normal erosion processes (fluvial processes) are also effective in the formation and development mechanisms of poljes.

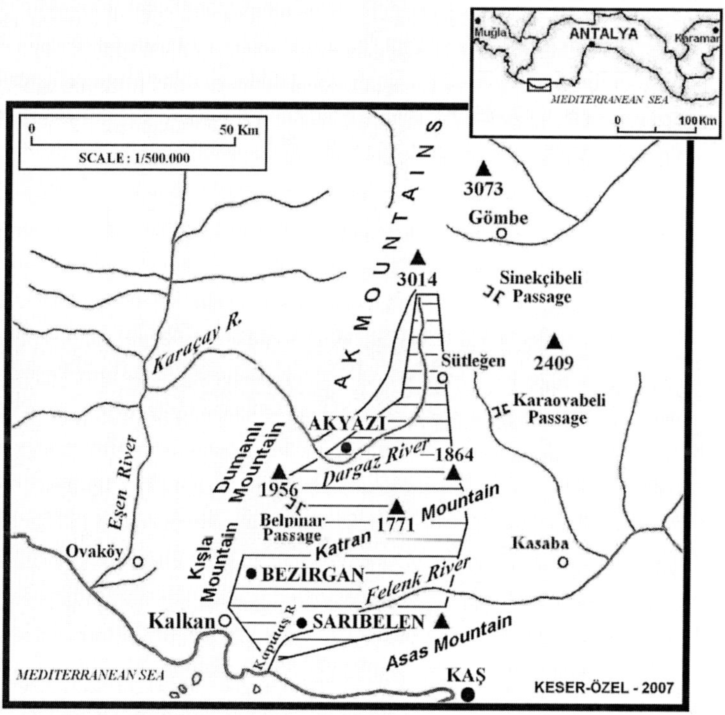

Fig. 14-1: The location map of poljes.

Structural Characteristics and Discontinuities: Polje bottoms exist at the transition zone of autocthonous and allocthonous rocks (drag zone) or in contact with limestone sequences and dissoluble or poor-soluble rocks, on the synclinal axis or along the faults. These lateral discontinuity lines of rocks having different lithology form, besides karstic decomposition, and structural weaknesses causing erosion due to external factors. The Akyazı polje has developed entirely on the drag zone of the autocthonous Bey Mountains formation and Yeşilbarak nappe (the Elmalı formation). This drag zone, comprising the contact between karstic and unkarstic rocks in the formation and development of poljes, must have formed the structural weakness belt. The northern parts of Bezirgan polje also exist on the drag zone of allocthonous rocks. Another important structural feature of this ploje is the normal faults bordering the depression from the east, west and south. This polje is on the descending block of the faults. The graben and drag zone, both of which are formed by these faults in the formation and development of the Bezirgan poljes, must have formed the structural weaknesses in which karstic decomposition started. The Sarıbelen polje, on the other hand, exists at the contact point of limestone sequences along the synclinal axis and dissoluble (claystone, siltstone) and is a less soluble (gritty, argillaceous limestone and conglomerate) formation. It is also bordered by normal faults from north-east and south-west (Fig. 14-2). The synclinal structure of the Sarıbelen polje that keeps the water on the rock surface during its location and development, the lithologic transition zone and faults have been the determinant and controlling factors.

Geomorphologic Characteristics

The area where the poljes exist in composed of such geomorphologic units as the Katran Mountains in the centre, the Ak Mountains in the north, the Dumanlı Mountains in the north-west, the Kışla Mountains in the west, karstic platoes in the south, and the Asas Mountains in the south-east. Poljes are found in these mountainous areas where the latitude decreases to the south in accordance with orographic lines. In this section, the geomorphologic features of the relevant poljes will be explained.

The Akyazı Polje. This polje extends in a northwest-southwest (NW-SW) direction with the Ak Mountain in the north (Uyluk Apex 3014 m), the Katran Mountain in the south (Eren A. 1864 m), the Dumanlı Mountain in the west (Eren A. 1956 m) and the Susuz Mountain in the east. The distance between the NW and SW points of the polje whose bottom lies at 900 meters is approximately 9 kilometers. Its widest place at a N-S direction is 3 kilometers, while the narrowest is 250 meters, and the whole bottom is 6 kilometers wide (Fig. 14-3). The polje is drained by the Dargaz Stream (a tributary of the Eşen Stream), which comes from karstic sources in the Ak Mountain. The bottom of the depression is typically in polje characteristics and covered with red soil. Also, there are sand dolines during the rainy season, due to leakages. The base appearing as a valley to the south is covered with gravel and sand. While the northern slope has a 30° angle of obliquity, the southern slope has 60° on average and at times an 80° angle of obliquity can be seen. This is due to the difference between the resistance against the erosion of the northern sides composed of clastic rocks (sandstone, claystone, siltstone) and of the rugged southern sides composed mostly of limestone; moreover, it developed in relation to perpendicular faults as well as the layer distributions connected with the anticlinal structure of the Katran Mountain. The perpendicular depth of the depression base has also developed as a result of doming along the anticlinal axis (the Katran Mountains) and of local risings with normal faults. Deformations of the erosion surfaces (Serravaliyen surfaces) -- which have been determined to be at different altitudinal levels although they belong to the same periods -- on Ak Mountain, Dumanlı Mountain and Katran Mountain surrounding the Akyazı polje have developed as a result of these local and wholesome risings. The base level of the area that changes depending on this tectonic activity has caused a periodical increase in the efficiency of Dargaz Stream, which runs through the depression. Along the slopes at the polje base exist many fluvial-karstic pockets. These are the base recessions, having been formed as a result of the hydraulic and corrosive effects of new and ancient rivers which reach to the depression along the sides. The bigger of these, in whose formation the erosion of rivers and dissolution of carbonated rocks have been effective, exist in front of the paleovalleys. The lateral extension of the polje base has mostly been in this way. The Arma uvala at the eastern point of the polje has also been seized by the Dargaz Stream, thus causing an extension in the length of the depression. This section of the polje also includes Quaternary figures. The Akyazı Polje, which was closed to external drainage at its origin, is a fluvial-karstic depression in whose today's geomorphological development rivers have had tremendous effect.

The Bezirgan Polje: It is surrounded by the Dumanlı Mountains (Kırgedik A. 1930meters) in the north, by Kışla Mountain (Göl A.1219meters) in the west, by the Katran Mountains (Yapıyol A. 1434 meters) in the south and east. The base of the polje is at 750m on the SN axis. The NS length is about 4.5 km, and the widest point at EW direction is 1750m. The width of its bottom is about 6 km^2 (Figure 14-3). There are two dolines- one active and the other fossilised- at the southern tip of the polje. The fossilised doline, about 10 m high and 8 m wide, is a collapse doline with a cave-like appearance on the limestone hillside. It became dysfunctional when it became blocked due to the movement of the live fault on which it had developed. The active doline exists 3 m lower than the fossilised one, and it has the characteristics of sand doline. The water of this doline reappears on the surface at the spring in İnbaşı, which is on the hillside 1 km east of Kalkan. There are also numerous suffusion dolines along the limestone cracks at the base of the polje where leakage occurs. There appears a seasonal lake on the south of the polje between December and March when there is a lot of precipitation. The bottom of the lake is covered with hardpan. At the places where the hillsides of the polje around the lake and this impervious base meet, corrosion notches and fluvial-karstic pockets exist. The Darboğaz Stream originates in Dumanlı Mountain and drains into the doline after going along the polje in N-S direction. The depression has developed within the limestones of the Bey Mountains formation except its northern section. The northern parts, mainly composed of clastic rocks, are covered with alluvions composed of the dissolusion of these rocks along the valley of Darboğaz Stream. The polje is surrounded by faults in all directions and by rugged limestone ramps with 450 m height and 70°-80° slopes. Along the ramps exist young fault rigidities with intense slope gravels in front of them, suspended fossilised or active water exites. These are geomorphological findings showing the fact that the faults surrounding the polje used to be live faults which were active until recently (Keser, 2004a).

The polje exists in the depression that was formed by the faults having developed in the late Pliocene period (Şenel, 1997). The Bezirgan Polje is a tectono-karstic polje whose position and development have been controlled by these faults. In the development of the depression, Darboğaz Stream has also been effective, and its impact has at times increased with a view to the base level which changes depending on the tectonic activity. There are deformed erosion surfaces on the high platoes around the polje dating back to Serravallian (on Dumanlı Mountain), Messinian (on Katran Mountain and Dumanlı Mountain) period, and quaternary figures along the Darboğaz Stream valley on the north. Fluvial processes are directly or indirectly effective in the erosion of the northern part of the polje, filling the base of the polje with alluvions and in the occurrence of decomposition along the slope that have contact with water of the lake in rainy seasons.

The Sarıbelen Polje: This depression, with the Katran Mountain (Tınaz A. 1771m) on its north and west, karstic platoes (approximately 1100m) on its south and the Asas Mountain (Tuzla A. 1366m) on its east extends in a NE-SW direction parallel to the Katran Mountain. Its bottom is at 750 meters altitude, and it has developed along the Sidek synclinal on the south of Katran Mountain anticlinal. The distance between the two opposite edges in the NE-SW direction of the polje base is 12 kilometers. Its widest point in N-S direction is 1500 meters, while the narrowest is 250 meters. It has a width of roughly 10 km^2 (Fig. 14-3). The base of the depression is covered with red soil, except from the stream basins. Small ravines exist in patches where leakage occurs. The rugged tops surrounding the polje are composed of limestone, claystone, siltstone and conglomerates. The parts in the north are formed of the Bey Mountains formation limestone, while those in the south are composed of the Felenk Mountain conglomerates. This formation, with its fast-decomposing carbonated cement, has had an important effect on the development of the polje.

In the beginning and development of this polje, the litholitic transition zone composed of these rocks and limestones formed the structural weakness line where the karstic decomposition started. The conglomerates, which have been effective in the current morphology of the polje, became the source of alluvial range besides being the reason for not only the altitude difference but also an asymmetry between the southern and northern slopes. While the slope value of the polje at the southern ramps is between 30°- 40°, this value is on average 50°-60° at the northern ramps. Whereas the east of the polje is bordered with the Yumrutaş fault, which is a normal fault, in a NW-SE direction, the west is surrounded by a normal fault in EW direction that cuts the valley of Kaputaş Stream in addition to the Bezirgan fault. Along the faults, rugged ramps with 70°- 80° slopes are formed with intensive gravels. At the hillsides of the polje, which are bordered with faults, are numerous suspended dry valleys and along the slope cracks are miniature waterfalls and springs. These are the findings that make one think that the live faults, as all through the area, developed at the end of the Pliocene have retained their effect up to now.

The polje is drained into the Kısık Stream, which is a branch of the Felenk Stream, in the east, and into the Kaputaş Stream in the southwest. Moreover, along the slopes, there are numerous small seasonal rivers that join these streams. With faults developed in the eastern and western parts of the polje (or with the reactivity of the alive faults), these rivers have been adapted to the bottom level and they have opened their way to the outside by covering the Sarıbelen polje. The Kaputaş Stream carries the water of the polje to the Mediterrenean Sea through its karstic limestone valley. In the current geomorphology of the polje which is a fluvial-karstic depression, the rivers have an important place. There are a lot of developed fluvial-karstic pockets along the polje slopes and there are also quaternary figures at the east of the bottom, along the Kısık Stream. On one hand, the rivers enable the polje to swell alterally via its pockets; on the other hand, they fill the bottom by flattening the slopes through the alluvions they carry.

Geomorphological Evolution

The Western Taurus, where the polje exists, is in such an area that has retained its tectonic activity since paleo-geographical periods. In this area, which is composed of authocthonous and allochthonous structural unites as a result of a compaction regime in a N-S direction that repeated in the geological past for a few times, the Bey Mountains authocthonous appeared above water for a few times in the Upper Lutetian and before the Burdigalian periods and were exposed to transgression. The nappe unites (Likya nappes), which moved to the area from the northwestern Menderes massive, overlapped on the Bey Mountains authocthonous in the Lower Langhian due to a repeating compression at the end of the Middle Eocene (Erakman-Alkan, 1986; Önalan, 1979; Şenel, 1983, 1986, 1994). The Likya nappes (The Dumanlı Mountain nappe and the Yeşilbarak nappe) continued their formation after the Langhian by overlapping of the old formations, which are viewed with angular dissonance, on the young ones with reverse faults (Şenel, 1989; Erol, 1990). In the area where a new tectonic regime (Neotectonics) started with the arrival of the nappes (Erinç, 1973; Şengör, 1980, 1985b, Erol, 1989a; Şenel, 1983, 1986), the current geomorphology of the research area was determined in this period. The Sidek synclinal, in which the Sarıbelen polje developed, and the structural gorge and the Katran Mountain anticlinal, in which the Akyazı polyje developed, are huge curvature systems developed in this period. At the end of the Middle Miocene the research area had totally become land (Şenel, 1994).

Soon after that, in the place where the river network had developed, Late Middle Miocene (Serravallian), Late Upper Miocene (Messinian) and Late Pliocene (Piacenzian) erosion surfaces (Erol, 1980, 1983, 1983a, 1984, 1989, 1990) began depending on the effective erosion processes and morphotectonic development that continued from the end of the Middle Miocene to the end of Pliocene (Ketin, 1959; Ardos, 1979; Uysal et all. 1980; Erol, 1983, 1990; Yalçınkaya ct all. 1986; Atalay, 1987, 1988, 1996, 2003; Şenel, 1994). The sediments at the polje base (while it was being carried to the Mediterrenean Sea via rivers and dolines) must have been

FIG. 2 : GEOLOGICAL MAPS OF AKYAZI - BEZİRGAN - SARIBELEN POLJES

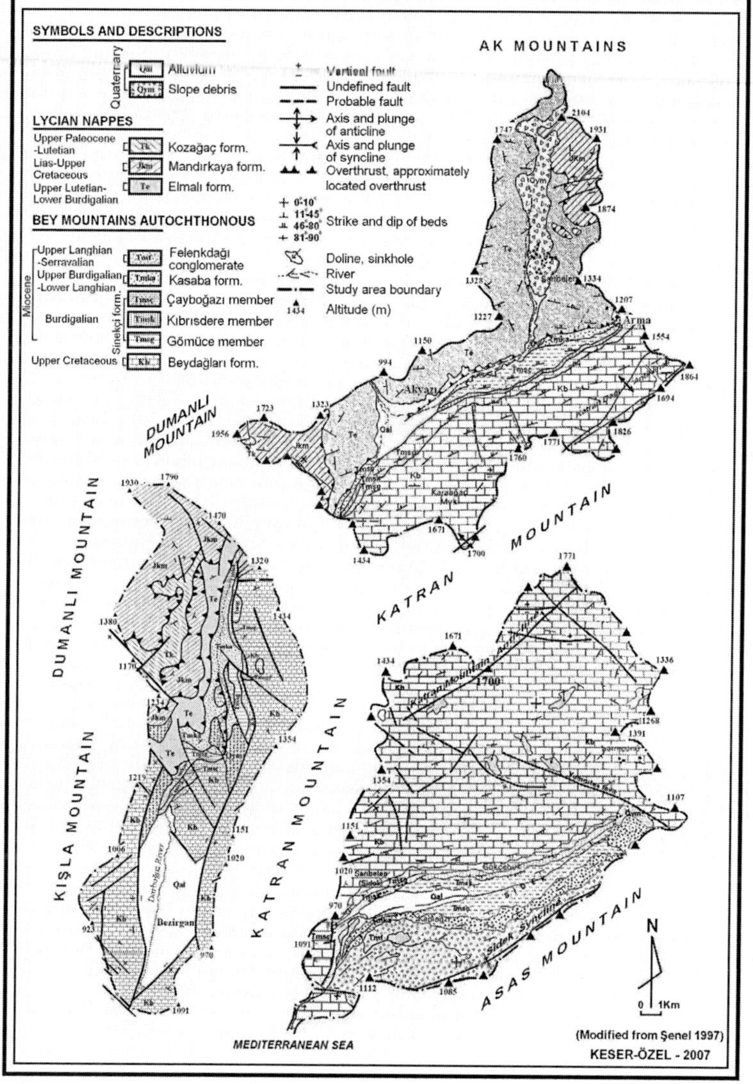

Fig. 14-2: Geological map of Akyazı-Bezirgân-Sarıbelen poljes.

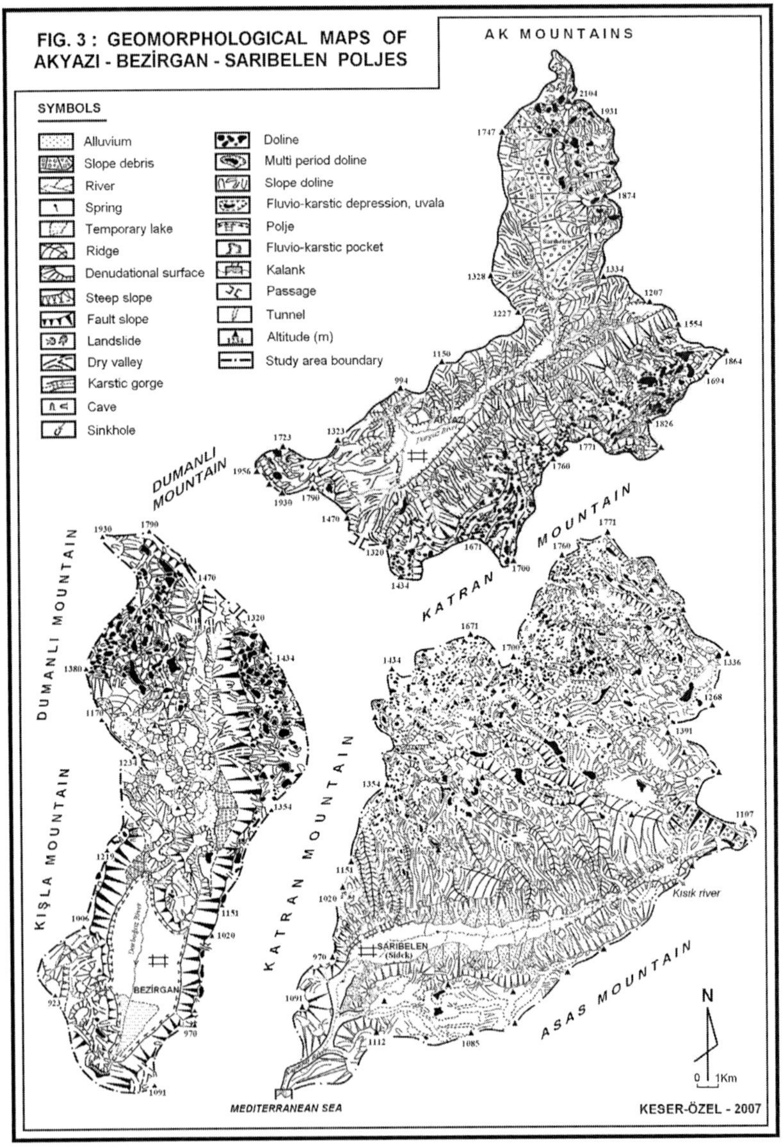

Fig, 14-3: Geomorphological map of Akyazı-Bezirgân-Sarıbelen poljes

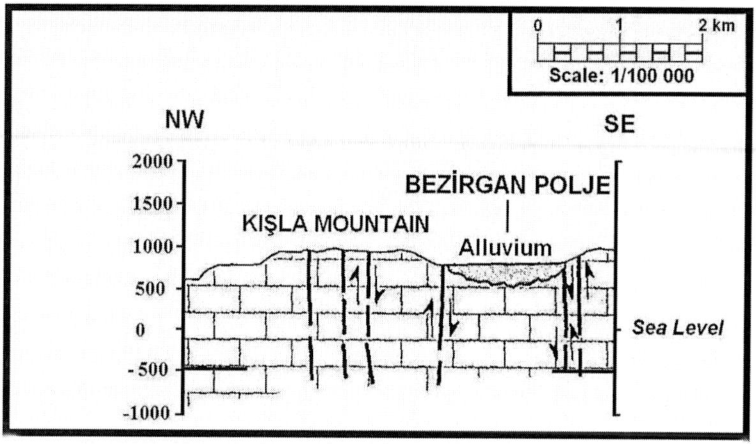

Fig. 14-4: Geological cross-section of Bezirgan Polje and its surrounding (modified from Şenel 1997).

the correlation of these erosion surfaces on the Ak Mountains, the Dumanlı Mountains (Serravallian surfaces) and the Katran Mountains (Serravallian and Messinian surfaces).

Unlike the subtropical climate conditions of the Upper Miocene, the climate was warmer and wetter at the beginning of the Pliocene (Erol, 1980, 1983, 1990), when river networks compatible to orogenic and tectonic lines in the area developed. At the end of the Middle Pliocene, unlike the Lower Pliocene, moisture started and rivers increased their impact (Erol, 1979). At the end of the Upper Pliocene, drought started, but the Pliocene basins started to be covered via fluvial erosion which tectonic activity rejuvenated. The network got bigger when the rivers got longer due to the merging of the basins (Atalay, 1987). The Akyazı polje continued to be shaped by fluvial processes, besides being karstified by the Esen Stream (the Karaçay branch), which increased its impact under the climatic and tectonic conditions of this period. The karstic strait of the Kara Stream, formed in limestone 3 kilometers northwest of Akyazı, was also developed in the climatic and tectonic conditions of this period. In this, period when suitable climatic and tectonic conditions were available for karstification, besides new karstic formations are being formed, the paleokarstic figures of the earlier periods also continued their development.

At the end of Pliocene, when the Akyazı and Sarıbelen poljes continued their development, with the tectonic activity effective all around the area (Colin, 1962; Ardos, 1979, Erinç, 1982; Koçyiğit, 1984; Şenel et al., 1989; 1994), a new geomorphological period started in the research area. The fracture systems in the area that had formed many normal faults, the Bezirgan polje and the Eşen graben, and the Yumrutaş fault also developed in this period. Some of the normal faults developed in this period, when the Bezirgan poljes also developed with its outline, are the live faults that have controlled the development of the poljes with their impact up to now. Dislocations developed along these faults and the fact that the Katran Mountain anticlinal has continued its ascent has been the major factor in enabling the poljes to get their current depth.

In the period until the Middle Pleistocene, when a hot and low-high precipitational climate prevailed with cool-rainy cycles (Erol, 1983, 1991a), karstification accelerated parallel to the ascents depending on the tectonic activities and the Pliocene river network turned into an underground drainage depending on the advancement of this karstification (Atalay, 1987). These paleovalleys, which have been effective in the recession of the polje slopes and in filling the bases with sediments, extend towards the poljes in NE-SW and NW-SE directions in the higher parts of the Ak, Dumanlı and Katran Mountains. These extremely karstified valleys are characterized with successive, deep and very important dolines and uvalas. At the bottom of some poljes, there are 15 meter deep water storage pools, in and under which there is terra-rosa and red-brown Mediterrenean soil. This soil was formed in a period when hotter and moister climate conditions prevailed in the Taurus (Erinç, 1965, 1971; Bener, 1965; Atalay, 1973, 1984, 1987, 1988).

With the effect of precipitation in the area where the elevation increases in patches, the efficiency of rivers and karstification increased as the tectonic efficiency that continued as Early-Middle Pleistocene faults (Erinç, 1970; Ardos, 1979; Koçyiğit, 1984; Atalay, 1987) rose and as the climate in the Middle Pleistocene became cooler and wetter than that in the Early Pleistocene (Nazik, 1992). The opening of the Sarıbelen polje to the exterior drainage through the Kaputaş and Felenk Streams (Kısık Stream) and the formation of the Kaputaş strait occurred under the climatic and tectonic conditions of the Middle Pleistocene. The young, dry valleys that end with fluvial-karstic pockets and alluvial range on the polje hillsides also belong to the rivers of this period. The captivation and inclusion of the Arma uvala in the east of the Akyazı polje by the Dargaz Stream also happened in this period.

Conclusion

1. The poljes evolved in contact with either Jura-Cretaceous limestones or sandstone, which have high potential of karstification to their litholigic characteristics, conglomerates and breccias, which are pebble rocks with low resistance to abrasion, or claystone and siltstone, which are rocks without dissolubility. Therefore, it is understood that not only karstic decomposition but also other erosional processes were effective in the formation mechanism of the poljes. It is also determined that the poljes are in a NE-SW direction and their longitudinal axis are compatible with tectonic and orographic lines. The parallel structure of the shape and longitudinal direction of the poljes to the contact point of the rocks show that the stratigraphic properties that are effective with the lateral and vertical continuities of the layers are also effective in the position and development of the depressions, apart from their lithology. Moreover, besides the chemical and petrographic characteristics of the rocks, their stratigraphic position that involves the slope and vertical values of the layers are also effective in the current geomorphological caharcteristics of the poljes.

2. Generally, the neritic limestones surrounding the poljes are authocthonous, while the dissoluble or poor-soluble rocks are allochthonous. Therefore, the existence of authocthonous and allochthonous structural units are important as a reason for lateral discontinuity where different lithologies adjoin, and this structural characteristic is a major factor determining the position of the poljes. Almost all the poljes in the Western Taurus which also includes our research area in Teke district developed at the same direction and on these tectonic lines where they show authocthonous and allochthonous properties.

3. The second structural factor that determines the position and developmental direction of the poljes is found to be their synclinal structure, which allows water to be held on the rock surface, which is the second requirement in karstic decomposition after the existence of pure limestone. This tectonic sluice not only enables the start of karstification with the effect of the water running along the slopes to the bottom, but also provides a suitable environment for the formation and development of the polje with the establishment of the drainage network and alluvial activity in the further stages.

4. The faults and the depressions they formed are found to be the third structural factor determining their position and developmental directions of the poljes. It is also understood from the geomorphological findings about the fact the Katran Mountain anticlinal still goes on its activity with numerous faults in the area such as Bezirgan and Yumrutaş that besides other morphodynamic processes, local risings with live faults and total risings along the other curving axis are effective in the depth of the poljes. The tectonic activity that evolves as vertical dislocation and doming in the area indirectly affected the development of the poljes by increasing the effects of the rivers with a change in the bottom level.

5. It is also understood from the existence of current geomorphological elements at the bottom of and around the poljes, alluvions, terraces, alluvial ranges, suspended and dry valleys, karstic springs, straits, fluvial-karstic pockets and rivers along the bottom that fluvial processes have been effective in the development mechanism of the depressions. The rivers that use the valleys at the bottom in the current climatic conditions has a corrosive effect on the limestone hillsides by covering the bottom which is made up of impervious materials like clay, milestone, siltstone or by creating a lake (most likely in the paleogeographical periods). Fluvial-karstic pockets formed depending on linear corrosion along the slope are geomorphological elements that show the bottom-development mechanism of the poljes. The rivers are understood to be the major factor of hydraulic and corrosive activities in the polje slope recession mechanism. Whereas the direction of the longitudinal axis of the polje shows compatibility to the structural lines explained above, it is observed that these rivers have been effective in the extension mechanism of the bottom. Furthermore, it is understood from the asymmetrical slope profile of the polje that results from further corrosion of the hillsides made up of unkarstic rocks that catbonated rocks have had a secondary effect in the bottom development.

References

Ardos, M., 1979. Neotectonics in Turkish Geomorphology. Istanbul Univ. Institute of Geography Publ. No:113, Istanbul.

Atalay, I., 1973. Toros Dağlarında Karstlaşma ve Toprak Teşekkülü Üzerine Bazı Araştırmalar. Jeomorfoloji Dergisi Sayı:5, Ankara. “Researches on Karstification and Sand Formation in the Taurus”

—. 1982. Toprak Coğrafyası'', Ege Üniv. Sos. Bil. Fak. Yay., No:8, İzmir. “Earth Geography”

—. 1987. Türkiye Jeomorfolojisine Giriş. Ege Üniv. Ed. Fak. Yay., No:9, İzmir. “Introduction to Turkish Geomorphology”

—. 1988. Toros Dağlarında Karstlaşma ve Karstik Alanların Ekolojisi. Jeomorfoloji Derg., Sayı:16, Ankara. “Karstification in the Taurus and the Ecology of Karstic Areas”

—. 1996. Karstification and karstic landforms in Turkey. Karen Landforms (Eds. Fornos and Gines). Universitat de les Illes Balears, pp. 325-334.

—. 2003. Effects of the neotectonic movements on the karstification in Anatolia, Turkey. Acta Carsologica, 32: 195-204.

Bener, M., 1965. ''Göksu Vadisi ve Çevresindeki Platolarda Karst'', İst. Üniv. Coğr. Enst.Doktora Tezi (Yayınlanmamış), İstanbul. "Karts in Göksu Valley and the Platoes around It" (Unpublished)

Colin, H.J., 1962. ''Fethiye-Antalya-Kaş-Finike (Güneybatı Anadolu) Bölgesinde Yapılan Jeolojik Etütler'', MTA Enst. Derg. Sayı: 59, S: 19-61, Ankara. "Geological Studies in Fethiye-Antalya-Kaş-Finike (Southeastern Anatolia)

Dumond, M. and Monod, D., 1976. Dipoyraz Dağ Masifinin Triyasik Karbonatlı Serisi (Batı Toroslar –Türkiye), MTA Enst. Derg., Sayı: 87, Ankara. "The Triassic carbonated Series of the Massive Dipoyraz Mountain"

Erakman, B., Alkan, H. 1986. ''Kalkan-Elmalı-Yeşilova-Acıpayam-Fethiye Arasının Jeoloji ve Petrol Olanakları'', TPAO Raporu (yayınlanmamış), No: 2190 Ankara. "Geological and Oil Possibilities of the Area Between Kalkan-Elmalı-Yeşilova-Acıpayam-Fethiye" (Unpublished)

Erinç, S., 1965. ''Türkiye'de Toprak Çalışmaları ve Türkiye toprak Coğrafyasının Ana Çizgileri'', İst. Üniv. Coğr. Enst. Derg., Cilt:8, Sayı:15, S: 1-39, İstanbul. "Earth Studies in Turkey and the Outline of Turkish Earth Geography"

—. 1970. ''Türkiye Kuvaterneri ve Jeomorfolojinin Katkısı'', Jeomorfoloji Derg., Sayı:2, S: 12-35, Ankara. "The Contribution of Turkish Quaternary and Geomorphology"

—. 1971. ''Jeomorfoloji II, (2. Baskı)'', İst. Üniv. Coğr. Enst. Yay., No: 23, İstanbul. "Geomorphology II"

—. 1973. ''Türkiye'nin Şekillenmesinde Neotektoniğin Rolü ve Jeomorfoloji- Jeodinamik İlişkileri'', Jeomorfoloji Derg., Sayı:5, Ankara. "The Role of neotectonics in the Formation of Turkey and the Relation between Geomorphology and Geodynamics"

—. 1982. ''Jeomorfoloji I'', İstanbul Üniversitesi Ed. Fak. Yayınları, No: 2931, İstanbul. "Geomorphology I"

Erol, O., 1979. ''Dördüncü Çağ (Kuvaterner) Jeoloji ve Jeomorfolojisinin Ana Çizgileri'', Ankara Üniv. Dil Tarih Coğr. Fak. Yay., No:289, Ankara. "The Outlines of the Geology and geomorphology of the Fourth Period (Quaternary)"

—. 1980. ‘‘Türkiye’de Neojen ve Kuvaterner Aşınım Dönemleri, Bu Dönemlerin Aşınım Yüzeyleriyle Yaşıt Tortullara Göre Belirlenmesi’’, Jeomorfoloji Derg., Sayı:11, S:1-22, Ankara. “The Neogene and Quaternary Erosion Periods in Turkey, The Determination of These Periods According to the Sediments of Erosion Period Ages”

—. 1983. ‘‘Türkiye’nin Genç Tektonik ve Jeomorfolojik Gelişimi’’, Jeomorfoloji Derg., Sayı:11, S: 1-22, Ankara. “Young Tectonics and Geomorphological Developments of Turkey”

—. 1984. ‘‘Geomorphology and Neotectonics of the Pluvial Lake Basins in the Taurus Belt and South Central Anatolia’’, Geology of the Taurus Belt International Symposium, 26-29 September 1983, Ankara, Proceedings, Page: 119-124, Printed in M.T.A., Ankara.

—. 1989a. ‘‘Batı ve Orta Torosların Neotektonik, Jeomorfolojik Gelişimini Araştırma Projesi Ara Raporu’’, TPAO Raporu, Ankara. “The Interim Report of the Research Project on the Geomorphological Development and Neotectonics of the Estern and Middle Taurus”

—. 1990. ‘‘Batı Toros Dağlarının Messiniyen Paleojeomorfolojisi ve Neotektoniği’’, Türkiye 8. Petrol Kongresi (16-20 Nisan 1990), Genişletilmiş Bildiri Özleri, S: 91-82, Ankara. “The Mesinian Paleogeomorphology and Neotectonics of the Western Taurus”

—. 1991a, ‘‘Geomorphological Evolution of The Taurus Mountains’’, Turkey., Zeitschrift für Geomorphologie, N.F. Supp. Bd.82, Berlin.

Folk, L.R., 1968. ‘‘Petrology of Sedimentary Rock’’, Univ. Texas Geology 370 K, 383 L+M, USA.

Koçyiğit, A., 1984. ‘‘Güneybatı Türkiye ve Yakın Dolayında Levha içi Yeni Tektonik Gelişim’’, TJK Bülteni, Cilt: 27, Sayı:1, S: 1-15, Ankara. “New Intra-Plate Tectonic Developments in the Southwestern Turkey and Adjacent Areas”

Keser, N., 2004a. ‘‘Bezirgan Polyesi ve Yakın Çevresinin Karst Jeomorfolojisi’’, Türk Coğrafya Derg. Sayı: 42, S: 11-45, İstanbul. “The Karst Geomorphology of the Bezirgan Polje and Its Vicinity”

—. 2004b. ‘‘Sarıbelen (Sidek) Polyesi ve Katran Dağının Karst Jeomorfolojisi’’, Marmara Coğrafya Derg. Sayı: 10, S: 19-52, İstanbul. “The Karst Geomorphology of the Sarıbelen (Sidek) Polje and Katran Mountain”

Ketin, İ., 1959. ‘‘Türkiye’nin Orojenik Gelişmesi’’, MTA Derg. Sayı: 53, Ankara. “The Orogenic Development of Turkey”

Monod, O., 1977. ‘‘Recherches Geologiques dans le Taurus Occidental au Sud de Beyşehir (Turquie)’’, These d’Etat Univ. Paris sud Orsay.

Nazik, L., 1992. ‘‘Beyşehir Gölü Güneybatısı ile Kembos Polyesi Arasının Karst Jeomorfolojisi’’, İst. Üniv. Deniz Bil.Coğr. Enst. Doktora Tezi

(Yayınlanmamış), İstanbul. "The Karst Geomorphology of the Area between the Southwest of Beyşehir Lake and the Kembos Polje"

Önalan, M., 1979. ''Elmalı-Kaş (Antalya) Arasındaki Bölgenin Jeolojisi'', Doktora Tezi, İst. Üniv. Fen Fak. Monografileri, No: 29, İstanbul. "The Geology of the Area between Elmalı and Kaş (Antalya)"

Oztürk, E.M., Dalkılıç, H., Ergin, A. ve Afşar, Ö.P., 1987. ''Sultan Dağları ve Güneydoğusu ile Anamas Dağı Dolayının Jeolojisi'', MTA Raporu, No: 8191, Ankara. The Geology of the Sultan Mountains and the Vicinity of Anamas Mountain"

Şenel, M., Serdaroğlu, M., Gözler, M.Z., 1983. ''Teke Torosları Güneydoğusunun Jeolojisi'', MTA Derg. No: 95-96, Ankara. "The Geology of the South-western of the Teke Taurus"

Şenel, M., Arbaş, A., Bilgi, C, Bilgin, Z.R., Dinçel, M.A., Durukan, E., Erkan, M., Karaman, T., Kaymakçı, H, Örçen, S., Selçuk ve Şen, M.A., 1986. ''Gömbe Akdağı'nın Stratigrafisi ve Yapısal Özellikleri'', Kaş-Antalya. Türkiye Jeoloji Kurultayı, Bildiri Özetleri, Ankara. "The Stratigraphy and Structural Properties of Gömbe Ak Mountain"

Şenel, M., Selçuk, H., Bilgin, Z.R., Şen, A.M., Karaman, T., Dinçer, M.A., Durakan, E., Arbaş, A., Örçen, S., Bilgi, C., 1989. ''Çameli (Denizli)- Yeşilova (Burdur)- Elmalı (Antalya) ve Dolayının Jeolojisi'', MTA Raporu, No: 9429, Ankara. "The Geology of Çameli (Denizli)- Yeşilova (Burdur)- Elmalı (Antalya) and Their Vicinity"

Şenel, M., Öztürk, E.M., Özdemir, T., Kadınkız, G., Metin, Y., Serdaroğlu, M., Örçen, S., 1994. ''Fethiye (Muğla), Kalkan (Antalya) ve Kuzeyinin Jeolojisi'', MTA Raporu, Ankara. "The Geology of Fethiye (Muğla), Kalkan (Antalya) and the Northern Parts"

Şenel, M., 1997. ''Fethiye-M-8 Paftası'', 1:100.000 Ölçekli Türkiye Jeoloji Haritaları, No:4, MTA Jeoloji Etütleri Dairesi, Ankara. "M-8 section of the Fethiye Map"

Şengör, A.M.C., 1980. ''Türkiye'nin Neotektoniğinin Esasları'', T.J.K Konferanslar Serisi Yayını, Ankara. "The Fundamentals of Turkish Neotectonics"

—. 1985b. ''Türkiye'nin Jeomorfolojik Evriminde Orta Miyosen'in Önemi'', Türkiye 9. Jeomorfoloji Bilimsel ve Teknik Kurultayı (13-15 Mart 1985), Bildiri Özleri, Ankara. "The Importance of Middle Miocene in the Geomorphological Evolution of Turkey"

Uysal, Ş., Dumont, J.F., Poisson, A., 1980. ''Batı Toros Platformları'', MTA Raporu (Yayınlanmamış), No: 6861, Ankara. "The Platforms od Western Taurus" (Unpublished)

Yalçınkaya, S., Ergin, A., Taner, K., Afşar, Ö.P., Dalkılıç, H., Özgönül, E., Ergin, A., 1986. ''Batı Torosların Jeoloji Raporu'', MTA Raporu

(Yayınlanmamış), No: 7898, Ankara. "The Geological Report of Western Taurus" (Unpublished)

CHAPTER FIFTEEN

AN OVERVIEW OF PLANT DIVERSITY AND LAND DEGRADATION INTERACTIONS IN THE EASTERN MEDITERRANEAN

MUNIR OZTURK, SALIH GUCEL, SERDAR SAKCALI, CIGDEM GORK, CELAL YARCI AND GUVEN GORK

Introduction

Geographically, the Mediterranean countries around the nearly enclosed Mediterranean Sea include the land masses of Europe on the northern, Africa on the South, and Asia to the East and northeast (Quezel, 1985; Ozturk, 1995). It extends over an area of nearly 2.3 x 106 km^2, which is equivalent to half the total area of the Mediterranean climate zone in the world. Various definitions of the limits of the Mediterranean climatic region, based on bioclimate, biogeography or floristics have been proposed, but there is no complete diagnostic set of criteria that can be used, although a delimitation based on bioclimatic criteria is most widely employed today. The crucial difference between the Mediterranean and adjacent arid climate zones is the mean annual rainfall. But summer is always the driest season (Akman, 1982, 1999). The range of the olive plant is also used to define the extent of the Mediterranean climate, like that of the holm oak. According to Dallman (1998), where any two species of olive, Aleppo pine, holm oak, and kermes oak are seen growing together, this clearly shows that it is a Mediterranean climate.

The Mediterranean area shows unique ecological and topographical diversity, with a striking alternation of complex stream eroded mountains, hills, alluvial plains, a very long coastline, and hundreds of islands. In some areas the high fold mountains are rising close to the shore. The bare and steep slopes, broad flood plains and torrential forms are the main kinds of land forms. The soils in the basin are a mixture of old and newer soils, but distinctive terra rossa soil is widely distributed on the lowlands (Atalay, 1994; Ozturk et al., 2002). The dry summers and the limited seasonal rainfall allow the development of a mosaic pattern of soils. Many different species are adapted to grow on these different soil types (Atalay, 1994; Ozturk, 1995; Efe, 1998).

The basin can be divided into two distinct regions - east and west on the basis of summer drought intensity. A great percentage of the eastern Mediterranean, which includes Turkey, Syria, Lebanon, Palestine, Israel and Jordan, is covered by arid lands, depicting a typical transition zone between the Saharo-Arabian desert biome and temperate climates (Heywood, 1995b, 2003a,b). The area shows a rich plant diversity (Dallman,1998), due to its unique position as a meeting area of the phytogeographical regions of the Mediterranean, Irano-Turanian, Saharo-Sindian and Sudano-Decanian; each of which have their own flora along with a large number of biregional and pluriregional species (Davis et al.1994; Ozturk et al.,1996a,b). An attempt will be made here to analyse the plant diversity of the eastern Mediterranean region in light of degradation processes.

Plant Diversity

A conservative estimate puts the total number of species presently existing on earth at around 12.5 million (WCMC, 1992). Heywood (1995a), based on several sources, lists some 1,750,000 described living species. One of the major centres of this diversity is the Mediterranean Basin, which covers some 2.3 million km^2 and contains about 10 percent of the world's flowering plants (Heywood, 1995b). It embodies more than 25,000 species of plants, containing 11 of the 231 centres selected for their global importance (Davis et al. 1994; Cowling et al.1996). The endemism is very high, estimated to lie at around 50 percent; with some areas being especially rich in endemics (Greuter, 1991). A high number of species are found in unit areas and the plant life is very rich. This region is also a major centre of crop origins and their relatives (Zohary and Hopf 1993). The bare deserts, hostile to any kind of life, make up a large part of the total surface area of the eastern Mediterranean.

A great variability is seen in the flora and vegetation of the eastern Mediterranean, which includes nearly 15,000 different plant species belonging to 1000 genera and 500 families (Heywood, 1995b). Agriculture, in the course of its long history, has contributed much towards enrichment in plant species. Many segetals and ruderals have been introduced and others have penetrated from adjacent regions. The eastern

Mediterranean includes two of the eight centers of origin of cultivated plants, the Mediterranean and southwest Asia (Vavilov,1951; Davis et al.1994). It is an area of megadiversity of important food crops, pasture and range land species, the native area of more than 160 species of cultivated plants, as well as a source of desirable genes for existing varieties and for the supply of new crops. The total gene pool of wild species and land races of cultivated plants has a great potential in the flora of this region. A large number of our food crops such as wheat, barley, lentils, chick peas, beans, cabbage, asparagus, artichokes, leeks, figs, olives, and almonds originated here and it is also a peripheral area of origin and home to fruit trees such as plum, almond, and pistachio (Harlan 1995).

According to Jana (1995) several factors have contributed to the region's rich plant diversity: (a) the political climate and royal interests in agricultural innovation; (b) stress-resistant alleles developed in response to the harsh climatic conditions of the region; (c) ecological-habitat diversity that contributes to adaptive diversity in indigenous crops and wild relatives; (d) genetic, such as dynamic interactions between natural and human factors which maintain crop genetic diversity; e) evolutionary processes such as geographical isolation and limited gene migration through seed exchange which have led to adaptive divergence and geographical differentiation of crop species; f) ecological conditions created by the topographical diversity within limited areas and g) its unique geographical position. The replacement of floras in the region at very short intervals and paleo-plant geography have brought many changes in floral composition since the Tertiary period, leaving behind many elements as relics in suitable habitats (Ketenoglu et al., 1993; Ozturk et al., 2004).

The total area under the forest, and secondary or subseral shrub communities like maquis and phrygana, is 12.9 million hectares, out of which 6.7 million hectares form Mediterranean type components (Fig. 15-2,15-3). The forests in this region are considered to be areas of great botanical importance for threatened species, for habitats, and for plant diversity in general, discrete sites which support one or more high-priority species and exceptional habitats. The total area of forest, maquis and phrygana in different countries of the region is 12.3 million hectares (Turkey), 0.4 million hectares (Syria), 0.1 million hectares (Lebanon) and 0.1 million hectares (Palestine). The Mediterranean component is 6.1 million hectares, 0.4 million hectares, 0.1 million hectares, and 0.1 million hectares, respectively, for these countries (Ozturk et al.2002). Turkey has more than 9,000 taxa of higher plants belonging to nearly 1000 genera. In

Syria and Lebanon this number lies at around 870 genera and in Paletsine together with Jordan the number is 718 genera and 2,250 taxa. The endemics are also abundant. The number of endemics varies as follows: 145 (Jordan), 165 (Israel), 311 (Lebanon), 395 (Syria), 1350 (Greece) and 3300 (Turkey) (Ozturk et al.1991).

The proximity of this area to Asia accounts for more than 300 plant species of Asian origin. The high and low maquis are composed of different vegetation formations. The typical representatives from garrigue (phrygana) are *Sarcopoterium spinosum* and *Euphorbia acanthothamnus*, which are native to the eastern Mediterranean (Dallman, 1998). There are several drought-deciduous, dimorphic and aromatic species like; *Euphorbia dendroides*, *Calycotome villosa*, *Rosmarinus officinalis*, *Lavandula stoechas* and *Origanum* species. Many of the trees, shrubs and herbs from the east Mediterranean are planted locally and elsewhere in the world. *Platanus orientalis, Pinus halepensis, P.pinea, Cupressus sempervirens, Cedrus libani, Abies cilicica, Ceratonia siliqua, Pistacia atlantica,* and *Rhus coriaria* are the well known afforestation trees (Ozturk, 1995; Dallman,1998). The area has also provided the gardens of the world with several ornamentals, such as *Adonis aleppica, Allium* spp., *Alnus orientalis, Anemone coronaria, Arbutus andrachne, Arundo donax, Buxus sempervirens, Cercis siliquastrum, Chrysanthemum coronarium, Cistus* spp., *Clematis vitalba, Colutea* spp., *Cotonaster racemiflora, Crocus* spp., *Glaucium* spp., *Gypsophyla* spp., *Hyacinthus orientalis, Iberis odorata, Jasminum officinale, Laurus nobilis, Lavandula stoechas, Ligustrum vulgare, Lilium candidum, Linum* spp., *Matthiola* spp., *Nerium oleander, Nigella damascena, Ornithogalum* spp., *Pistacia lentiscus, Reseda odorata, Retama raetum, Rhamnus alaternus, Ruscus aculeatus, Ruta chalepensis, Salvia triloba, Spartium junceum,* and *Tamarix* spp. (Ozturk et al., 2000)

Palestine, as a part of this region, comprises plants from most of the Old World. There is a peculiar northward migration of the tropical and Saharan floras along the coastal plain and the Jordan Valley. Important characteristics of the flora are: a lack of tropical elements, the presence of a fairly large number of trees and shrubs that find their southern limit of distribution in the country; the presence of a large number of northern (Auxinian) plants as relics of past humid periods; the presence of alpine and sub-alpine mountainous flora and a high percentage of endemism (11%) (Juneidi, 1994). Out of 26,000 hectares of forested area on the West Bank, more than 20,000 hectares are natural forests, and just over 3,500 hectares are man-made forests. Historical and recent data indicate that this area was probably naturally planted with juniper, carob, and oak species, as well as olives and fruit trees. The forests tend to be diverse, composed of *Pinus halapensis*, *Pistacia* spp., *Ceratonia siliqua, Cuppressus sempervirense*, and *Quercus* spp. According to Zohary (1973) and Daninn (1988) approximately 2,500 species of plants exist in the West Bank and Gaza, representing a high degree of diversity. Among the most important indigenous plant groups adapted here are *Acacia, Artemisia, Prosopis, Ziziphus, Populus* and *Platanus* species. The indigenous forest trees are *Arbutus andrachne*, *Laurus nobilis*, *Pinus pinea*, *Pistacia* spp., *Populus euphratica*, *Quercus* spp., and *Tamarix* spp.

Rhamnus palaestinus, Quercus calliprinos, Cyclemen persicum, Cuppressus horizontalis, and *Phlomis pungens* also exist in some forests. The ecological sites of natural forest also contain *Opuntia ficus-indica.* In the southeast, degraded rangelands of *Thymelaea hirsuta* are very common. In the southwest is a forested area, probably important in the maintenance of the mountain ecosystem that covers agriculturally important areas, as well as biologically important habitats for *Pinus* and other species. The area extending down to the Jordan Valley shows native species like *Paronychia argentea, Majorana syriacum, Gundelia tournefortii, Aruma palestinum* and *Chamomile*. The area around Mar Saba has *Capparis spinosa*, a species of high importance. The north side contains important ecological resources including *Thymus bovei, Asphodelus ramosus, Phlomis* spp., some *Trifloium* and *Medicago* species. The valley of Wadi Nar is, biologically, relatively fertile. The area just behind Mount Olives and Mount Scopus has relatively old forest and is the center of origin of *Nicotiana glucua*. Wadi Hajla has a beautiful landscape with a high abundance of *Ziziphus spini-christe, Eucalyptus,* and *Pinus halapensis* together with Date palms, *Cuppressus, Rhamnus alternus, Echium palaestinum, Onosoma gigentea, Notobasis syriaca, Ballota undollata*. Some sporadic native trees found in the Jericho area are *Calotropis procera, Acacia albida* and *A. tortilis*. *Artimisa sieberi, Haloxylon persicum, Hammada negevensis* and *Balanitis aegyptiaca* are distributed further north. The area towards the Dead Sea contains savannah-like vegetation, including grass species. In the Ramallah district much of the land is agricultural or built up. A large wooded area is found in the north of Turmus Aya. The Nablus District too is an area with high plant diversity. The pine and olive forest to the west side of Tulkarem District is a highly sensitive area, including old forest lands. The Jenin District is an area of high ecological significance as it includes a rich vegetational cover of rangeland species with a high diversity of *Anemone* spp. and *Tuilpa* spp., *Cyclamen persicum, Podonsoma syriaca, Notobasis syriaca,* and *Centaurea iberica*. In the east we find *Retama raetem,*

Gundelia tournefortii, Pistacia sp., Calligonum comosum, Ziziphus lotus, Ephedra, and *Mercurialis* spp. The Aqqaba area has a beautiful rangeland. The Raba forest to the northeast of Aqqaba is a highly sensitive area of pine, oak and cuppressus trees. The area just before Raba has a great potential for rangelands, olive trees and plants like *Pimpinella, Ridolfia segetum, Achillea critica*. Natural forests in Palestine are found more than 700 metres above sea level and are concentrated mainly on the central highlands being represented by Aleppo Pine and Evergreen Oak Maquis.

Israel is characterized by a wide variety of flora (Shmida and Darom, 1990,1992). Its geobotanical position has endowed the country with a rich plant diversity, including some 2,780 plant species. The climatic regime is Mediterranean in the north and desert in the south. The central part of the country is a transition area between these two biogeographical regions, where desert biota is gradually replaced by Mediterranean biota. Species widely distributed over the entire Mediterranean climate region reach their southern limit of distribution in Israel, while Saharan or Asian desert species reach their northernmost limits in this country.

The total number of plant species in Jordan is estimated to be between 2,300 and 2,500, of which 2,000 are vascular plants (Juneidi, 1973). Also, there are several vegetation types. The forests occur mainly within the Mediterranean region, while steppes are confined to the Irano-Turanian region. *Acacia* woodlands occur in the sub-tropical region. Hammad vegetation occurs predominantly in the Saharo-arabian region.

A major part of the Syrian territory is desert to semi-desert, with a volcanic plateau in the south, but a large fertile basin formed by the Euphrates River lies in the north. The steppe ecosystems occupy more than half the land area (Al-Jundi et al. 1999, 2001; Mirreh, 1999). There is a rich forest of deciduous trees of *Quercus cerris* ssp. *pseudocerris*, in Forloq. This ecosystem is quite distinctive for Syria and is composed of an ecotone of unique assemblages of species, which are remnants of the ancient climate reign in Syria and mingled with Mediterranean and Irano-Turanian species. The northeastern corner of Syria supports remnant forest/steppe associations which represent the nearest living examples to Neolithic sites along the Euphrates. Key tree species here are *Pistacia khinjuk* and *Pistacia atlantica*. Another geologically, geo-morphologically and biologically important area is Abou Qubies. It possesses unique flora due to micro-climatic conditions. The number of species in the area is estimated at 350, with perhaps six endemics, and 25 rare or endangered species. The steppes in Palmyra support large wildlife.

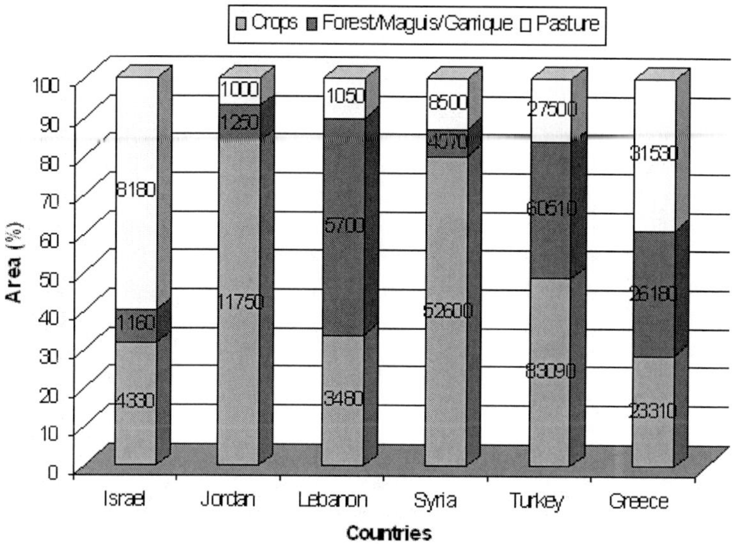

Fig. 15-1: Percentage cover of crops, forest-maquis-garrigue and pastures in the eastern Mediterranean countries (histogram numbers represent area in km^2).

Asian Turkey includes one of the largest areas of Mediterranean climate and shares a great number of its species with Iran, Afghanistan and Central Asia (Ozturk et al. 1996a,b). A rough estimate of the number of plant taxa in the flora of Turkey is more than 9,000 (Davis 1965; 1988; Davis et al., 1988; Davis. 1998). A feature of the flora is the presence of giant genera with hundreds of species; so as a result, the country could be considered as being a main or secondary centre of speciation. There are different relics, enclaves, endemisms and disjunctions in the flora. The Turkish Mediterranean region covers 198.165 kilometers and includes states along the Aegean and Mediterranean Sea coasts with a high endemism (25%), comprised of 1,768 endemic taxa; the same being true for the number of tree species, especially Rosaceae and rich alpine flora (Ozturk et al., 1991; Ozturk, 1995). The families with the highest number of taxa are: *Compositae* (225), *Labiatae* (190), *Fabaceae* (184), Scrophulariaceae (154), *Caryophyllaceae* (135), *Brassicaceae* (112), *Liliaceae* (110), *Umbelliferae* (83), *Boraginaceae* (65), *Campanulaceae* (45), *Poaceae* (42) and *Ranunculaceae* (22) (Fig.15-3). The distribution of the endemic phytogeographical elements is as follows: eastern Mediterranean 966 taxa (80 %), Irano-Turanian 203 taxa (17 %), Euxine

29 taxa (2 %), Mediterranean 12 taxa (1 %), and Euro-Siberian 5 taxa. The number of taxa with unknown phytogeographical origin is 553 (31.27 %) and those with known origin is 1,215 (68.72 %). The life forms of different endemic taxa are distributed as follows: hemicryptophytes 1,007 (57 %), geophytes 283 (16 %), therophytes 254 (14 %), chamephytes 192 (11 %) and phanerophytes 32 (2 %).

According to the data published by Cetik (1976), Akman et al. (1978;1979a,b), Ozturk et al. (1995, 2002,2006) and Mazzoleni et al. (2004), the natural vegetation of the area reflects typical Mediterranean features. It includes a major part of the 20.2 million hectares of the total forest area of Turkey. The forests generally occupy high altitudes, but half of these are composed of maquis vegetation and are regarded as low productive land. The six vertical zones (Atalay, 1994) from the coastline to the highest peak in this region are:

Altitude and Salient Vegetation Type

- 0-500 meters, Maquis, Phrygana, Plantations (Mediterranean);
- 500-1200 meters, *Pinus brutia* (Upper Mediterranean)
- 1200-1600 meters, *Pinus nigra* subsp. *pallasiana* (Oro Mediterranean)
- 1600-2100 meters, *Abies cilicica, Cedrus libani, Juniperus excelsa* trees (Subalpine)
- 2100-2600 meters, Tragancanthic plant cover (Alpine)
- 2600-3600 meters, Wet meadows, Lower plants,Geophytes, Lithophytes (Alpine)

The coastal zone includes wetlands, psammophytic and halophytic vegetation zones located in the form of a narrow belt. *Eryngium maritimum, Ipomoea stolonifera, Cakile maritima, Salsola ruthenica, Cyperus capitatus, Plantago maritima, Elymus farctus, Trachomitum venetum, Arthrocnemum fruticosum, Halimione portüacoides, Holocnemum strobilaceum, Sporobolus virginieus, Juncus maritimus, Salicornia europaea* and *Zygophyllum album* are the most abundant plant species occurring on sandy and saline areas (Guvensen and Ozturk, 2003; Ozturk et al. 2006).

The phrygana covers coastal areas in general, with low thorny plants which are usually leafless in the dry season. Major representatives are: the species of *Cistus, Phlomis, Teucrium, Fumana, Origanum, Calicotome villosa, Coridothymus capitatus, Euphorbia acanthothamnos, Erica manipuliflora, Genista acanthoclada, Micromeria nervosa, Salvia fruticosa, Sarcopoterium spinosum, Satureja thymbra,* and *Thymbra spicata.* This formation occupies a mainly stony and arid thermo Mediterranean belt at places where maquis and *Pinus brutia* forests have been destroyed (Ozturk, 1995).

The maquis zone includes evergreen and dwarf shrubs with hard leathery leaves, widespread from sea level to 700-800 meters. Mixed forest remnants of *Pinus brutia* and *Pinus halepensis* in maquis vegetation show this vegetation is of secondary origin, representing a secondary succession of the thermo-Mediterranean. The species like *Styrax officinalis, Cercis siliquastrum, Arbutus andrachne, Myrtus communis, Quercus infectoria, Q. libani, Nerium oleander, Ceratonia siliqua, Vitex agnus-castus* indicate the existence of humid habitats whereas, *Quercus coccifera, Pistacia terebinthus, Rhamnus oleoides ssp. grecus, Paliurus spina-christi, Berberis cretica, Cistus creticus, C. salviifolius, Olea europaea, Calicotome villosa, Erica manipuliflora, Cotinus coggyrea, Phyllyrea latifolia* represent aridity and poor habitats (Ozturk et al. 2002).

Forest formation (Akman et al. 1978; 1979 a, b; Atalay, 1994) in the western and southern Anatolian regions can be divided into Eu-Mediterranean and Oro-Mediterranean parts on the basis of climatic and altitudinal characteristics. Out of these, *Pinus pinea* forests occupy a special place. These occur in the Kozak Yayla (Bergama), and on the sand beaches between Side and Serik (Antalya). They prefer south-facing slopes with generally low productivity and poor undercover (Atalay,1994; Ozturk,1995).

Pinus brutia forms the climax forests in the area beginning from sea level and reaching up to 1600 meters in the south, 600 meters in the north and 800 meters in the west. The floristic composition and the physiognomy of the Red Pine forests depend on the physico-chemical characteristics of the parent material (Ozturk, 1995).

Pinus nigra ssp. *pallasiana* forests are found between 800-2000 meters but extend up to 2250 meters in the humid oro-Mediterranean part of the Taurus mountains and are mixed with *Cedrus libani* or *Abies cilicia*. The understory cover of these forests is poor floristically. Maquis elements frequently appear on the open areas, while other species occurring in such forests mainly belong to the genus *Quercus*. Wet valleys and shady slopes are generally covered by *Fagus orientalis, Carpinus orientalis, Juglans regia, Platanus orientalis, Sorbus torminalis* and *Ostrya carpinifolia.* The destroyed *Pinus nigra* forest areas are occupied by the species of *Juniperus, Astragalus, Asphodeline taurica,* and *Euphorbia kotschyana* (Atalay, 1994; Yucel and Ozturk, 2000).

Abies cilicica ssp. *cilicica* and *Cedrus libani* forests lie in the oro-Mediterranean zone, distributed between 800-2000 m. on the Taurus mountains, but between 1400-1500 m. on the slopes facing south (Cetik, 1976). Major companions are *Quercus, Juniperus, Fraxinus, Celtis, Pinus nigra* ssp. *pallasiana, cer hyrcanum* ssp. *tauricolum, Sorbus torminalis, S. umbellata, Ostrya carpinifolia, otoneaster nummularia, Styrax officinalis, Fagus orientalis, Carpinus orientalis* ssp. *orientalis, Populus tremula, Salix alba, Ulmus minor* ssp. *canescens, Cornus mas, Pistacia terebinthus* ssp. *palaestina, Tilia rubra* ssp. *caucasica, Rubus canescens* and some Euro-Siberian herb elements. *Juniperus excelsa* and *Juniperus foetidissima* are common on the Taurus mountains at places where Cedar and Black Pine forests have been cleared. Such forests can be accepted as a regression in the succession.

Oak forests occur between thermo and oro-Mediterranean belts from 800 to 1200 meters, commonly in the western and the eastern parts of the Taurus. These are represented by 4 taxa of genus *Quercus*. *Q. pubescens,* which dominates in places where the continental climate is effective. *Quercus vulcanica* is an endemic oak found on the Taurus Mountains in karstic depressions, together with *Sorbus torminalis, Tilia rubra* ssp. *caucasica, Fraxinus excelsior* ssp. *excelsior, Ulmus glabra* and *Ostrya carpinifolia,* all of which are Euro-Siberian elements (Gemici, 2000).

Fagus orientalis forests are found in small groups at the Amanos mountains in the mist covered zone along with *Euonymus latifola, Cedrus libani, Corylus avellana* and *Staphylea pinnata,* but *Carpinus orientalis* ssp. *orientalis* forests lie on its north facing slopes. The latter are mixed with *Fraxinus ornus* ssp. *ornus, Quercus* spp., *Daphne sericea, Cornus mas, Ilex colchica, Buxus balearica, Taxus baccata, Ulmus glabra, Juniperus drupacea, Eriolobus trilobatus,* and *Pinus nigra* ssp. *pallasiana* (Akman, 1978, 1979a,b).The small stands of *Castanea sativa* humid forests are associated with the species of *Cornus, Quercus, Salix alba, Rhus coriaria, Phillyrea latifolia, Arbutus andrachne, Sorbus torminalis, Corylus avellana* and *Tilia argentea* (Gemici, 1995). The Alpine and Sub-alpine Zones embody perennial herbaceous and semi-woody cushion plants from 2250 meters to the peaks. Major species found here are *Acantholimon, Astragalus, Thymus* and *Convolvulus,* as well as *Berberis crataegina, Cerasus prostrata, Cotoneaster nummularia, Rosa pulverulanta* and geophytes like *Allium, Fritillaria, Orchis, Crocus, Ornithogalum* and *Colchicum* (Gemici, 1995).

The upper parts of the Taurus are sparsely covered by thorny cushion herbs like *Acantholimon* and *Astragalus* (Ozturk et al. 1991), but middle parts principally embody Sub-alpine and Irano-Turanian herbaceous elements. *Festuca* and *Astragalus* are dominant in the Sub-alpine vegetation of the

Bolkar and Ala mountains between 2300-2400 meters altitudes (Gemici, 2000).

Land Degradation: Land degradation means a process of soil impoverishment and vegetation cover loss that may lead to a progressive and irreversible reduction in biological or economic productivity (Brandt and Thornes, 1996; Geeson et al., 2002; Feoli et al. 2003). Drought and desertification threaten the livelihood of millions of human beings in more than 100 countries globally. Human dominance and activities are destroying the soil system, a resource which is not permanent and which is renewable only over a long period of time, and which supports all living and nonliving beings and is a depot and filtering agent for our waters. It is the natural source of our forests, food and clothing, allowing sustenance of desired conditions of lands and other communities, as well as economic enterprises that depend on land. Our life on this planet is especially dependent upon fertile soils and it is highly essential for our well-being and to life itself. We thus need to understand the nature of soils so that we can support and carry out the practices that will help preserve them. The sustainable use of our soils is a question of paramount importance and a prudent path to pursue for the stewardship of biological communities.

The eastern Mediterranean shows a typical transition zone between the Saharo-Arabian desert biome and temperate climates. The plant cover in this zone has suffered much from biotic interference in the past. Thousands of years of destruction have set in motion a chain of feedback processes leading to land destruction. The areas covered by tragacanthic vegetation have been continuously changed into agricultural lands for immediate benefits (Ozturk, 1995; Ozturk et al., 1999, 2002). The main drivers of degradation are socio-economic factors, climatic fluctuations, water scarcity, food security, urbanisation, agricultural policies and demographic changes.

The advent of man in the east Mediterranean took palce in the Early Pleistocene and evidence proves that hunters and gatherers were in the Villefranche period in the Jordan Valley. Egypt, the Middle East, and Turkey were among the earliest places settled by agricultural societies and a cradle of civilization for thousands of years, with a great cultural heritage that developed with the mixing of people, architecture, food and religion. A great percentage of lands in these countries is arid with an average annual precipitation of 50-200 millimeters per year. These arid lands have a great variability in their flora and vegetation due to many factors, some of them are biotic; others abiotic. Most countries in the region are facing progressive degradation of their vegetation cover due to increasing anthropic pressure, leading to improper use of resources. The pollen diagrams from the area reveal early and widespread forest clearance and grazing and cultivation, even at high altitudes. The fire effects are 10.000 years old and periodic fires continue to destroy the woody and herbaceous vegetation, leaving the soil exposed to erosive effects. It is regarded as a powerful factor in the evolution and differentiation of flora, as well as landscapes in the eastern Mediterranean (Naveh 1991). Agriculture and animal husbandry began earlier than 10,000 B.P, whereas fruit crops appeared during the Bronze Age. The breeding of sheep and goats began 12,000-8,000 B.P. The transformation of different types of plant communities and remnants of forests into agricultural lands, or urbanization on fertile soils add to these processes. Overgrazing and indiscriminate urbanization are common factors contributing to the impoverishment of the Mediterranean forests and grasslands, hence, increasing environmental aridity (Vertovec et al., 2001; Sakcali and Ozturk, 2003). Deforestation began as early as 8000 B.P., with many negative consequences due to the growth of urban societies. This led to a greater demand for wood for fuel and construction material, which in turn have profoundly influenced the landscape and vegetation. The demographic explosion has led to a decrease in the rural inhabitants, but urban ones have increased (Naveh, 1991; Trabaud, 1981). Nearly 300 million people live in the Mediterranean Basin. Population growth is causing the deterioration of monuments and buildings, as well as of biodiversity. The ratio of agricultural soils covered by the settlement areas varies between 35–70 %. The agricultural soils are continuously being transformed into tourist resorts (Ozturk et al., 1999). The last two centuries have seen an increasingly severe destruction of vegetation, and today timberless vegetation in the area is included in deserts or semi-deserts by many researchers. Overgrazing and over-exploitation of the steppe vegetation and the drainage of marshes, estuaries and deltas have led to the disappearance of many ecosystems, in particular, of agroecosystems. Sedentary people have proved to be more destructive than nomadic ones. The areas covered by tragacanthic chamaephytes like *Artemisia, Astragalus* and *Acantholimon* have been changed into agricultural lands for immediate benefits. The present pressure of grazing by herders in the region affects the biomass and decreases the rangeland potential. Only *Iris palestina, Asphodelus ramosus* and *Sarcopoterium spinosum* exist in such places. During the last 5 decades drastic changes have taken place in the range conditions. They show much lower potential than their productive capacity. This has a direct effect on the land and an increased demand for livestock products increases the pressure placed upon grazing resources. There has been a huge increase in the number of sheep on the range. Every year thousands of animals grazed by the villagers generally follow a grazing pattern that involves moving animals to the lower and less abundant pastures during winter and spring, and driving them back to the relatively more fertile highlands in summer and autumn.

Three main kinds of land forms are characteristic of Mediterranean lands: bare and steep slopes, broad flood plains and torrential forms. The risk of landslides, floatation and avalanche occurrence increases because of uncontrolled overgrazing (Mendizabal and Puigdefabregas, 2003). The biotic interferences have set in motion a chain of feedback processes leading to heavy soil erosion. A retrogressive succession is visible in the *Pinus nigra* and *P.brutia* forests. The present plant cover has developed adaptations such as surface reduction ranging between 40 to 90 percent, which can be observed as microphylly, aphylly or sclerophylly. Trabaud (1981) and Le Houerou (1981) postulated that all these disturbances are largely responsible for the rich floristic and ecological diversity in the region. These interferences have led to the formation of many secondary or subseral shrub communities such as maquis, phrygana, matorral, garrigue; forming the aesthetic landscape of the Mediterranean (Quezel,1985; Ozturk, 1995). The climatic variations, particularly during the last few decades in conjuction with the global patterns of climate change, have had significant effects on plant life. A large number of exotic species have become established in the region and many of the Mediterranean species are noxious weeds in other parts of the world (Heywood, 1989). The important feature of agriculture in this area, particularly in the drier parts, is extensive misuse and over-exploitation of the natural resources, which are a) limited, due to the fact that there are few areas of good soil with proper rainfall or irrigation water, and b) prone to degradation due to such factors as irregularity of rainfall and erosivity (or "aggressivity") of the climate towards soils, which are often on slopes and fragile.

The eastern Mediterranean area is considered by Davis et al. (1994) as including two of the eight centers of origin of cultivated plants. These are the Mediterranean and southwest Asian centers. According to Vavilov (1951) the region is the original home of more than 160 species of cultivated palnts. These plants are at the same time still growing wild in the region. The total gene pool of wild species and land races of cultivated plants has great potential in the flora of the eastern Mediterranean region. Despite the scarcity of water and cultivable land in these semiarid and arid areas, the region is and was one of the major food producing regions of the world. It was termed as the granary of the Roman Empire (Cowling et al. 1996). However today, many plant species occur in small fluctuating and poorly dispersed populations, as a result of habitat fragmentation and face extinction; the abandonment of terracing; the movement away from the land to the towns and cities; crop substitution both in terms of individual crops or of whole agroecosystems - such as the replacement of dryland cultivation of almonds, olives, vines, figs and carob in the coastal areas with subtropical American species or horticultural quality crops under plastic greenhouses; the introduction of new crops and intensive commercial horticulture; an increase in the spread of alien and invasive species; the invasion by exotics that were introduced as living hedge plants, displacing other local species as they became naturalized; the devastating effects of agricultural, industrial and urban pollution and the endangerment of plant genetic resources by genetic erosion and genetic pollution (Jaradat 1995). The degradation of natural habitats, intensification of the cultivation of arable lands, the expansion of cultivation into marginal areas and remnants of forests, the replacement of land races by new cultivars and the overgrazing of natural pastures and range lands is resulting in the erosion of plant genetic resources. Yet most of the plant species have managed to survive, although in small populations. Nonetheless, the continuing pressure on these habitats makes the survival and management of remaining populations of many plants and wild relatives a complex and demanding task. Anthropomorphic pressure is active on coastal ecosystems as well, due to their aesthetic appeal (Ozturk et al., 2008). The Mediterranean shore zone exhibits a striking alternation of complex stream eroded mountains and hills, and alluvial plains. Moderate to steep slopes, with thin soils and much bare rock, are commonplace. Erosional features in different parts of the Mediterranean suggest its location as being a climatic borderland. The Mediterranean today presents a mosaic of old and newer soils. A distinctive soil of rather wide distribution on the lowlands is the terra rossa (soil developed from limestone).

The plant diversity in transitional and fragile ecosystems has declined dramatically and some species have become totally extinct from the wild. In the Turkish Mediterranean area, the number of endemic taxa under threat is 355 (20.1%), out of which 63 (3.56%) are facing a low threat and 292 (16.52%) are under great threat. More than 200 species are rare or endangered in the West Bank. Some of these include important rangeland forages such as *Coleutea, Artemisia herbaalba, Haloxylon persicum, Acacia albida*. As many as 40 plant species are reported to be endangered and about 150 are known to be rare. Several plant species are threatened by unregulated collecting. These include medicinal and ornamental plant taxa from the genus *Tulipa, Narcissus*, *Lupinus* and *Cyclamen* and hundreds of specimens are illegally smuggled yearly either in the form of living or dry herbarium specimens. The most important endangered endemics in the Turkish Mediterranean are: *Achillea phrygia, Ajuga bombycina, Alkanna phrygia, Alyssum aurantiacum, Anthemis dipsacea, A. wiedemanniana, Aristolachia hirta, Asperula lilaciflora* ssp. *phrygia, Astragalus karputanus, A. petilodes* var. *petilodes, Aubrieta canescens* ssp. *canescens, Campanula lyrata* ssp. *lyrata, C. tomentosa, C. raveyi, Centaurea calolepis, Cirsium sipyleum, Consolida phrygia* ssp. *phrygia, Crocus olivieri* ssp. *balansa, Cyclamen mirabile, Dianthus cibrarius, D. lydus, Digitalis cariensis, Erysimum pusillum, Euphorbia anacampseros* var. *anacampseros, Ferulago humilis, Fritillaria bithynica, Galium brevifolium* ssp. *brevifolium, Gladiolus anatolicus, Gypsophyla tubulosa, Haplophyllum myrtifolium, Hieracium patentissimum,Lotus macrotrichus, Malope anatolica, Minuartia juressi* ssp. *asiatica, Nepeta nuda* ssp. *lydiae, N. pilinux, Papaver wirchowii, Parakaryum auchei, Ranunculus reuterianus, Scabiosa reuteriana, Scrophularia floribunda, Sedum samium* ssp. *samium, Sideritis leptoclada, Trigonella cretica, Tulipa armena* var. *lycica, Velezia hispida, Velezia pseudorigida, Verbascum donii, V. mykales, V. parviflorum, V. symirnaeum, Ziziphora taurica* ssp. *cleonioides, Aethionema capitatum, Allium gayi, Astragalus antiochianus, Centaurea haradjiani, C. iycopiflia, C. plosimopappoides, C. solsitalis* ssp. *carneola, Cephalaria taurica, Chamaecytisus dreponolobus, Euphorbia rhridosperma, Hyacinthella hispida, Hypericum polyhyllum* ssp. *polyphyllum, Lathyrus laxiflorus* ssp. *angustifolia, Leucocyclus formosus* ssp. *amanicus, Lamium garganicum* ssp. *nepetifolium, Phlomis longifolia* var. *bailanica, Scabiosa kurdica, Scutellaria glaphyrostachys, Stachys annus* ssp. *cilicica, Thlaspi densiflorum, Trifolium davasii, Verbascum caesareum* and *V. linearlobium.*

Conclusions

The plant diversity of the Mediterranean has been a major source for humans for millennia. Today it is at risk on an unprecedented scale. As Greuter (1991) comments '... the Mediterranean ... is threatened by the very culture it engendered, a culture which has become dangerous and destructive.' If we are to continue to profit from the cornucopia of plant diversity, we must make wise choices when establishing priorities for conservation and sustainable use. It is among the 4 most significantly altered, out of 34 biodiversity hotspots on Earth. The environmental problems in the region and its vulnerabilities and risks, are challenging human security, food security, health security, inducing migration, posing livelihood challenges, triggering violent social consequences and thus becoming a social and regional security challenge. The driving forces are markets, agricultural policies, climatic factors and demographic changes, together with water security, food security and urbanisation (Brauch, 2003). Soil productivity is decreasing, although we know that our life on this planet is highly dependent upon this factor. It is essential for our well-being and to life itself. We need to understand its nature so that we can support and carry out the practices that will help preserve it. The potential carrying capacity of the soil is relatively constant, but cultural-social demands change it over time. We need to know what society wants from the soil in the face of different pressures and develop a system that reflects the wants and needs of society. The protection of soils is important not only from an economic, but also from ecological and sociological points of view, in order to grow food for humans. The natural vegetation covering our soils plays an important protective role, which is lost on transformation. Deforestation to obtain grazing and arable lands induces an imbalance in the precarious morphoclimatic equilibrium. Overgrazing and over-exploitation of plant cover, drainage of marshes, estuaries and deltas is leading to the disappearance of many ecosystems.

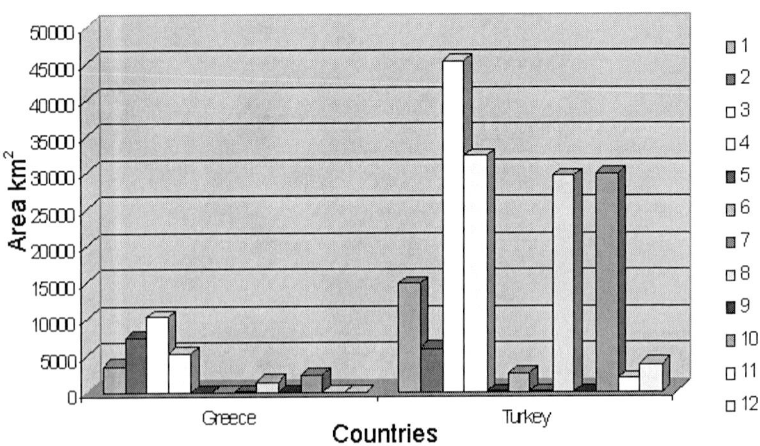

Fig. 15-2a: The cover of important plant species in km^2 in Turkey and Greece

The eastern Mediterranean has a megadiversity of important food crops, pasture and range land species. The land resource is both slender and fragile, and the conservation and maintenence of its plant diversity and productivity is of crucial concern. The plant genetic resources are being eroded through the degradation of natural habitats; intensification of the cultivation of arable lands; the expansion of cultivation into marginal areas and remnants of forests; the replacement of land races by new cultivars and the overgrazing of natural pastures and range lands. These degradative forces pose a serious threat to development. During the last three decades nearly one million people have been affected by social turmoil in the eastern Mediterranean. Out of 100 events, 5 were directly related to drought and famine. Combating this problem is a major environmental, developmental and security task in the region. It needs pro-active policies by states and international organisations. The subject has been accorded greater priority in recent years with the growing awareness that land scarcity, coupled with unprecedented population and economic growth, threaten to deplete natural resources and open space landscapes (Briassoulis et al., 2003).

Efforts for the conservation and management of natural resources must be derived from a set of clear objectives, a mechanism for action and a commitment from all stakeholders. Seven target groups such as industry, energy, transport, tourism, agriculture, the urban sector, and biodiversity can be addressed. Discussions should be conducted within a roundtable framework, with the participation of all stakeholders. Apart from halting the process of degradation and species loss, specialized solutions and an understanding of ecological processes are required. Protecting biodiversity does not merely involve setting aside areas as reserves. Instead, all the ecological processes that maintain biodiversity need to be understood for its proper management. This requires comprehensive in-service and external training of nationals in the areas of management, wildlife, socioeconomics, community participation and extension. A landscape approach to conservation where protected and non-protected areas are integrated through significant protection measures, at both state and the community levels and through the improvement of cooperative lands surrounding the biodiversity reserve, should be one of the points in this connection.

To promote community integration we need to introduce the participation of rurals populations in scientific fieldwork, identify and establish a demand driven extension approach for the population living in the area, prepare a communication strategy to reach local communities, and provide concrete services and income generating activities to locals. Key stakeholders, such as farmers and co-operative representatives, should be trained in the use of resource management techniques, informal resource mapping, ranking and living conditions. A mechanism to establish dialogue and communication is critically important. Therefore, communication materials such as flipcharts and leaflets should be prepared to deliver extension messages. This will prove effective in raising awareness. The building of trust quite often happens when concrete services are provided to target groups. Community demands should be met by conducting training programs aimed at income generation. A holistic approach to resource management and the involvement of all stakeholders is crucial for long term sustainability. Therefore, the steps to be taken are capacity building of the staff in agriculture and forestry services; environmental and socioeconomic monitoring; methodologies and techniques of rehabilitating degraded lands; wildlife reintroduction; biodiversity surveys and community participation for the conservation and management of natural resources.

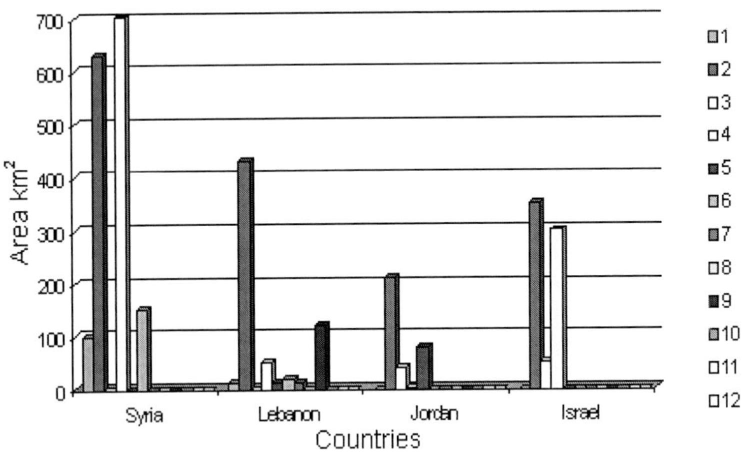

Fig. 15-2b: The cover of important plant species in km^2 in the eastern Mediterranean.

(1-*Abies* spp., 2-*Quercus coccifera,Q.coccifera* ssp.*calliprinos*, 3-*Q. cerris,Q.infectoria,Q.pubecens*, 4-*Juniperus phoenicea*, 5-*Cedrus* spp., 6-*Cupressus* spp., 7-*Pinus nigra,* 8-*P.pinea*, 9-*Fagus* spp., 10-*Castanea sativa,* 11-*Juniperus excelsa,* 12-*J.thurifera*

There is no other region in the world like the eastern Mediterranean, where the development of ecosystems has been so intimately associated with humans for so long. A demographic explosion took place 4000 years earlier in the east than in the west. Over-exploitation has led to the disappearance of many ecosystems and agroecosystems. Forested areas are decreasing at a rate of 2 percent per year. There is an immediate need for increasing the number of active and passive germplasm collections, herbaria, gene parks and genetic reserves to allow for the regional exchange of plant materials; to maintain a database and information network; to promote regional cooperation and coordination; to organize workshops and training activities; and to disseminate information. One of the most important points is to produce a comprehensive taxonomic survey of the flowering plants of the region, since our information on the demography, population structure, reproductive biology and genetic diversity of most species is limited or virtually non-existent.

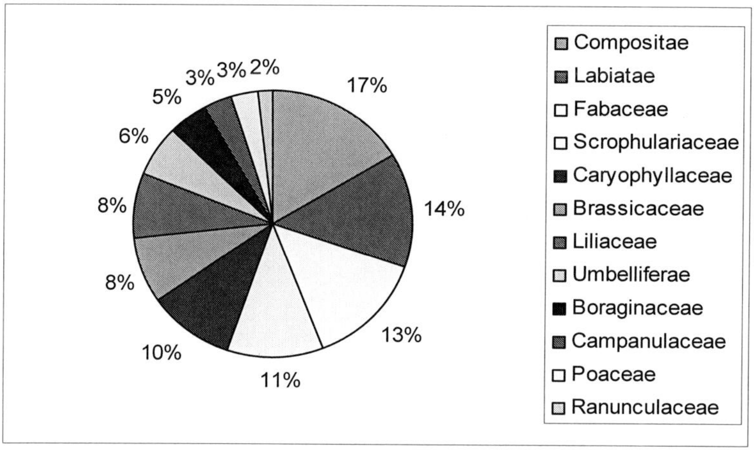

Fig. 15-3: Percentage distribution of major families of endemic taxa in the Turkish Mediterranean zone.

The pool of genetic resources and vegetation biomass in the area is declining. Therefore, investigations should be undertaken on the ecological status of plant species. Harlan (1995) pointed out that the importance of the Mediterranean regarding crop diversity, can be judged by the fact that about a third of the foodstuff used by humankind comes from the Mediterranean climatic region. There is a need for an integrated multi-disciplinary approach covering genetics, plant breeding, population biology, population genetics, demography, phytochemistry, molecular biology, taxonomy, ecology, sociology, legislation, seed physiology, gene bank management, protected area management, and conservation biology to solve the problems of the conservation and sustainable use of these plants (Skoula et al. 2002; Tress and Tress, 2003). For most of the endangered plant species, no conservation measures have been taken and current reserve networks are inadequate. Additionally, knowledge on the use of plant genetic resources by traditional societies is now threatened with extinction. Landscape surveys and evaluations will provide planners with the necessary background information to ensure that development will not destroy the ecosystem, wildlife and landscape features of the rivers. The integration of such considerations as the sensitivity and vulnerability in development is expected to help secure biodiversity and to preserve visual resources. One of the pressing environmental problems in the region is the reclamation of swamps and the diversion of rivers. Most of the wetlands have been drained and there is a great need for ecological and environmental surveys.

Drastic socio-economic changes have taken place in the last fifty years and the general attitude of people has been to seek healthy, clean and green surroundings. This unification of people from different economic, social and political backgrounds on such an aspect is an important event, which clearly supports the far-sightedness of ecologists who for decades have been suggesting that economic development should be considered in parallel with nature conservation. There is now a great need for framing the interrelationships between man and nature under the principles of environmental protection, if we think a little about our future generations. Sustainable use of our soils is a question of paramount importance ecologically, economically and socially and a prudent path to pursue for the betterment of human-biological communities. If we do not protect our lands based on ecologically sound applications, our future generations will not succeed as our earlier civilizations once did.

The stabilisation of population growth, until equilibrium is reached, is a must. For erosion control, native or adapted grasses, close-growing biennial or perennial crops, shrubs or a mixture of grasses, shrubs and trees should be used. Widespread education for the public and special and technical education for farmers is needed for protection of the soil. Politicians and mayors need to follow strict rules, with a scientific foundation, for the sustainable use of soils and strategies should never be associated with their short term political gains. These could include the proper evaluation of soil groups, environmental education, social and economic security for rurals peoples, application of family planning, and agri-business based on eco-agriculture.

References

Akman, Y., Barbero, M., Quezel, P.1978. Contribution a l'etude de la Vegetation Forestiere d'Anatolie Mediterraneenne. Phytocoenologia 5(1): 1-79.

Akman, Y., Barbero, M., Quezel, P. 1979a. Contribution a l'etude de la Vegetation Forestiere d'Anatolie Mediterraneenne. Phytocoenologia 5(2): 189-276.

Akman, Y., Barbero, M., Quezel, P. 1979b. Contribution a l'etude de la Vegetation Forestiere d'Anatolie Mediterraneenne. Phytocoenologia 5(3): 277-346.

Akman, Y. 1982. Climats et Bioclimats mediterraneens en Turquie. Ecologia Mediterranea VII (1/2).

—. 1999. Climate and Bioclimate (Bioclimatic Methods and Climates of Turkey). Kariyer Matbaacılık, Ankara, 350 pp.

Al-Jundi, A., Khaddour, H., and Mirreh, M. 1999. Range monitoring in the Syrian Steppe.Technical working papers 2&3. Range Rehabilitation and Establishment of a Wildlife Reserve in the Syrian Steppe: Project GCP/SYR/003/ITA Palmyra

Al-Jundi, A., Khaddour, H., Mirreh, M. 1999. Range monitoring in the Syrian Steppe.Technical working paper 5. Range Rehabilitation and Establishment of a Wildlife Reserve in the Syrian Steppe: Project GCP/SYR/009/ITA Palmyra.

Atalay, I. 1994. Vegetation Geography of Turkey. Ege University Press, Izmir-Turkey, ISBN 975 95527 8 7, pp: 352.

Brandt, C.J. and J.B. Thornes. 1996. Mediterranean Desertification and Land Use. John Wiley and Sons, Ltd., Chichester, England.

Brauch, H.G. 2003. Security and Environment Linkages on the Mediterranean Space: Three Phases of Research on Human and Environmental Security and Peace, pp. 35-143. *In* Security and Environment in the Mediterranean, Conceptualising Security and Environmental Conflicts (Eds. Brauch, Liotta, Marquina, Rogers, Selim). Springer-Verlag, Berlin.

Briassoulis, H., Junti M., and Wilson G. 2003. Mediterranean Desertification, Framing the Policy Context. Directorate General for Research, Sustainable Development, Global Change and Ecosytem, EUR 20731, Brussels. 35 pp.

Cetik, R. 1976. The Phytosociological and Ecological Studies of the *Cedrus* Woodland Vegetation of Caykara and Bucak at Elmalı.Comm. Fac. Sci. Univ. Ankara, C 2: 20.

Cowling, R.M.,Rundel, P.W., Lamont, B.B., Arroya,M.K., Arianoutsou, M. 1996. Plant Diversity in Mediterranean-Climate Regions. Tree 11:362-366.
Dallman, P.R. 1998. Plant Life in the World's Mediterranean Climates. University of California Pres, California,USA, pp:258.
Danin, A. 1988. Flora and vegetation of Israel and adjacent areas. p. 129-157. The Zoogeography of Israel. Junk, Dordrecht.
Davis, P.H., 1965-1985, *Flora of Turkey and East Aegean Islands*, Vols. I-IX, Edinburgh University Press.
Davis, P.H., Mill, R.R., and Tan, K., 1988. *Flora of Turkey and East Aegean Islands*, Vol. X, Edinburgh University Press.
Davis, P.H. 1998. Flora of Turkey and the East Aegean Islands (Supplement)). Vol. 10. Edınburgh Unıversıty Press, Edınburgh.
Davis, S.D., Heywood, V.H., Hamilton,A.C.(eds.) 1994. Centres of Plant Diversity. A guide and strategy for their conservation. Volume 1:Europe, Africa, South West Asia and the Middle East.xıv+578.WWF- IUCN. IUCN Publ.Unit, Cambridge –UK.
Efe, R. 1998. Natural Environmental Characteristics of Ermenek River Basin. FA.U publication no. 1. Istanbul. pp. 210, ISBN 975-303-001-0
Feoli, E., Giacomich, P., Mignozzi, K., Ozturk, M., Scimone, M., 2003. Monitoring desertification risk with an index integrating climatic and remotely-sensed data: An example from the coastal area of Turkey. Management of Environmental Quality: An Intern. Journal, Vol. 14, No.1, pp. 10-21.
Geeson, N.A., C.J. Brandt, J.B. Thornes (Eds.). 2002. Mediterranean Desertification, A Mosaic of Processes and Responses. John Wiley and Sons, Ltd., West Sussex, England. 440 pp.
Gemici, Y. 1995. Three important areas on the Bolkar mountains for the in-situ conservation of plant biodiversity. Karaca Arboretum Magazine, 3 (2) : 55-78.
—. 2000. Plant diversity in the Bolkar Mountains (Southern Anatolia, Middle Taurus), Second Balkan Botanical Kongress, 14-18 May, Istanbul.
Greuter, W.1991. Botanical Diversity, Endemism, Rarity, and Extinction in the Mediterranean Area: An Analysis Based on the Published Volumes of Med-Checklist. Bot. Chron. 10:63-79.
Guvensen, A., Ozturk, M. 2003. Halophytic Plant Diversity of South Aegean Coastal Zone in Turkey. Pakistan Journal of Botany, 35,5,853-864.
Harlan, J.R. 1995. The Living Fields: Our Agricultural Heritage. Cambridge University Press, Cambridge.

Heywood, V. H. 1989. Patterns, Extents and Modes of Invasions by Terrestrial Plants. In: Drake, J.A. et al.(eds.), Biological Invasions: A Global Perspective, 31-51. SCOPE 37. John Wiley, New York.

—. 1995a. The Mediterranean Flora in the Context of World Diversity. Ecologia Mediterranea 21: 11-18.

—. 1995b. Global biodiversity assessment. Cambridge University Press, Cambridge, UK. 1140 pp.

—. 2003a. The future of floristics in the Mediterranean Region. *Israel Journal of Plant Sciences*, 50, S-5-S-13.

—. 2003b. Mediterranean Plant Collections: needs and options. Setting the scene: what we have inherited. *Bocconea*, 16(1), 283-287.

Heywood, V. H., Iriondo, J. M. 2003c. Plant conservation: old problems, new perspectives. In: Iriondo, J.M. and Heywood, V.H. (eds) *Plant Conservation Biology: Emerging tools and strategies. Biolgical Conservation*, 2, 113, 321-335.

Heywood, V. 2004a. Modern approaches to floristics and their impact on the region of SW Asia. *Turkish Jour Bot*, 28, 7-16.

Heywood, V.H. 2004b. Invasives. In: Prance, G. and Nesbitt, M. (eds.) *The Cultural History of Plants*: Routledge/Taylor & Francis Ltd, 377-386.

Jana, S. 1995. Factors contributing to crop diversity in the Mediterranean. Diversity 11:15.

Jaradat, A.A. 1995. Modern Agriculture Endangers Rich Fruit and Nut Reservoir of the Mediterranean Basin. Diversity 11: 127-128.

Juneidi, M. 1973. Natural Plants of Jordan and their ecological distribution. Amarzyan Publ. Co. Amman.

—. 1994. Wild plants of Palestine & their medicinal values. Intern. Engin. & Printing CO (IEC) Amman.

Ketenoglu, O., Akman, Y., Quezel, P., Kurt, L. 1993. Analyse syntaxonomique des forets de *Liquidambar orientalis* en Turquie. Ecologia mediterranea XIX (1/2): 49-57.

Le Houerou, H.N.1981.Impact of Man and His Animals on Mediterranean Vegetation. In: Di Castri, F. et al.(eds.),Mediterranean-Type Shrublands, 479-517. Elsevier Scientific Publishing Company, Amesterdam.

Mazzoleni, S., Di Pasquale, G., Mulligan, M., Di Martino, P., Rego, F. (Eds.), 2004. *Recent Dynamics of the Mediterranean Vegetation and Landscape* John Wiley & Sons, 306 pp.

Mendizabal,T.,Puigdefabregas,J.2003.Water and soil erosion in the Mediterranean, 687-701; "NATO Security through Science Security

Issues of Desertification in the Mediterranean Region Debated at NATO Workshop 2-5 December 2003." http://www.nato.int/science/news/2003/n031202a.htm

Mirreh, M. M. 1999. An illustrated booklet of project activities and progress: Project GCP/SYR/003/ITA, Range Rehabilitation and Establishment of Wildlife Reserve in the Syrian Steppe. Palmyra.

Naveh, Z. 1991. The role of fire in the Mediterranean Vegetation. Bot.Chron. 10: 385-405.

Ozturk, M., Gemici, Y., Gork, G., Secmen, O. 1991. A general account of high mountain flora and vegetation of mediterranean part of Türkiye. Ege Univ. Journal of Faculty of Science 13(1-2): 51-59.

Ozturk, M.A., 1995.Recovery and rehabilitation of mediterranean type ecosystem- A case study from Turkish maquis. In: Evaluating and NATO-ARW (Eds. D.Rapport, C. L. Gaudet, P.Calow), Springer-Verlag, Berlin, pp: 319-332.

Ozturk, M., Secmen, O., Gork, G., (Eds.) 1996a. Plant Life in Southwest and Central Asia. Vol. I, Ege Univ. Press Izmir, Turkey, pp:499.

Ozturk, M., Secmen, O., Gork,G., (Eds.) 1996b. Plant Life in Southwest and Central Asia. Vol. II, Ege Univ. Press Izmir, Turkey, pp: 1093.

Ozturk, M. 1999.Urban Ecology and Land Degradation. In: Perspectives in Ecology (Editor: Almo Farina). Backhuys Publishers, Leiden, NL,115-120.

Ozturk, M., Ozcelik, H., Aksoy, A., Akcicek, E. 2000.Gardens in Turkish culture and botanical gardens in Anatolia. III. Conf. Int. Assoc. of Bot. Gardens- Asia Division, 26-29 August, Urumqi, P.R.China.

Ozturk, M.,Celik, A.,Yarci, C., Aksoy,A., Feoli,E., 2002. An overview of plant diversity, land use and degradation in the Mediterranean region of Turkey. Environmental Management & Health, 13,5,442-449.

Ozturk, M., Parks,C. R., Coskun, F., Gork,G., Secmen,O., 2004.Vanishing Tertiary Genetic Heritage in the East Mediterranean,. *Liquidamber orientalis* Mill. Environews, 10,4, 6-8.

Ozturk, M., Guvensen, C. Gork G. Gork, G., 2006. An overview of coastal zone plant diversity and management strategies in the mediterranean region of Turkey. In: Biosaline Agriculture& Salinity Tolerance in Plants (Eds.Ozturk et al.). 89-100, Birkhauser Verlag (Springer Science), Basel.

Quezel, P.1985. Definition of the Mediterranean Region and the Origin of its Flora. In: Gomez Campo,C.(ed.), Plant Conservation in the Mediterranean Area 9-24, W.Junk, Doderccht.

Sakcali, M. S., Ozturk, M. 2003. Eco-physiological behaviour of some Mediterranean plants as suitable candidates for reclamation of degraded areas. Journal of Arid Environments , 57, 141-153.

Shmida A., Darom, D. 1990. Handbook of Wild Flowers of Israel. 3 vols., Keter Publishing House Ltd., Jerusalem.

Skoula, M., Rakiz, Z., Boretos, N., Heywood, V.H., Johnson, C.B. The MEDUSA Information System: a tool for the identification, conservation and sustsianable use of Mediterranean plant diversity. *ICSEB VI "Biodiversity in the Information Age"Abstracts*, Patras, 2002, 62.

Trabaud, L.1981. Man and Fire: Impacts on Mediterranean Vegetation. In: Di Castri,F. et al.(eds.),Mediterranean-Type Shrublands, 527-537. Elsevier Scientific Publishing Company, Amesterdam.

Tress, B. and Tress, G., 2003. Shaping Future Landscape: The Scenario Approach, in: *Agricultural Impacts on Landscapes: Developing Indicators for Policy Analysis*, Proceedings from NIJOS/OECD Expert Meeting on Agricultural Landscape Indicators, W. Dramstad and C. Sogge, eds., Norwegian Institute of Land Inventory, ISBN 82-7464-308-9, pp. 68-81.

Turkmen, N., Duzenli, A., Cakan,H., Ozturk, M. 1996.Impact of environmental problems on the natural vegetation in the east Mediterranean part of Turkey. Int.Conf. on Land Degradation, Adana, pp:245-253.

Vavilov, N.I. 1951. Phytogenetic Basis of Plant Breeding: The origin, Variation, Immunity and Breeding of Cultivated Plants. Chronica Botanica 13: 1-366.

Vertovec, M., Sakcali, S., Ozturk, M., Salleo, S., Giacomich, P., Feoli,E., Nardini, A., 2001.Diagonosing plant water status as a tool for quantifying water stress on a regional basis in Mediterranean drylands. Ann. For. Sci. 58, 113-125.

WCMC,1992. Global Biodiversity: Status of the Earth's Living Resources. Chapman & Hall, London

Yucel, E., Ozturk, M. 2000. A geobotanical survey of *Pinus nigra* ssp. *pallasiana* forests in Turkey."Plant Life in South-West & Central Asia, Vth Intern. Symp.",pp. 196-200,Tashkent, Uzbekistan.

Zohary M. 1973. Geobotanical Foundations of the Middle East. Gustav Fischer Verlag Vol. 1-2. Stuttgart.

Zohary, D., Hopf, M.1993. Domestication of Plants in the Old World. Ed.2, Clarendon Pres, Oxford.

Chapter Sixteen

The Analysis of the Tornadoes in Alanya On 16–20 October 2006

Ülkü Eser Ünaldi

Introduction

Turkey is a country where, excepting for the tropical cyclonic storms, all kinds of natural hazards based on atmosphere such as strong winds, heavy rain, floods which may turn into disasters and most of which emerge as a result of storm systems can be observed. "Tornadoes" which are very dangerous systems and can be seen on land and sea are one of these hazards. A large number of these natural hazards including tornadoes are induced by frontal cyclonic storm systems and physical geographical conditions. Because Turkey is a temperate-zone country takes place in the belt of westerlies where the air is the most complex one. Turkey is surrounded by seas on three sides. And it lies between hot and cold mainlands. Its average altitude is very high and it has a very different topographic-orographic structure. Hence, all these features enable the conditions which make natural threats including tornadoes possible.

As it is known, tornado is the general name given to short-term, rotating wind storms which occur suddenly in a highly limited area. These are columns of air rotating around a small area (low pressure center) in the shape of funnel, trunk or rope in contact with the ground. Tornadoes are named according to their vertical structures, causes of formation, the place of occurance and the degree of effcct (such as supercell, nonsupercel, Gustnados, waterspouts, landspouts). Some are harmless or cause little damage whereas others are very destructive and are among the most devastating powers of the nature ever known. The formation mechanisms of the tornadoes are not known exactly. But it is certain that most of them are directly correlative to strong thunderstorm systems shaped by emergence of cumulonimbus clouds (Cb).

Because, it is agreed on that tornadoes stem from different characteristics of the winds between down and upper levels of troposphere and between strong updrafts and downdrafts in thunderstorm cloud. For this reason, the main prerequisite for their occurrence, as in Alanya, is firstly the emergence of Cb's and then depending on it the existence of thunderstorms.

However, it is doubtless that each thunderstorm has not the potential to produce a tornado and only 1% of them can produce a tornado. The reason of this that some atmospheric conditions must by all means occur for tornadoes to emerge. These are as follows:

- The existence of warm and moist air in the levels of atmosphere close to the ground and
- Putting into action a mechanism which will help this warm and moist air move upward, that is, provide a strong convection.

To this effect, the followings are necessary:

- Convergence in the low levels of troposphere,
- Speedy updrafts,
- Divergence in the upper levels of troposphere,
- In parallel with the height, a change in the direction and/or in speed of the wind (wind shear),
- The existence of variable weather conditions,
- Appropriate topographic conditions.

The existence of all or a few of them has a direct effect on the formation and severity of these storms.

According to thunderstorms they produce, tornadoes are divided into two as:

1. Supercell Tornadoes
2. Nonsupercell Tornadoes.

Tornadoes occurring in the ground are nonsupercell tornadoes. The difference of them from supercell tornadoes is that rotating mesocyclone and wall clouds do not develop in cumulonimbus clouds they form, as do during the supercell tornadoes. Nevertheless, there exists a mesocyclone rotating from the base of the cumulonimbus cloud to the earth in these tornadoes. This mesocylone occurs as a cloud in the shape of a funnel or as a rotating dust or debris cloud from the ground. In addition, when compared to supercell tornadoes they are less speedy and are shorter-termed.

Nonsupercell tornadoes generally develop over the region under variable atmospheric conditions such as the cold air and a related cold trough in the upper parts of the atmosphere and a convergence on the earth. The formation of these are generally correlated with the "nonsupercell thunderstorms" developing from very moist and warm air masses and with the storms called "ordinary cell thunderstorms".

The speed of rotating wind in nonsupercell tornadoes is 80 km/h or slower. They move slowly and they last only 10-15 minutes.

This kind of tornadoes can develop both over the land and the sea and they are named landspouts and waterspouts accordingly. The "Waterspouts" seen offshore of Alanya, which is the subject of our study, is the name given to the air column rotating over large water masses. Its facilitative conditions are the moist and warm air over the water and the atmospheric variability occurring as a result of the cold air in the upper layers.

In other words, if the air is variable and if cumulus clouds develop, it means that there are suitable conditions for waterspouts. Whereas some of them occur with weak thunderstorms, most of them occur in relation with cumulus congestus clouds.

A large number of waterspouts in Turkey have happened over the Mediterranean and in the Mediterranean coastline according to the data from the General Directorate of State Meteorology Affairs. These are the waterspouts occurring offshore of İskenderun, Anamur, Alanya, Manavgat and Datça and causing loss of lives and properties.

Study area

The field of study is in the east of Antalya Gulf, in Adana division of the Mediterranean Region (See Fig. 16-1). Within the scope of this field, the place where the first tornadoes were observed on 16-20 October 2006 was Okurcular town, in the 30-35 km west of Alanya. And the place where the second tornado has occurred is within 14-15 km offshore. They have reached the shore in the Kargıcak town of Alanya.

Material

The probable causes of formation of the tornadoes observed on sea in the field have been studied in terms of synoptic. To this effect, the following documents obtained from the General Directorate of Turkey States Meteorological Organization have been used:

- Surface charts ($00.^{00}$GMT),
- 500 hPa upper level charts ($00.^{00}$GMT),
- Radiosondes Observations of Isparta, İzmir and Adana. (RAOB),
- Satellite images (METEOSAT- Channel 9),
- The findings of observation of Alanya Meteorological Station.

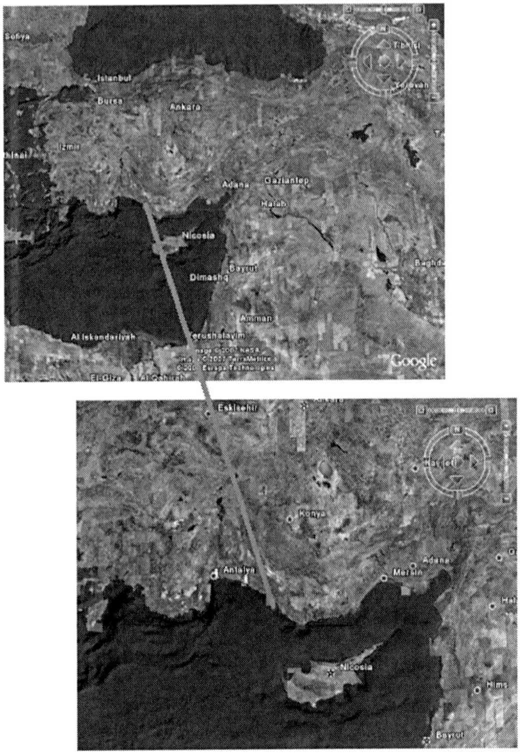

Fig. 16-1: The position of the study area (www.google.earth).

Discussion

A series of waterspouts have been observed on sea in Antalya Gulf between 16 and 20 October 2006.

The first one of the tornadoes has occurred in Okurcalar town, in the 33-35 km west of Alanya on 16 October between 6:00 pm and 6:45 pm and 9 tornadoes following one another have been determined. Two of them have reached to the shore and so caused damage on the tourism facilities though not on a massive scale.

The second tornado has been observed again on sea within 14-15 km offshore of Alanya. The second tornado has reached to the shore in Kargıcak town which is 15 km from the west of downtown Alanya and brought about damage on cultivated lands and greenhouses.

In the stated dates, according to the ground and high level maps, cold-core Siberian Anticyclone formed in Siberia and hanging down from the country to the Mediterranean is effective in a big part of Turkey while warm-core Basra Cyclone is effective in the other parts of the country (Fig. 16-2).

Hence, Turkey is under the influence of cold and dry Continental Polar air mass (cP) and warm and dry Continental Tropical air mass (cT).

As a consequence of Siberian anticyclone, cyclones develop in the upper troposphere over the regions where Siberian anticyclone is effective and it has cold air and troughs. On the other hand, in the upper parts of the regions where Basra cylone is effective, a ridge and warm air dominates as a consequence of the low pressure (Fig. 16-3).

The Mediterranean Region is rather warm as a result of the fohn wind forming by the tropical air and north drafts. Thus, according to the measurements of Alanya Meteorology Station, the ground temperature varies between 23-25 °C and the seawater temperature varies between 26.1-26.3 °C.

In the meteogram reporting the weather forecast of Alanya between 14-24 October 2006, the ground temperature was forecasted to be above 20 °C, 850 hPa temperatures to be above 10 °C and the relative humidity was to be between 90-100%. According to these data, a rather hot and moist air dominates in the area where the temperature is above the season average (Fig. 16-4).

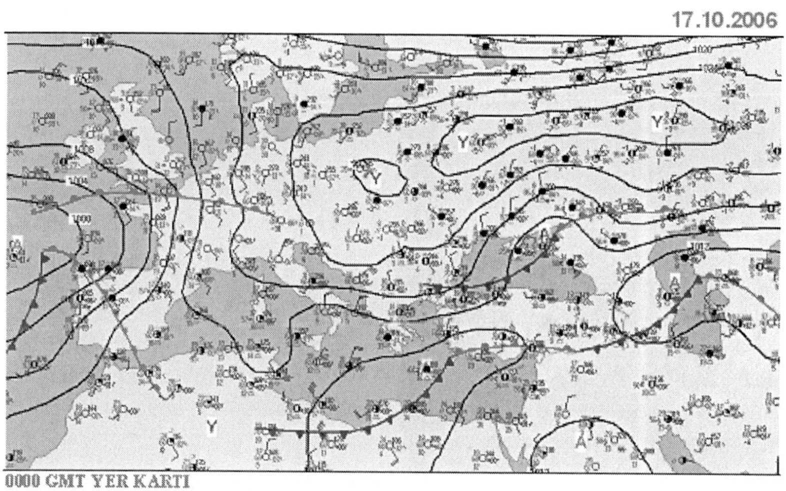

Fig. 16-2: Surface Chart on 17.10.2006 (DMIGM).

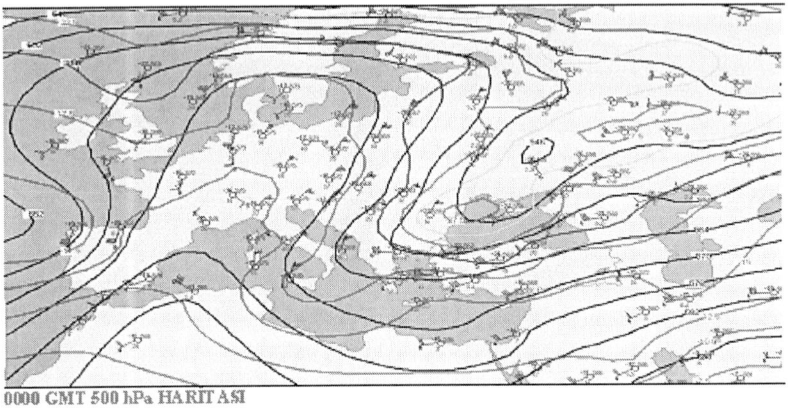

Fig. 16-3: 500 hPa Upper Level Chart on 17.10.2006 (DMIGM).

Under these conditions, in a big part of our country, particularly over Antalya Gulf, there is a strong variability influenced by the 500 hPa cold air and the 26 °C-seawater. The Radiosonde Observations in İzmir, Adana and particularly Isparta ($00.^{00}$ GMT), verify this. Thus, according to the variability indexes, the variability in the ground is at sufficient level (Table 16-1).

Furthermore, the "level of free convection (LFC)" in Isparta temp, which can represent Alanya best, is rather low (792, 849, 903 hPa). This means the easy ascent of the moist and warm air which is near to the ground and its condensation from LFC by continuing to ascend.

It is already understood from the measurement values and satellite images of Alanya Meteorology Station that the northern parts of Antalya Gulf is covered with Cb clouds which have the potential to transform to thunderstorm clouds. In addition, in the ground level, there are weak front systems but warm fronts are more effective over the Mediterranean (Fig. 16-5 and 16-8).

According to these data, between the stated dates, it is indispensable that strong convective activities such as thermic, orographic and frontal ones occur intensely as a result of the cold air in the upper troposphere and the strong warming on the ground and it is possible that Cumulus congestus develop locally at any time over the gulf and they transform into Cb clouds and Cb clouds can transform into the clouds having the potential to create tornadoes particularly waterspouts and they can even transform into thunderstorm clouds.

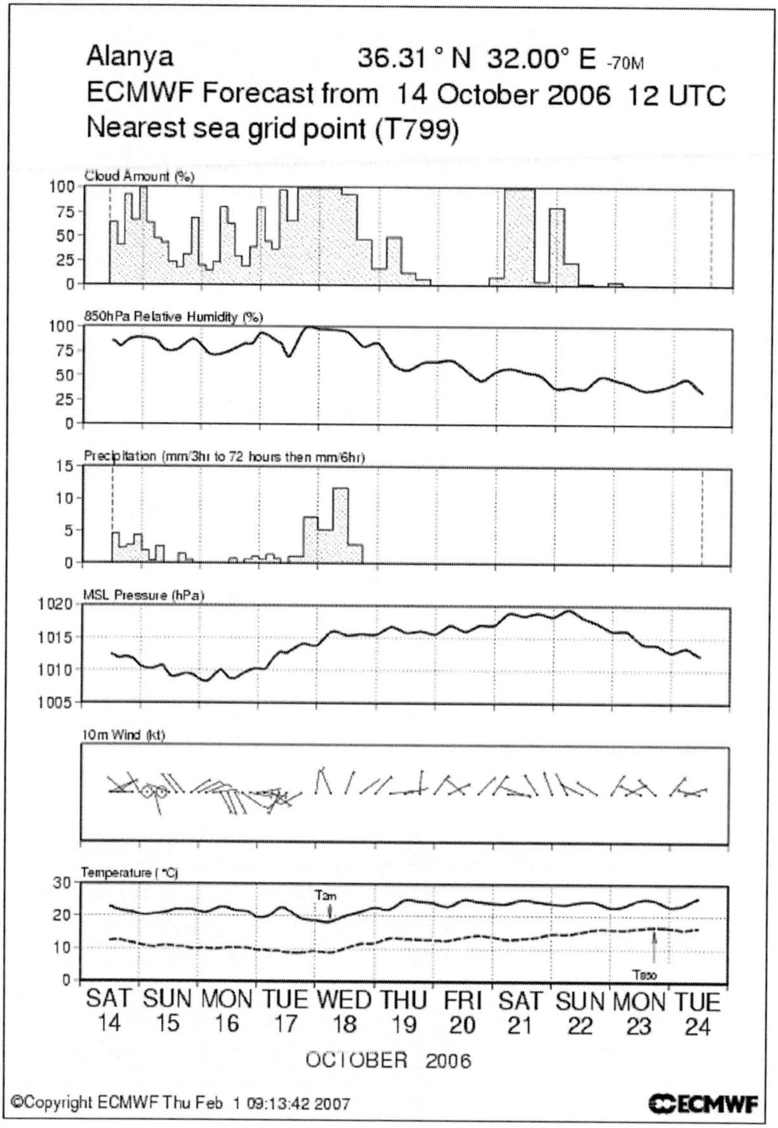

Fig. 16-4: Meteogram belonging to Alanya. Date: 14-24 October 2006 (DMIGM).

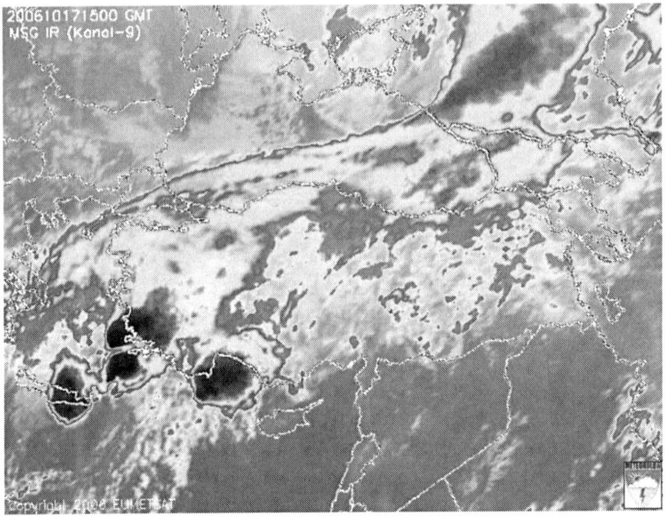

Fig. 16-5: The image (VIS) taken from the Meteosat satellite (17 October 2006, Time: 3.00 pm GMT) (DMIGM).

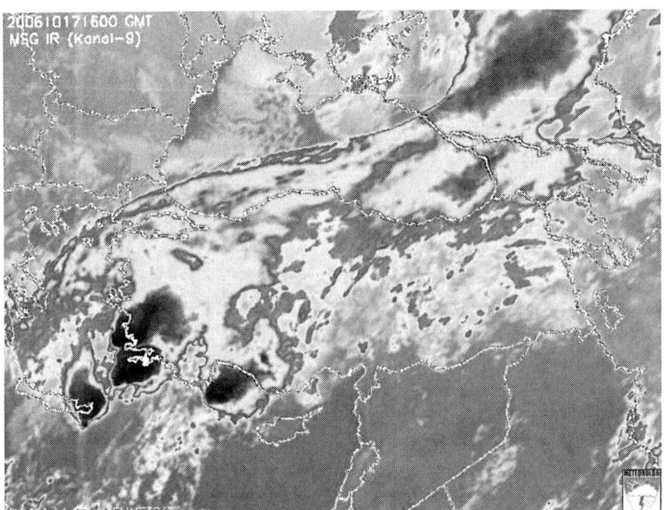

Fig. 16-6: The image (VIS) taken from the Meteosat satellite (17 October 2006, Time: 4.00 pm GMT) (DMIGM).

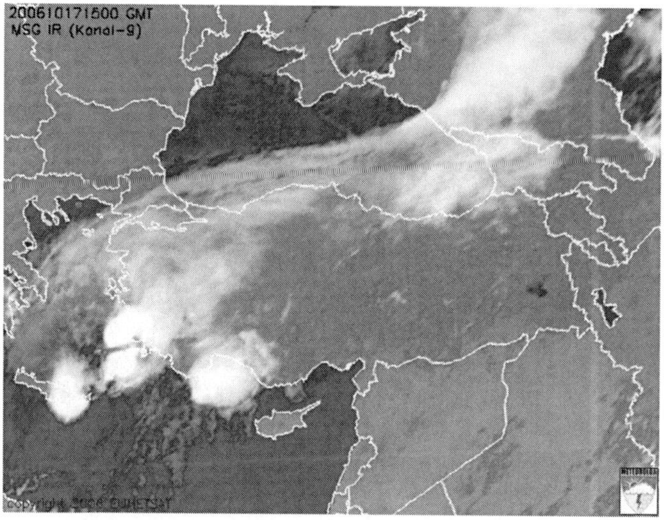

Fig. 16-7: IR image taken from meteosat satellite (17 October 2006, Time: 3.00 pm GMT) (DMIGM).

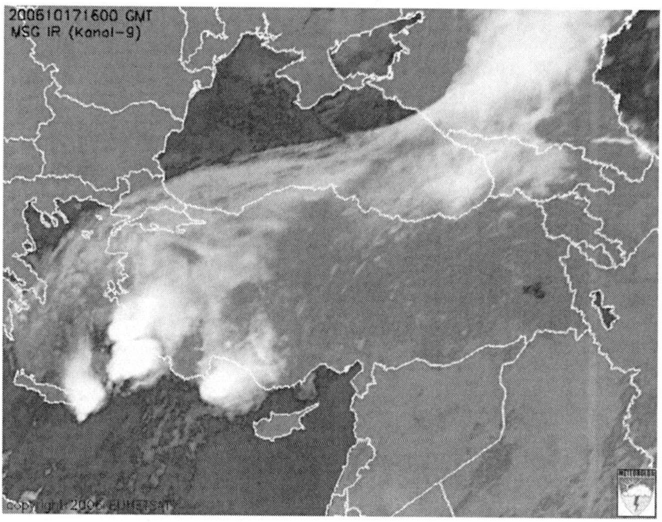

Fig. 16-8: IR image taken from meteosat satellite(17 October 2006, Time: 4.00 pm GMT) (DMIGM).

Table 16-1: Temp of Isparta (DMIGM).

Weight	Parameters	Weak	Moderate	Strong
1	Surface Dewpoint (C)	10.8		
1	850 mb Dewpoint (C)	5.3		
1	850 mb Wind Speed (kt)	15		
1	500 mb Wind Speed (kt)		35	
1	200 mb Wind Speed (kt)	49		
1	LFC - Level of Free Convection (mb)			849
1	LI - Lifted Index	-.5		
1	SI - Showalter Index		2.7	
1	TI - Thompson Index		30	
1	JI - Jefferson Index		30	
1	K Index	30		
1	TT - Total Totals		48	
1	SWEAT Index	216.1		
1	CAPE (B+)	174		
1	srH - storm-relative Helicity (0-3 km)		155	
1	EHI - Energy Helicity Index	.2		
1	CAP Strength	.6		
1	BRN - Bulk Richardson No.		3	
1	BRN Shear (m/s)	7.8		
1	WBZ - WetBulb Zero Hgt (ft,AGL)			5765
1	Hail (inches)			
1	Windex (kt)	25		
1	T2 Gust (kt)			
1	KO Index			
1	Boyden Index			
1	S Index		47	
	Weighted Category Totals:	12	8	2

Because almost every condition given below and required for the formation of each tornado (supercell, nonsupercell, Gustnados, Landspouts, and Waterspouts) is present in the ground:

- Seawater temperature exceeded 26 °C. (This is the value that tropical cyclonic storms bring about.)
- There becomes a warmer and more moist air over the sea when compared to the general season conditions.
- In the level 500 hPa, a cyclone (cyclonic rotation) and -when compared to the general season conditions- a very cold air and trough exist.

- The temperature in the earth is 25 °C, and -18 °C in 500 hPa.
- According to these values, there is a difference of temperature between the lower and upper parts of trposphere, which is 43 °C.
- Thus, there becomes a strong variability in the atmosphere.
- In the ground, convergence zones develep because of the front moves, topographical structure and thunderstorm clouds.
- There is a wind shear –even at a small amount- because of the atmospheric mobility.

Fig. 16-9: A view from the waterspouts around Alanya (DMIGM).

As a result of these conditions, it is indispensable that a tornado develops and so there became a series of tornadoes over the sea on 16-20 October 2006 (Fig. 16-9).

The tornadoes in question occurred, in relation with fronts and particularly hot front transitions, on the convergence lines which developed from frontal lines or other reasons. The formation of tornadoes develops in the following stages:

- The air which is over the sea and warmer than the average season conditions ascended fast on convergence lines depending on the variable atmospheric conditions.

- As a consequence, Cumulus (Cu), Cumulus congestus (Cu congestus), Cumulonimbus (Cb) and maybe even thunderstorm clouds (tower clouds) formed respectively.
- Formed clouds were moved to the areas where inverse winds meet and rotate clockwise on convergence lines.
- With this rotation, the updrafts rotating spirally toward the cloud formed the waterspouts in the ground.

The formation of tornadoes in a short time, their low speeds, short lives and their numbers –more than one- make us think that these are the multi-core nonsupercell waterspouts occurring on the variability line.

Conclusion

On 16-20 October 2006, in Antalya Gulf, a series of waterspouts was seen on the sea. The first group of these was determined to be 9 in number, on 16 October in Okurcular Town which is located 30-35 km far away from Alanya. Two of them caused damage on buildings by reaching the coast. The second group of tornadoes was seen 14-15 km offshore on 19 October and they caused great damage on the plantation and greenhouses by reaching the shore of Kargicak Town, Alanya.

Therefore, we aimed at examining the tornadoes in the study. For this reason, the causes of their formation and their types were discussed. While explaining the causes of their formation, “synoptic observations and maps”, “Radiosonde observations” and hourly “satellite images” obtained from The State Meteorological Affairs Department were used.

According to these data, the seawater temperature in the ground reached the temperature needed for the formation of tropical cyclonic storms and it was measured as 26 ^{o}C. Therefore, the air over the sea was warm and moist. In the upper parts of the atmosphere (500 mb) there was a much colder air (about -18 ^{o}C) when compared to the seasonal average and troughs in relation with this. As a result, these conditions created variable air and Isparta and Adana Radiosonde observations verified this situation. There became a convergence zone and a weak frontal system over the sea surface when mP and cT met. Because of the atmospheric mobility, there appeared a small wind shear.

All of these conditions formed a series of tornadoes in the ground. Formed tornadoes are probably multi-core nonsupercell tornado complexes occurring under variable atmospheric conditions of cold air and a relative trough in the upper parts of the atmosphere and of a convergence in the earth.

References

Ahrens, Donald, C. 2007. Meteorology Today: An Introduction to Weather, Climate, and the Environment, ISBN 0-495-01162-2, USA.

Lutgens, F. K., and E. J. Tarbuck. 2006. The Atmosphere and Introduction to Meteorology. Tenth Edition, ISBN -10 0 13187462, USA.

Oliver, E. J., and J. Hidere. 2002. Climatology an Atmosphorie Science. Second Edition, New Jersey.

DMİGM, Surface Charts dated 16-20 October 2006 in Turkey and its surroundings.

DMİGM, 500 hPa Upper Level Charts dated 16-20 October 2006 in Turkey and its surroundings.

DMİGM, MM5 Meteograms dated 16-20 October 2006 in Isparta and Adana.

DMİGM, Meteosat-7 Visible Satellite Images dated 16-20 October 2006 in Turkey and its surroundings.

DMİGM, Meteosat-7 Infrared Satellite Images dated 16-20 October 2006 in Turkey and its surroundings.

DMİGM, Temp Diagrams of the Stations of Isparta, İzmir and Adana dated 15-19 October 2006.

PART II

CULTURAL ENVIRONMENT IN THE MEDITERRANEAN REGION

CHAPTER ONE

SOCIAL CONFLICT IN WATER RESOURCE MANAGEMENT AND ITS ENVIRONMENTAL IMPACTS IN SOUTH-EASTERN TUNISIA

PAOLA MINOIA AND FRANCESCA GUGLIELMI

Introduction

The water scarcity issue is highly relevant in Tunisia, as well as in the whole Mediterranean region. The causes exist due to the natural physical scarcity of water that is typical in arid environments, together with the presence of anthropogenic pressures, such as overexploitation of aquifers, the contamination of fresh and groundwater resources, increasing water demand by the agricultural, domestic, tourist and industrial sectors, and difficult institutional regulatory settings and law enforcement related to water management. The agricultural sector is the greatest water consumer, amounting to 80% of the total consumption of the Southern Mediterranean countries (Chrik, Ghorbel and Zouari, 2005). Paradoxically, the highest consumption rates occur when the seasonal scarcity is worse: in summer, the water demand presented by the agricultural and tourist sectors dramatically increases. The situation is particularly sensitive in coastal areas, where tourist resorts are mainly concentrated. Intensive water use produces marine ingression in the aquifers, thus worsening the quality of water used for drinking, irrigation and ecosystem functionality.

The linkage between environmental and social degradation is particularly evident in poor and marginal areas. Poor communities increase their vulnerability if they depend on a depleted source. Therefore, the need to protect a natural resource has to be coupled with the need to sustain the local sources of income, e.g. by supporting the acquisition of technical options to rationalise the water demand or by providing alternative solutions to those relying on the endangered resource. A cooperative approach between communities and institutions is a fundamental prerequisite for a sustainable development of the target areas. In all cases, decisions have to be made through processes involving stakeholder participation. In contrast, the application of top-down decisions and partial or specific sectoral solutions can constitute a cause for social and institutional conflicts.

Local governance mechanisms play a key role in the integrated management of water resources as well as in human development, by enhancing negotiation dynamics with all relevant players, acceptance of decisions made, and support towards the groups that might be at risk because of the efforts concentrated on the preservation of the natural resource. During the past decade, some Mediterranean countries introduced new regulations (e.g. Morocco, 1995; Tunisia, 1999) introduced new regulations on water management, including provisions for decentralisation of decision-making. However, the local authorities' capacity to involve local stakeholders hasn't always been effective, as in the case of the overexploited Djeffara aquifer of Northern Gabès in Tunisia.

The aim of this paper is to assess how water management policies have been applied in Tunisia, particularly by analysing a case of water stress in the south- eastern region of the country, showing that institutional decisions targeting water conservation issues, when not integrating social and governance aspects, can promote conflict with important target groups.

Water resource management in Tunisia

In Tunisia, during the decade 1991-2000, a national strategy of water resource mobilization was set up in order to satisfy the growing water demand from the agricultural, domestic, industrial, and tourist sectors. The proposed solutions were to increase the exploitation ratio of the overall resources to 95% by the year 2010 (Benabdallah 2003) and to increase the utilisation of non-conventional resources, in particular from treated sewage waters, that grew from 37 Mm^3 in 1988 to 200 Mm^3 in 2000, from the domestic sources and, to a minor extent, from artificial groundwater discharges and desalinisation plants (Ministère de l'Agriculture/ DGRE 2003). This strategy included new engineered infrastructures, but also demonstrated a progressive integration of demand management policies, aimed at moderating water allocation per hectare in agriculture, reduce losses, protect the quality of water resources and obtain greater volume for other growing sectors. These decisions were in line with international policies addressing water issues at the global level, such as those introduced by the Dublin and Johannesburg Conferences, respectively in 1992 and 2002 (GWP 2000; Scanlon, Cassar and Nemes 2004). Tunisia has since approached the principles of Integrated Water Resource Management (IWRM). A primary principle is related to the *integrated* and *demand-side* approach to water management, meaning that institutional, technical, economic, environmental and social aspects have to be integrated in order to rationalise water consumption and ameliorate the balance between limited water quantities and the increasing human demand. Another aspect of the IWRM principles refers to the need to ensure a level of *governance* sufficiently close to local communities, in order to allow for their meaningful participation (Wallace *et al.* 2003).

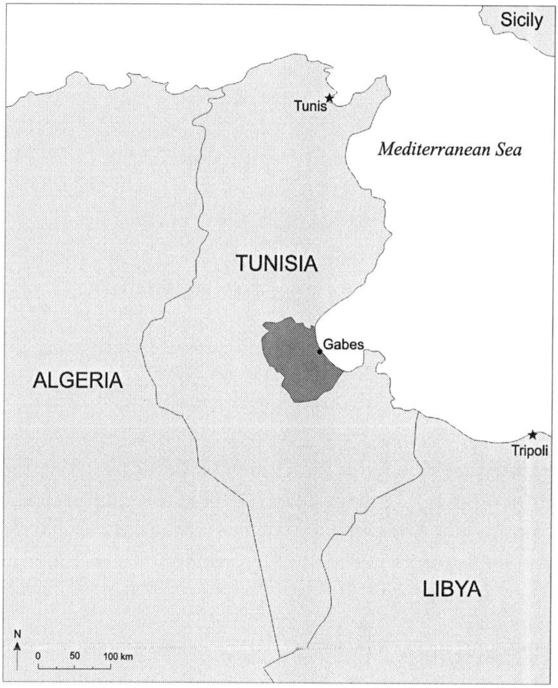

Fig. 1-1: Location of Gabès Governorate

As for the first principle mentioned, the public policies undertaken in Tunisia follow, in practice, a "weak approach", focussed on reducing the losses and providing incentives to consumers for virtuous behaviour by means of pricing, subsidies, management tools and sensitization activities; while a "stronger approach", that would also address the inefficiencies in water allocations and distribution among the sectors, and would apply monitoring, control and sanctions interventions for non-compliant behaviours, is not sufficiently applied (Romagny et al 2004; Chrik, Ghorbel and Zouari 2005).

In line with the local governance principle, the Governorates and the regional *Commissariat Régional au Développement Agricole* (CRDA) of Tunisia have started to decentralize some tasks to local associations of water users. In 1999 (Law n° 99-43) the pre-existing groups related to farmers activities (*Associations d'Intéret Collectif* - AIC) were renamed as *Groupements de Développement Agricole* (GDA) and assumed more responsibility in steering the members towards better agricultural practices, for soil, water and environmental protection, particularly with respect to pollution, irrigation and overexploitation control. However, not all GDA carry out these tasks because of internal weaknesses, like those identified in the case study of the Gabès Governorate.

Water deficits in Gabès and the institutional responses

The water resources utilized in the Gabès Governorate (Fig. 1-1) are mainly from deep aquifers, followed by phreatic aquifers. The volume of available and consumed water resources in Gabès are summarized in Table 1-1.

Four main aquifers supply the Gabès Governorate: Djeffara of Gabès (serving North Gabès, South Gabès, El Hamma Chenchou and Oglet Mertaba); Continental Intercalaire; Turonien Matmata; and Crétacé de Sidi Mansour. The total volume of available groundwater is 156.6 mm^3 and the volume utilized is 123 mm^3, allocated to the following sectors (Abidi and Rahali 2004):

- Agriculture: 90.1 mm^3/yr (73% of total use);
- Tourism and domestic use: 28.9 mm^3/yr (24%);
- Industry: 4.1 mm^3/yr (3%).

The main part of the exploited water resources comes from the Djeffara Aquifer, with a volume of 115.80 Mm^3/yr and a utilised amount of 99.9 mm^3/yr; the other aquifers show less important volumes of available and consumed water resources. A rapid increase in consumption rates has been recorded since 1998 (Abidi and Rahali 2004).

The intensive exploitation has caused various kinds of environmental impacts on the inshore Djeffara aquifer. In particular, there has been a decrease in the piezometric level, a deterioration of the water quality due to salinity and serious marine intrusion. In particular, the Djeffara of Gabès North is a coastal and phreatic aquifer, supported by sands of the Miocene which consist of two surfaces separated by a semi permeable sandy-clay layer of about 40 meters thickness (Abidi, 1999). Because of this relatively thin layer, the area is naturally at risk of marine intrusion. Other pressures are caused by local farmers, who dig wells to irrigate their fields, thus increasing the dynamics of sea water intrusion and decreasing the piezometric level of the aquifer.

These environmental damages induced the governorate to react by imposing measures to prevent further degradation. Having assessed the relationship between the intensive exploitation of the inshore aquifer and irrigation uses, the response was the implementation of restrictive rules of water uses, based on the protection principles enunciated in the Tunisian *Code des Eaux* of 1975 (Chkir et al., 2005).

The first delimitation constituted a "protection zone", defined in 1985 (decree n° 85-1105) but appeared too limited (Fig. 1-2), and therefore in 1987 (decree n° 87-480) was extended to the area of "interdiction" defined by the *oued* Akarit in the north-east of Gabès Governorate, Teboulbou-El M'dou, south-east of Gabès city, and including part of Eastern and Western Gabès, Gabès Medina, Ghannouch, Metouia and a section of El Hamma (Fig.1-2). According to Tunisian regulations, an "interdiction area" is defined by the *Arrondissement Ressource en Eau* of the CRDA of Gabès Governorate, responsible for aquifer monitoring, whenever the following conditions are present: 1) overexploitation of the aquifer; 2) decrease of the piezometric level of the aquifer; 3) water salinity increase because of marine intrusion; and 4) localization in a coastal zone. In the interdiction area, farmers are forced to use fixed volumes of water and need permission from the *Arrondissement Ressource en Eau* of the CRDA of Gabès.

Table 1-1: Water resources in Gabès Governorate
Source: Data from Office du Développement du Sud 2004

	Available resources mm^3/yr	Consumed resources mm^3/yr
Groundwater	180.87	146.37
Phreatic Aquifer (<50 m)	24.27	23.37
Deep Aquifer (>50 m)	156.6	123
Surface water	44	6.50
Treated sewage water	5.69	1.20

Conflict analysis in the Oued Akarit-El Bsissi plains

Some aspects of the social conflicts originating from the restrictive rules and their environmental impact can be better understood by focusing on a specific zone of the interdiction area, the agricultural plain of Akarit-El Bsissi. It is a coastal plain located in the delegation of Metouia, bordered to the north by the *oued* El Akarit, to the east by the Mediterranean Sea, to the south by the *oued* El Melah and to the west by the national road connecting Gabès to Sfax (Fig. 1-3).

This is a marginal area with agricultural fields but only scattered settlements, as the farmers mainly come from other areas of the Metouia or Ghannouch delegations. In summer, the crops cover 830.5 hectares while in winter these cover 904.5 hectares. Other cash crops for local and regional markets, especially tobacco, occupy 194 hectares during the entire year (CRDA Gabès 2003). Small family companies make up the basic economy for the local communities, who have increased their vulnerability after the institution of the interdiction area.

The installation of common wells allowed by the CRDA in other zones of the interdiction area, to be managed by local GDA, could not apply in El Bsissi area, where the agricultural land is not intensively exploited but rather presents scattered cultivated plots. The distances between the parcels are too wide to allow for their grouping and connection by pipes from a common well. Therefore, the solution was only to dig several private wells. Theoretically, even in this case, their use is limited by fixed volumes authorized for the irrigation of their perimeters.

However, the reality is that there is a lack of control. The high number of private wells in this area poses great difficulties to the CRDA officers to effectively monitor the water consumption rates and consequently respect the interdiction law.

Although justified by the need for environmental protection of the resource, the interdiction was not accepted by the rural populations that perceived the allowed amounts as being scarce compared to their expectations and felt deprived of "their own water resource". By the interdiction law of 1987, they suffered due to the limitation of water volume compared to the extension of their perimeters. Still today, twenty years after the decision, some farmers do not understand why they have to pay to obtain a limited volume of water for their agricultural activities when they could directly extract the needed water through a simple well located on their parcel (Chkir et al., 2005).

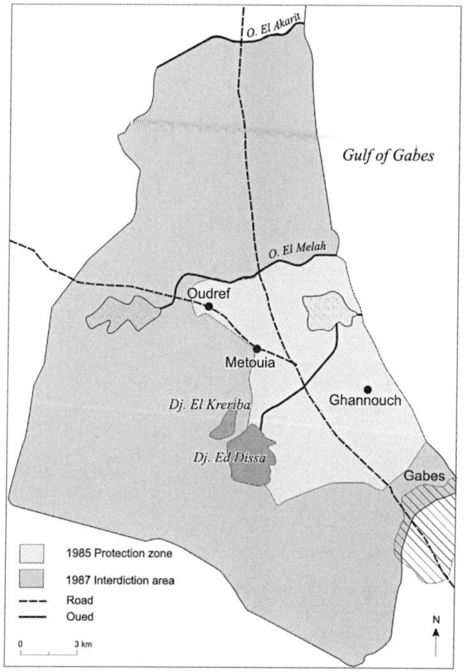

Fig. 1-2: Protection and Interdiction areas

Interviews of local farmers carried out in February and March 2006 revealed that before the delimitation of the area, the public institutions did not involve the affected communities by using awareness campaigns or negotiation activities. Nor have these parties benefited from any compensation measures for the limited volume of water to be utilised in their agricultural production. It is thus evident that the interdiction has established a conflict between the regional authorities of Gabès Governorate and the CRDA, on one side, and the farmers on the other. The way the interdiction law has been imposed, by a top-down instead of a participatory and negotiation process, has created a critical relationship between the public authorities and the local users, and as a consequence, has increased the number of illegal wells, posing serious constraints to the achievement of the very goals of the interdiction.

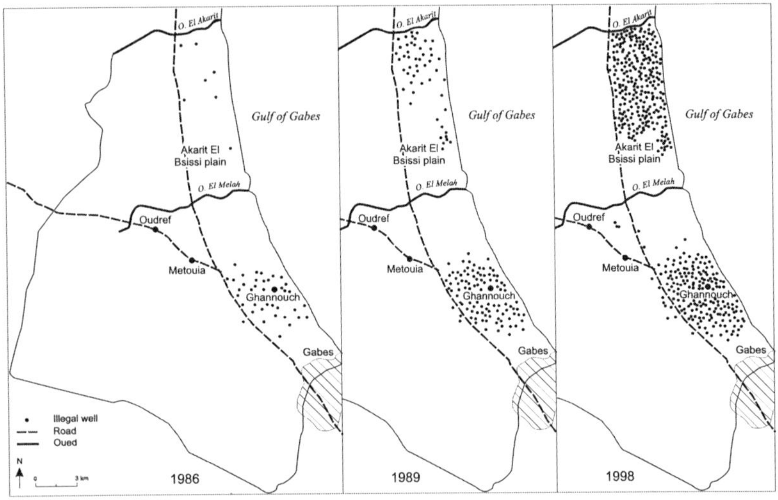

Fig. 1-3: Expansion of illegal wells in the Akarit-El Bsissi and Ghannouch plains

The limited supervision and the low depth of the aquifer of Gabès North (less than 50 meters) have induced the farmers to increase the wells, without submitting notification or permission requests to the CRDA. The discovery of these violations only brings administrative sanctions, which are not experienced as effective disincentives for the illicit farmers. However, the relations between the regional institutions and the local water users have become tense, with social, economic and environmental implications. In fact, the farmers were left alone with the new constraints to the local productivity, and as a reaction found a solution in digging illegal wells. Therefore, the overexploitation of Gabès North aquifer – particularly in the area of the *oued* Akarit-El Bsissi - continues, with the consequences of a further decrease in the piezometric level of the groundwater and marine intrusion in the inshore aquifer.

A survey compiled by the *Arrondissement Ressources en Eau* (CRDA Gabès 2003) shows the proliferation trend of illegal wells which have been discovered by the public authorities, including the following:

- 1986: 45 illegal wells, 7 in El Bsissi and 38 in Ghannouch;
- 1989: 150 illegal wells, 51 in the *oued* El Melah-El Bsissi-*oued* Akarit and 99 scattered between Ghannouch and Metouia;
- 1996: 285 illegal wells, 197 in El Bsissi and 88 in Ghannouch;
- 1998: 354 illegal wells: 264 in El Bsissi (105 artesian, 107 engine driven, and 47 abandoned) and 190 in Ghannouch.

Figure 3 shows the intense proliferation of illegal wells, mainly concentrated in the area of the *oued* Akarit-El Bsissi, and to a minor extent, in Ghannouch. Their depths have progressively increased, generally going beyond 50 meters, to reach the layers at 70-80 m and even at 130 m, thus crossing the phreatic aquifer to reach the deeper one. The increase of engine-equipped wells has evident impacts, especially in terms of diminution of the artesian flow, that is passed from 10-25 l/s to 4 l/s, with average values of 0.5 and 1 l/s. As a result, the diffusion of mechanised pumping has reduced the use of artesian wells (CRDA Gabès, 2003).

In the 1970s and 1980s, the piezometric level of the inshore aquifer of Gabès North was lying on isopieze between 20 and 25 meters. In 1998, as a consequence of the increased number of illegal wells, the piezometric level dropped by 10 meters, which was the threshold value to avoid sea intrusion (CRDA Gabès, 2003).

The need to solve this situation, as the legal measures have proven to be insufficiently effective, has recently motivated the CRDA to create a new GDA in the *oued* Akarit-El Bsissi. Some information and awareness raising meetings were conducted by the CRDA, the *Arrondissements en Ressources en Eau* and *Génie Rural*, in order to get the partnership of local farmers and their representatives from the Regional Farmers Union and from the delegations of Ghannouch and Metouia. The GDA cannot directly manage irrigation from common wells, but is supposed to provide support to farmers, through economic subsidies by the governorate and through technical advice, particularly on the application of water saving techniques for irrigation. From their side, the farmers pay for the water withdrawals and ordinary maintenance and are transparent and legally compliant in terms of the number of wells. The GDA is responsible for awareness-raising and agricultural extension activities. This is clearly contributing to a slower well proliferation pattern than the one that had occurred during the previous decade (personal communication by the head of the *Arrondissements en Ressources en Eau* of Gabès, 1st March 2006). Nevertheless, further efforts are needed, e.g. to strengthen the educational activities offered by the CRDA and the GDA, as the proposed ones are not fully incisive or understood by the farmers. From their perspective, the interdiction decision has jeopardized their economic activities, and some of them see the GDA institution as a further imposition by the regional authorities, and not a farmers' representative institution. These are the main constraints upon the effective role of the GDA to act as a bridge between the regional institutions and the local farmers, who are not fully aware of the need to protect the water resources and to ensure their sustainable use (Chkir, Ghorbel and Zouari 2005).

The situation that has arisen can be described as an institutional and intersectoral conflict (Minoia and Camuffo 2006). The institutional conflict regards in this case the relations between the local authorities and the civil society. The institutions are putting constraints upon the already poor economies and cultural habits of local farmers, as they do not have the means to utilise water saving techniques. Farmers suffer from the interdiction decision, because it brings a change in their lives and working patterns. The interdiction decision, though based on environmental considerations, is judged as being a cause of social problems.

Intersectorial conflicts appear when some water users perceive that their use is endangered by extractions done for other purposes. While this type of conflict has not been explored in the El Bsissi area, it is more and more apparent in other parts of the interdiction area, where tourist and industrial sectors are developing. Although agriculture is the major consumer of water, in the southern part of the area near the town of Gabès, tourism and industry also use the Djeffara waters (besides treated waters). However, the restriction does not apply to them because their withdrawal seems not to cause the overexploitation of the aquifer. The interdiction helps to emphasize the economical and social differences existing between the productive sectors, as tourism and industry are richer sectors than small and medium sized farming. Tourism and industry are still relatively small, but are rapidly expanding sectors, obtaining public support and facilitation by government, whereas the farmers feel that the water tariff increases, based on consumption rates, together with the interdiction, constitute a penalization of the agricultural sector. The competition among sectors is felt by the agricultural producers as being caused by a national policy that gives support to other economic activities, despite their environmental impact.

Concluding remarks

Through the case of Akarit El Bsissi, this paper has presented a particular situation of conflict, originating from an "interdiction" of irrigation practices from around a depleted aquifer. The reaction is expressed by thc affected farmers through non-compliant behaviour, rather than open conflict. Although the quantitative limitations of water abstractions imposed on the farmers are justified by urgent conservation needs, the decision-making process reveals different shortcomings of the policy institutions. Some adjustments for the future have to include the following strategies:

1) The establishment by institutions of effective cooperation with the stakeholders through awareness raising programmes about the environmental impact of their irrigation patterns;

2) The development and proposal of a negotiation process with farmers, to decide on modalities for the application of the interdiction rule, in order to permit adaptation by the vulnerable groups to take place, and to compensate the affected farmers for their social and economic losses;

3) The enforcement by inspectors of their monitoring system of water withdrawal, and with sanctions against farmers responsible for illegal boreholes.

Although it is clear that there is a willingness to avoid open conflicts that would worsen the situation, a coherent approach with respect to the interdiction rule has to be kept, to promote clarity in the legal environment and to facilitate compliant behaviour.

This is clearly a case that calls for effective *local governance* tools and for a complex approach that takes into consideration environmental protection needs, as well as the social and economic development of the area and its communities. If there is a lack of public participation in water resource management and the environmental policies fail and remain ineffective, then sustainable development will not be achieved (FEEM 2006). Good governance of water relies on the compliance of people from the political institutions, and on the capacity of these institutions to manage social problems and to gain the general consensus through agreements and compromise (Agricultural Research Institute, 2006).

A first step for the area of the *oued* Akarit-El Bsissi would be to convince the members of the *GDA* that the decision of interdiction will not constrain the socio-economic development of the region (Chkir et al., 2005). On the other side, the civil society has to become aware of the environmental impact of their actions, the relationship that local actions have to global ones, and the impact that present activities will have on the future (Tamas, 2002).

In conclusion, there is much to do to create a collaborative relationship with the *Groupement de Développement Agricole* of El Bsissi and with the *Commissariat Régional au Développement Agricole* of Gabès. So far, constraints have been in the paternalistic attitude of both scientists and policy-makers towards the public stakeholders, who were considered as an audience to be informed and made aware, rather than a group from which information could be gathered and to fully involve during the whole decision-making process. The challenge will be in enhancing the capacities of the GDA to become a facilitator, able to effectively overcome the traditional mistrust existing between the government and the public, and make use of scientific outputs in order to reach shared policies.

The participatory approach and involvement of the most relevant stakeholders in decision-making regarding the best uses of local resources, is highly needed to reach a concrete development of the agricultural sector and to protect water resources.

Acknowledgement

The authors wish to thank Dr. Najiba Chkir of the Laboratoire de Radio-Analyses pour l'Environnement, ENIS, and M. Brahim Abidi of the CRDA Gabès, for their scientific cooperation; Maria Vuori and Jouni Suominen, University of Helsinki, for drawing the maps. They also acknowledge the financial support of the EC through the "Nostrum-DSS" Co-ordination Action and of the Italian Ministry of University and Research through the FIRB project "Logos-Med".

References

Abidi, B. (1999). "Gestion des ressources en eau d'une zone d'interdiction. Cas des nappes souterraines de la region O. Akarit – El Bsissi (Gabès Nord), *Journée des Ressources en Eau,* INAT 3 juin 1999, Ministère de l'Agriculture, Direction Générale des Ressources en Eau, Tunis.

Abidi B. and Rahali M. (2004). "Arrondissement de Gabès", in Direction Générale des Ressources en Eau (by), *Annuaire de l'exploitation des nappes profondes*, pp. 333-345

Agricultural Research Institute (2006): "Comparative assessment of decision-making processes, regulations and laws in Mediterranean Countries", *Nostrum-DSS* (http://www.feem-web.it/nostrum/doc/d6-2_rev.pdf)

Benabdallah, S. (2003): "La réutilisation des eaux usées traitées en Tunisie", Bonn (http://www.diegdi.de/.../0/fa66ed1ab3b8efd6c1256ce100431e37/$FILE/reutilisation%20eaux%20Tunisie%20Benabdallah.pdf)

Chkir Ben Jemâa N., Ghorbel Zouari S., Zouari K. (2005): "Statistical information at country level. Assessment of water management practises, use and needs. Tunisia", *Nostrum-DSS National Report, Part I* (http://www.feem-web.it/nostrum/intranet/cd/docs/tun_1.pdf)

Chkir Ben Jemâa N., Ghorbel Zouari S., Zouari K. (2005). "Tunisia. Groundwater management of a protected area: Jeffara aquifer-Southern Tunisia", *Nostrum-DSS, National Report, Part II*

Chkir Ben Jemâa N., Ghorbel Zouari S., Zouari K. (2005): "Report on water uses in agriculture in the Mediterranean countries", *Nostrum-DSS* (http://www.feem-web.it/nostrum/doc/d2-1_rev.pdf)

CRDA Gabès (2003). *Gestion partecipative des nappes phreatiques. Cas du périmètre irrigue sur puits de surface: région d'El Bsissi-oued Akarit (Gouvernorat de Gabès). Rapport d'Evaluation*, Ministère de l'Agriculture, de l'Environnement et des Ressources Hydrauliques, Gabès

FEEM-Fondazione Eni Enrico Mattei (2006): "Thematic report on Dss and stakeholders' participation", *Nostrum-DSS* . (http://www.feem-web.it/nostrum/doc/d6-7.pdf)

Gaaloul N. and Zouari K. (2004): *Harnessing salty water to enhance sustainable livelihoods of the rural poor in Tunisia*, International Center for Biosaline Agriculture, Dubai

Global Water Partnership Technical Advisory Committee (GWP-TAC) (2000): "Integrated Water Resources Management", TAC Background Papers n°4, Stockholm. (http://www.gwpforum.org/servlet/PSP)

Ministère de l'Agriculture (1998): *EAU 21–Stratégie du secteur de l'eau an Tunisie à long terme 2030*

Ministère de l'Agriculture, Direction Générale des Ressources en Eau (2003): *Etude des nappes aquifères de Sfax* - lot n° 8 - Définition des besoins hydriques pour eau potable, d'irrigation et d'industrie, Sfax

Minoia P. and Camuffo M. (2006): "Thematic report on social aspects of conflicting water uses", *Nostrum-DSS* (http://www.feem-web.it/nostrum/doc/d6-6.pdf)

Office du Développement du Sud (2004). *Le Gouvernorat de Gabès en chiffres*, Gabès, Ministère du Développement et de la Coopération International

Romagny, B., Guillaume, H., Ben Ouezdou, H., Palluault, S. (2004). Ressources en eau, usages et concurrences dans la Jeffara tunisienne, *Série Usages, appropriation, gestion des écosystèmes. Documents de recherché n°1.* Laboratoire Population–Environnement–Développement Unité Mixte de Recherche IRD–Université de Provence 151, 38 p. (http://www.up.univ-mrs.fr/wiupenv/labo/d_lpe/publications/docs-recherche/lped-usages2.pdf)

Scanlon J., Cassar A. and Nemes N. (2004): "Water as a Human Right?", *IUCN Environmental Policy and Law Paper n°51*

Tamas, P. (2003): “Water Resource Scarcity and Conflict: Review of Applicable Indicators and Systems of Reference”, *UNESCO-IHP*, Paris (http://unesdoc.unesco.org/images/0013/001333/133307e.pdf)

Wallace J. S., Acreman, M. C. et al. (2003) “The sharing of water between society and ecosystems: from conflict to catchment based co-management.” *Philosophical Transactions: Biological Sciences* 358 (1440), pp. 2011-2026.

CHAPTER TWO

WATER RESOURCES AND ENVIRONMENTAL CHANGES IN MEDITERRANEAN HEADWATERS

ENRIQUE MORÁN TEJEDA
AND ANTONIO CEBALLOS BARBANCHO

Introduction

Several studies published in the last few decades demonstrate the concerns of the scientific community about the effects climate change could have on the availability of water resources (Loaciaga et al., 1996; IPCC, 2001; Milly et. al., 2005). The climatic variables and the physical characteristics of water catchments are the most important natural factors in the hydrological cycle. An understanding of these factors is essential for the development of planning and management strategies in regions where the availability of water is a problem.

From the climatic perspective in Europe, researchers relate the variability of river flows to the fluctuations that the NAO index denotes (Arnell, 1999; Trigo et al., 2004). Studies have indicated that the Mediterranean is predicted to be one of the areas most affected by the impact of climate change. What is projected to be particularly problematic is the availability of water (Parry et al., 2000). Indeed, Mimikou et al.(2000) in Greece and Ayala-Carcedo & Iglesias (2002) in Spain have detected a generalized decrease in precipitation, which in turn is provoking a reduction in river flows. An increase in the variability of both precipitation and river flow has also been demonstrated.

On the other hand, the physical characteristics of particular catchments determine the capability of each catchment to generate runoff. In the last few decades, an important relationship between changes in forest areas within catchments and water yield has been demonstrated. Growth in both the area and forest density intensifies processes such as interception and evapotranspiration, producing a reduction in the generated runoff (Boch &

Hewlett, 1982; Calder, 1998; Zhang et al., 1999; Andressian, 2004). Mountainous regions are most sensitive to such changes, as not only are they natural reservoirs, they also have the highest rainfall (Beguería et al., 2003). Therefore, headwaters found in mountainous regions play a key role in these processes. In Spain, the recuperation of forest cover, observed in the mountainous areas since the middle of the last century, has become a recent focus of scientific interest. Gallart & Llorens (2004) in the Ebro headwaters, and García-Ruíz et al. (2001) in the Pyrenees, have highlighted the expansion of the forest cover as one of the non-climatic factors which are critical to the reduction of river flow in these areas.

The main objective of this work is to analyze the evolution of water discharges in relation to the trend of the climatic variables and the changes of forest cover in a representative area of Mediterranean mountains. Three headwaters have been selected in a mountainous region in central Spain of such a size that allows us to precisely analyze the climatic and hydrological variables as well as land-use changes. Our findings have been discussed with reference to similar studies conducted in other parts of Spain, with the objective of gaining a better understanding of the water resources in our region.

Study Area

The three headwaters selected for this study are situated on the north face of the Central System, a mountain range in central Spain supplying water to the Duero Basin (Fig. 2-1). The headwaters are as follows (from east to west): Tormes Catchment (gauge station in *Hoyos del Espino*) with an extension of 83 km^2; Moros Catchment (gauge station in *El Espinar*) with a surface area of 37 km^2; and Riaza Catchment (gauge station in *Riaza*) with a surface area of 30 km^2. The altitudes vary from 1,100 m to 2,200 m above sea level.

The lithology within the region changes from granite, derived from hercynian plutonism in the nucleus of the Central System in the *Tormes* area, to gneiss in the *Riaza* area.

The climate is Mediterranean with continental characteristics, due to altitude. The annual precipitation varies from 660 mm to 840 mm. Throughout the months of July and August rainfall is scarce, producing the characteristic summer drought of the Mediterranean. The average annual temperature is 11.4 ºC; January being the coldest month with an average of 3.8 ºC, and July being the hottest with an average of 20.8 ºC.

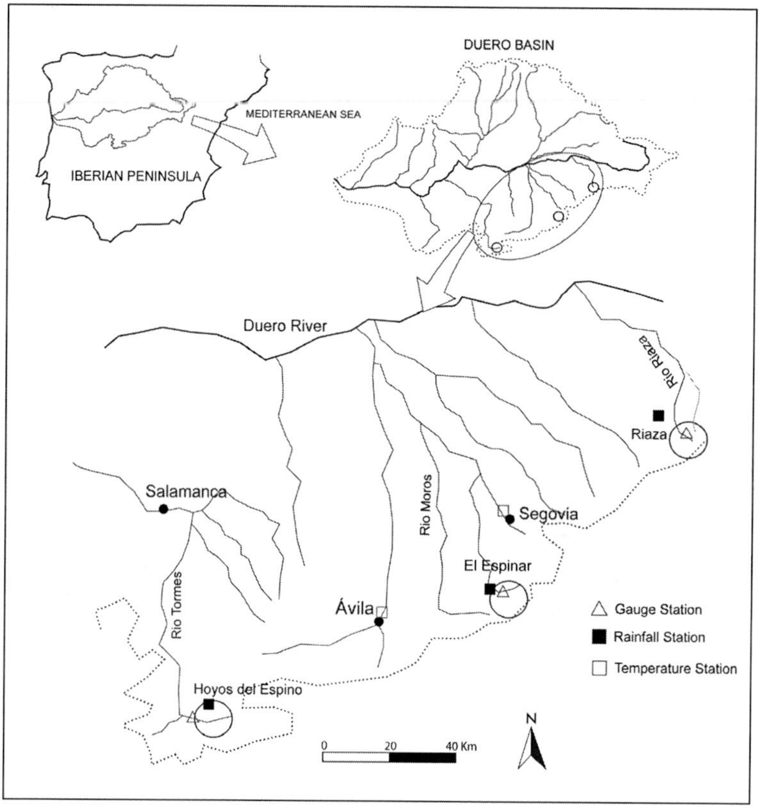

Fig. 2-1: The location of the Three Headwaters Selected for the Study.

The altitudinal gradient determines the distribution of the vegetation characteristic of the region. The forests of *Quercus pyrenaica* and *Pinus sylvestris*, with shrubbery of *Cytisus multiflorus, Genista cinerea*, occupy the supra-Mediterranean level. The oro-Mediterranean shrubs (*Cytisus* sp. and ericaceous) and the mountainous pasture occupy the highest zones.

Methods

In order to carry out the hydro-climatic analysis, data from five meteorological stations and three gauge stations was obtained. The metadata is shown in Table 2-1.

Rainfall and water discharge data were collected separately in each catchment area, enabling us to conduct individual analysis of these variables. Temperature however was recorded from two locations and the average of the two series was taken. This was possible due to a high correlation between the series: Pearson's coefficient (r^2) = 0.59 ($p<.001$).

The annual data has been graphically represented, showing the evolution of the variables over time. If the variables showed a trend, we plotted a line of regression to quantify the variation over the designated period of time. For temperature, the thermal anomalies were calculated by subtracting the median value of each year from the average value of the last normal climatic period. (CLINO 1961-1990). The moving average over five years was plotted, to observe possible partial trends. To study inter-annual variability of rainfall, we used a methodology proposed by Scian and Donnari (1997), who defined the "dry year" as that with annual rainfall values lower than the median minus the standard deviation, and the "wet year" as that with annual rainfall values higher than the median plus the standard deviation.

To determine whether the tendencies of the annual and monthly data were significant at a high confidence interval, (>95%) the Spearman test was conducted (Sneyers, 1992) for the whole series and the detected sub-periods.

Table 2-1: Metadata of Meteorological and Gauge Stations.

Name	Type	Coordinates	Altitud (m)	Data series
Hoyos del Espino	Rainfall Station	40°21'25"N 5°10'27"W	1,416	1956-2002
El Espinar	Rainfall Station	40°43'00"N 4°14'57"W	1,180	1956-2004
Riaza	Rainfall Station	41°16'40"N 3°28'47"W	1,180	1956-2004
Ávila	Temperature Station	40°39'00"N 4°42'00"W	1,126	1961-2004
Zamarramala	Temperature Station	40°58'00"N 4°08'07"W	1,000	1961-2004
Hoyos del Espino	Gauge Station	40°20'47"N 5°11'47"W	1,391	1953-2004
El Espinar	Gauge Station	40°44'59"N 4°09'28"W	1,250	1953-2004
Riaza	Gauge Station	41°16'55"N 3°28'09"W	1,040	1953-2004

Due to the fact that we were only able to record precipitation in one point within the catchment area, it was not possible to carry out a water balance, which would have allowed us to establish the relationship between runoff and precipitation. We therefore used different approaches as recommended by Beguería et al. (2003), to determine the dependency of the flow on precipitation, and the participation of other factors within the system:

- -The variables for the period (in hydrological years) 1961-62/2003-04 were normalised. Thus, the annual value was subtracted from the mean of the whole period (M), and dividing the result by the standard deviation (Sd). The result is a regional index that allowed us to compare the evolution of the different variables on the same scale.
- -For each catchment stepwise regressions were carried out with water discharges as the dependent variable and precipitation, temperature and time as the independent variables. This allowed us to calculate the degree of correlation of the water discharges with respect to the other variables, and to interpret the resulting residue of each model.

To study the evolution of the river regimes over the time period, the means of the monthly flow for the first half of the period (1956-1979), have been graphically represented, together with the means of the second half of the period (1980-2002).

Finally, the changes undergone by the forest cover in each catchment have been studied from the comparative analysis of aerial photographs (American Flight) of the region in 1957 (previously orthorectified), with the ortho-image (*Junta de Castilla y León*) of 2002.

Results

a) Hydro-climatic analysis

Analysis of the pattern of temperature change over time (Fig. 2-2), showed a change point in the early 1970s from a negative to a positive trend. From 1972 to 2004 the positive tendency (Spearman´s *Rho* = 0.67) was significant with a confidence interval of 99%. A linear regression carried out for this sub-period showed an increase (although not sustained), of 1.3 ºC, giving a ratio of 0.04 ºC year $^{-1}$.

During the aforementioned sub-period, the means for winter, spring and summer showed a positive trend (*Rho* = 0.38, 0.62 and 0.44 respectively), significant at the 95% confidence interval. March and June being the months to experience the most notable and significant rise (Table 2-2).

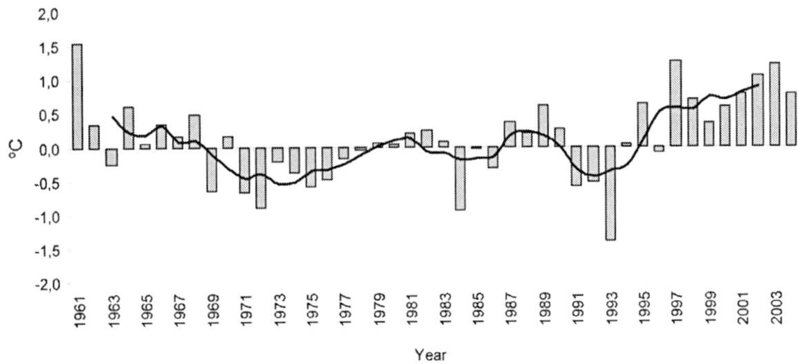

Fig. 2-2: Thermal Anomalies From the Average of *Ávila* and *Zamarrala* Stations for the Period 1961-04. Bold Line: Moving Mean Over 5 Years

Table 2-2: Spearman Rho Values and Confidence Intervals for the Monthly and Annual Temperatures Series, for the Sub-Period 1972-2004.

	Jan	Feb	Mar	Apr	May	Jun	Jul	Aug	Sep	Oct	Nov	Dec	Year
Rho Value	0.19	0.28	0.55	0.32	0.35	0.43	0.23	0.28	0.00	0.29	0.04	0.23	0,67
sig level	0.30	0.11	0.00	0.07	0.04	0.01	0.20	0.12	0.99	0.11	0.84	0.21	0.00

The trends over time for the precipitation were different for each catchment station (Fig. 2-3), for the time period 1957-2004. Table 2-3 shows the values of Spearman's Rho and the confidence interval statistically significant for the monthly and annual series of the variable.

In *Hoyos del Espino*, precipitation showed a negative trend, although not statistically significant. A linear regression revealed a decrease of 17.6%, across the whole time period, which can be explained by the decrease of precipitation during the winter and spring months (Table 2-3). From the perspective of the inter-annual variability, in Figure 2-3 we can see two different periods before and after 1979. The variation coefficient increases from 21.2% in the first sub-period (1957-79) to 36.3% in the second sub-period (1980-02). The frequency of dry years (M-Sd) also increased over time: 4.5% of the years were dry in the first sub-period, and 18.2% in the second.

In *El Espinar*, one can appreciate a slight increase in precipitation. However, this is not significant. Figure 2-3 shows how the frequency of dry years decreases over time, in contrast to the increase of the wet years in the last two decades. The high variability that one can observe in this graph is confirmed by the high variation coefficient (30%) that the series shows.

Table 2-3: Spearman Rho Values and Confidence Intervals for the Monthly and Annual Precipitation Series, for the Period 1957-2004.

	Jan	Feb	Mar	Apr	May	Jun	Jul	Aug	Sep	Oct	Nov	Dec	Year
Hoyos del Espino													
Rho Value	-0.06	-0.37	-0.24	-0.15	-0.03	-0.28	0.11	0.24	0.02	-0.02	-0.12	-0.07	-0.22
sig level	0.68	0.01	0.11	0.33	0.84	0.03	0.49	0.11	0.88	0.92	0.42	0.65	0.14
El Espinar													
	Jan	Feb	Mar	Apr	May	Jun	Jul	Aug	Sep	Oct	Nov	Dec	Year
Rho Value	0.12	0.00	0.02	0.07	0.09	-0.22	0.11	0.00	-0.13	0.19	0.07	0.10	0.10
sig level	0.43	0.99	0.91	0.63	0.56	0.13	0.50	0.99	0.39	0.21	0.66	0.50	0.49
Riaza													
	Jan	Feb	Mar	Apr	May	Jun	Jul	Aug	Sep	Oct	Nov	Dec	Year
Rho Value	-0.19	-0.26	-0.48	0.10	0.01	-0.47	0.20	-0.07	-0.26	0.07	-0.07	-0.16	-0.48
sig level	0.21	0.08	0.00	0.49	0.77	0.00	0.11	0.64	0.08	0.65	0.64	0.29	0.00

In *Riaza*, precipitation experienced a decline of 41% (*Rho* = -0.48; $p<0.001$) between 1957 and 2004. Table 2-3 illustrates how the majority of the months show a negative trend, only significant at the 99% confidence interval in March and June. The variability of the series is also high with a variation coefficient of 25%. In Figure 2-3, one can see a change in the behaviour of the variable from 1982. Until that year, 70% of the values of precipitation were between the normal range (M ± Sd), and the other 30% were above M + Sd ("wet years"). From 1982, the majority of the cases outside the normal range (30%), were below M – Sd ("dry years"). There was only one year with rainfall above M + Sd.

The evolution over time of water discharges is represented in Figure 2-4 for the period 1953-54 to 2003-04. As with precipitation, we found different trends in each case.

In *Hoyos del Espino* the variable showed a negative trend across the series, although not statistically significant. The linear regression showed a decrease of 28%. The variation coefficient for the period was 40.8%, which confirms the high variability that Figure 2-4 shows. Monthly analyses showed that water discharge between February and May (Table 2-4) followed a negative trend, although they were not significant at a high confidence interval. The change in river regime is represented in Figure 2-5. The average value of runoff for the second half of the period (47.4 hm^3) has notably decreased with respect to the first half (66.3 hm^3). Furthermore, the snow peak that used to be in April has disappeared, and now the maximum monthly flow occurs in December.

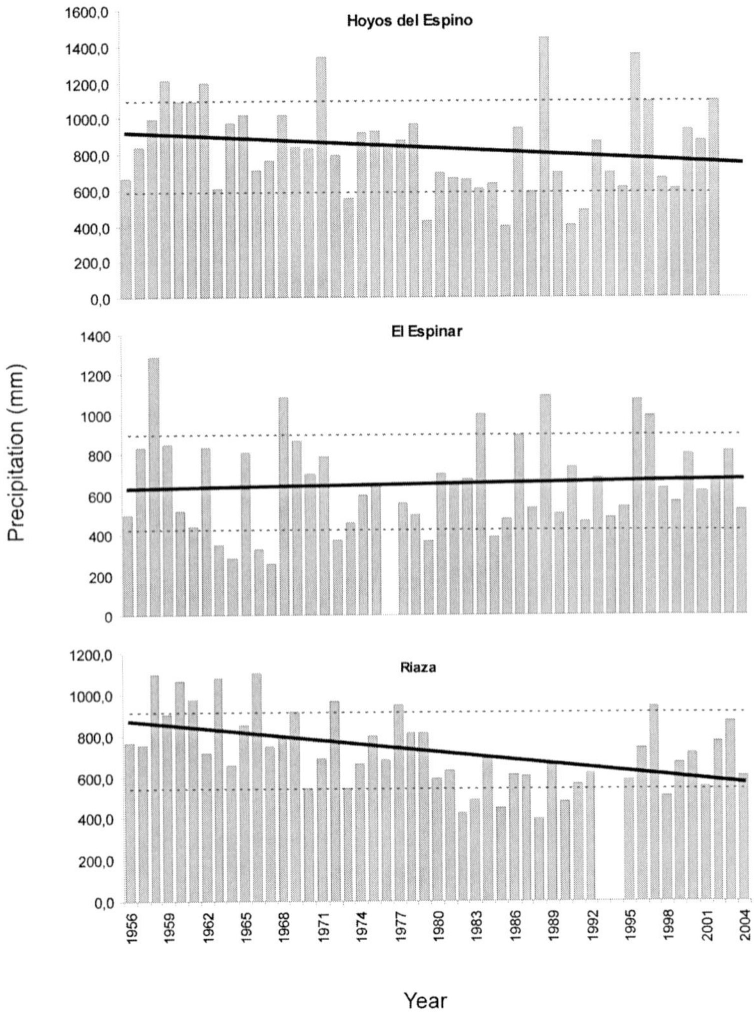

Fig. 2-3: Annual Precipitation in *Hoyos del Espino, El Espinar* and *Riaza* Stations. Bold Line: Tendency; Dotted line: M ± Sd

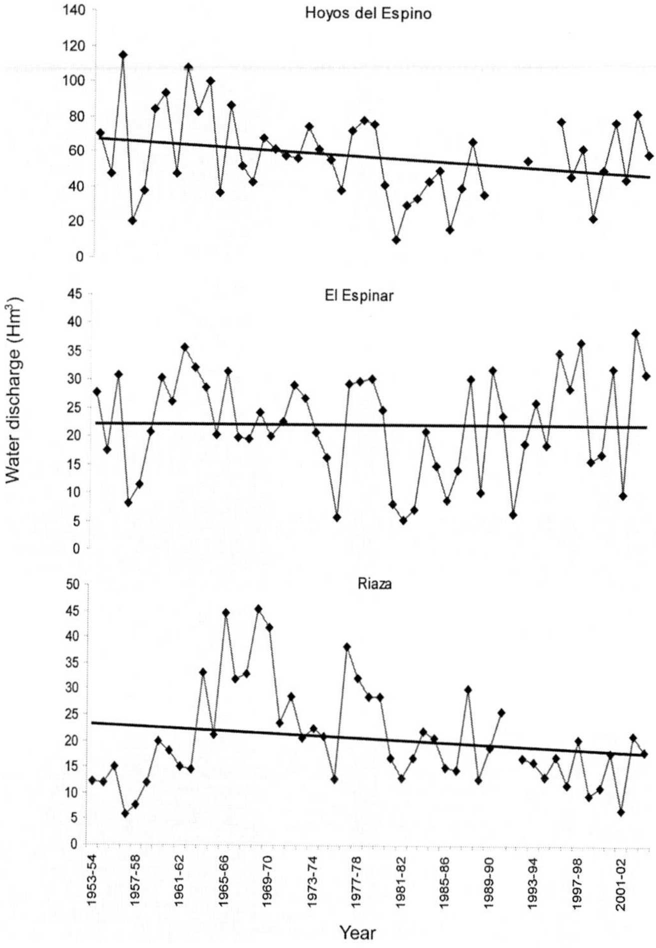

Fig. 2-4: Annual Water Discharges in *Hoyos del Espino, El Espinar* and *Riaza*. Bold line: tendency.

Water discharges in *El Espinar*, did not show any specific trend across the series. The inter-annual variability was high, with a variation coefficient of 41.4%. The monthly tendencies were not significant, although one can appreciate a change in the river regime with the disappearance of the snow peak in March, and an important decrease in water discharges from January to May.

In *Riaza* the water discharges showed a negative trend, although not significant for the period studied. February to June showed a negative trend, statistically significant in March, at a 95% confidence interval. In Figure 2-5, one can appreciate a change of trend from negative to positive during 1967-68. Before this year the variable experienced an important increase, and afterwards it underwent a significant decrease (*Rho* = -0.58) at a 99% confidence interval. The losses of water discharge for that sub-period totalled 60%. The river regime also experienced changes over time. The maximum value due to melting snow that occurred in April for the sub-period 1953-54/1979-80 disappeared, and the peak for the two last decades occurred in January. The stream flow in spring has decreased notably, and the drought for summer seemed to begin earlier, in the month of June.

Table 2-4. Spearman Rho Values and Confidence Intervals for the Monthly and Annual Water Discharges Series, for the Period 1953-54/2003-04.

Hoyos del Espino													
	Oct	Nov	Dec	Jan	Feb	Mar	Apr	May	Jun	Jul	Aug	Sep	Year
Rho Value	0.07	-0.09	0.00	0.10	-0.16	-0.23	-0.09	-0.13	-0.01	0.05	0.20	0.12	-0.23
sig level	0.68	0.01	0.11	0.33	0.84	0.03	0.49	0.11	0.88	0.92	0.42	0.65	0.14
El Espinar													
	Oct	Nov	Dec	Jan	Feb	Mar	Apr	May	Jun	Jul	Aug	Sep	Year
Rho Value	-0.04	-0.06	-0.01	-0.06	-0.02	-0.28	-0.10	0.07	0.03	0.06	-0.11	-0.17	0.00
sig level	0.76	0.74	0.96	0.70	0.90	0.05	0.49	0.62	0.84	0.69	0.47	0.23	1.00
Riaza													
	Oct	Nov	Dec	Jan	Feb	Mar	Apr	May	Jun	Jul	Aug	Sep	Year
Rho Value	0.04	0.04	0.06	0.02	-0.11	-0.31	-0.12	-0.02	-0.07	0.00	-0.10	-0.10	-0.10
sig level	0.77	0.76	0.66	0.88	0.43	0.03	0.42	0.89	0.63	1.00	0.50	0.50	0.50

b) Forest Growth

The forest areas within the catchments have experienced expansion during the last five decades (Fig. 2-6).

In the Tormes Catchment, P*inus sylvestris* forests occupied 982 hectares by the middle of the 20th century (11% of the catchment area). At the beginning of the 21st Century it filled 15% of the catchment´s surface, with 1,271 hectares, an increase of 30%. This growth was due to reforestation of small areas by local administration. The forest perimeter has not experienced any notable change.

In 1957, forest within the El Espinar Catchment occupied 55% of its surface, a total of 2,069 hectares. In 2002, it increased to 2,991 hectares, occupying 81% of the catchment area. This enlargement of 44.6% was due mainly to the natural growth of *Pinus sylvestris* (introduced in the 1960s) into the highest areas.

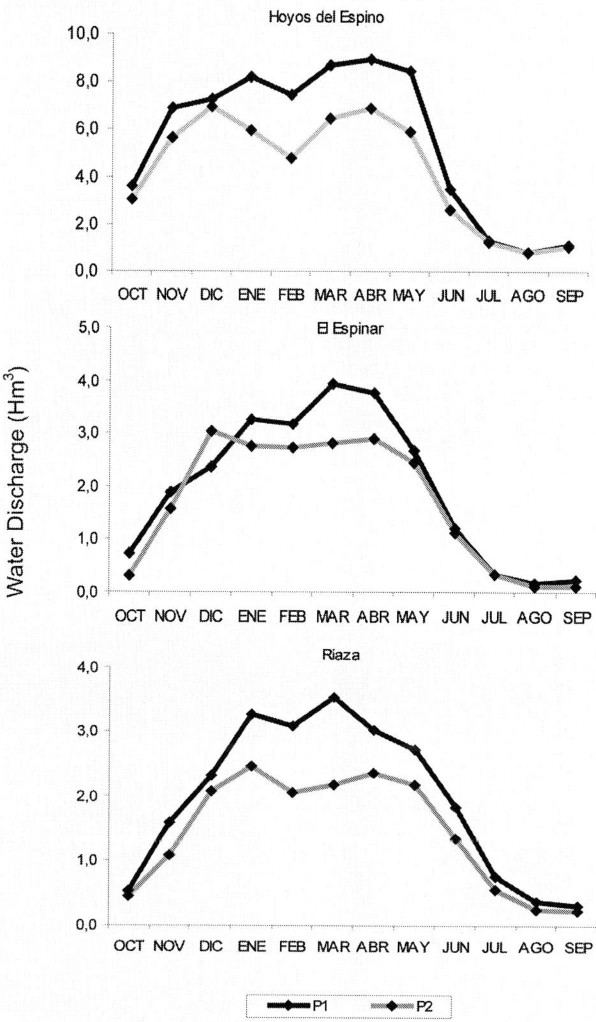

Fig. 2-5: River Regime in Average Values for the Sub-periods 1953-54 to 1979-80 (P1) and 1980-81 to 2003-04 (P2).

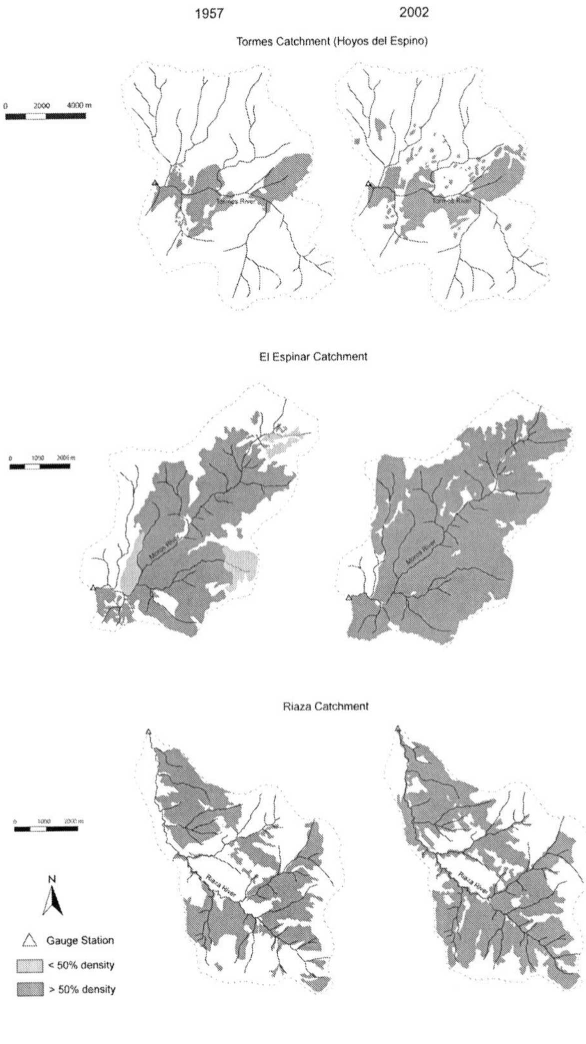

Fig. 2-6: Forest Areas Within the Studied Catchments.

The natural forest comprised mainly of *Quercus pyrenaica* has also increased in density.

In the Riaza Catchment, the forest area in 1957 measured 1,259 hectares, representing 42% of the catchment area. Reforestation, mainly by *Pinus sylvestris* in the highest areas, and the growth in density of *Quercus pyrenaica* and *Fagus sylvatica* varieties, increased this value to 1,513 hectares in 2002, occupying 50.4% of the catchment area.

c) Relationship between the evolution of water discharges and the environmental changes

In Figure 2-7, the normalized variables studied in earlier chapters have been represented graphically. To determine the degree of correlation between the variables, multiple regression analysis has been carried out for each case.

A step-wise regression model showed that precipitation explains most of the variance ($r^2 = 0.70$) of water discharges in *Hoyos del Espino*. Time and temperatures as independent variables do not increase the correlation in the regression. Figure 2-7 shows how the two variables follow a parallel evolution over the series.

In *El Espinar*, correlation between water discharges and precipitation is not significant ($r^2 = 0.16$) for the whole period. It is believed that this was due the construction of a small reservoir (1 hm^3 of capacity) and its regulation following its completion in 1975. Once the regulation was over, from 1983 onwards, the evolution of the two variables show a high correlation ($r^2 = 0.68$). The variables of temperature and time within the model are not statistically significant.

In *Riaza*, there is a weak correlation between water discharge and precipitation ($r^2 = 0.32$). The addition of time as an independent variable increases the r^2 value to 0.45, which suggests that a factor dependent on time may be responsible for a 13% decrease in the water discharge. This factor came into play from 1981-82, with the variables starting to display different trends: precipitations began to increase (99% significant confidence interval) whilst the water discharge continued to decrease. This suggested that forest growth in the last few decades is contributing (in the terms explained above) to a decrease of water discharge in the catchment area.

It is important to note at this point that a high percentage of the variation of water discharge still remains unexplained by the variables included in the regression model. Therefore we have to take into account the hypothesis that other factors are involved in the system.

Discussion and conclusions

The temperature increase across the series showed a similar pattern to those observed by other authors at different spatial scales. The same change point was detected by Morales et al. (2005) in their study for *Castilla y León*, and by the IPCC (2001) at the European and planetary levels. The increase was observed mainly in the mean values for winter and spring months, whilst in summer and autumn no significant trends were identified. From the hydrological perspective, the increase in temperature during winter and spring has triggered changes in the river regime. The disappearance of the snow peak is believed to be a result of both a reduction in the accumulation of snow, and melting beginning earlier. This process has also been observed in other mountainous areas in Spain, such as in the Pyrenees (García Ruiz et al., 2001).

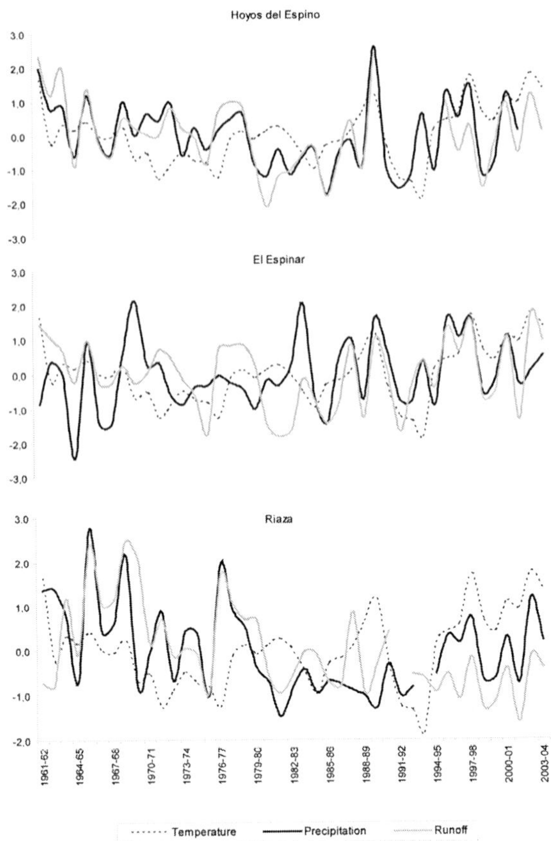

Fig. 2-7: Temperature, Precipitation and Runoff Indexes in Normalised Standard Deviations.

Discovering different precipitation trends over a relatively small distance (approximately 200 kilometers), is not unusual in Mediterranean regions, where rainfall often varies significantly from one place to another (González-Hidalgo et al., 2001). De Luis et al. (2000) demonstrated that precipitation trends showed a high spatial variability throughout the series 1961-90, in their study carried out with 97 rainfall stations in the *Comunidad Valenciana*. Together with this observation, the study also highlighted other characteristics common to rainfall within the Mediterranean region: high values of inter-annual variability (variation coefficient between 25% and 35%) and notable contrasts between “wet years” and “dry years”. Such factors make it difficult to find strong statistical trends. The negative trends of winter and spring precipitation in *Hoyos del Espino* and *Riaza*, are thought to be responsible for the decrease in water discharges observed in both catchments. Soils in winter are generally saturated with water due to autumn rainfall, reducing their capacity to absorb water. Subsequently, rainfall in winter converts easily into runoff (Ceballos et al., 2002). Therefore, if this rainfall diminishes, stream flow will suffer.

Water discharges clearly decrease across the series, except in *El Espinar*, where no trend was detected. The disappearance of the snow peak within the monthly river regime was the most clearly detected process in the hydrological analysis. The variability of precipitation explained most of the variance in the water discharge, although still a high percentage remains unexplained by the variables included in the model. These residuals are thought to be due to different processes that may induce alterations in the water output, such as modifications in the channel (in the case of *El Espinar*), or changes in the monthly distribution of precipitation. The latter may introduce new elements into the analysis, such as the infiltration capacity of soil or evapotranspiration. These require additional analyses that would help to achieve a better understanding of the hydrological behaviour of the catchments.

The abandonment of the countryside that has taken place in Spain over the last few decades has caused changes in the landscape, such as the natural recuperation of forest resources (Gallart and Llorens, 2002). Furthermore, reforestation plans managed by the government have helped to accelerate this process.

The growth of forest areas observed in this study has not had a clear effect on the water resources of the catchments studied. Our findings allow us to speculate on the hypothesis that if forest expansion is such that it allows forest cover to cross the threshold of approximately half the catchment area, its effect on the water resources become significant. In

Hoyos del Espino the forest cover never exceeded 50% of the catchment's area; in *El Espinar* on the other hand, in both years analysed (1957 and 2002), the forest had already covered more than 50%. Consequently in these two catchment areas no notable effect of the forest on water discharge was found using multiple regressions. Only in *Riaza* did the forest area cross the hypothesised threshold (42% to 50.6%), and in this catchment, a slight effect of the forest on the reduction of water resources was detected. Further analyses are therefore needed to verify the existence of such a threshold, and to confirm the hypothesis. Beguería et al. (2003) found evidence that forest expansion played an important role (30%) in the decrease of water discharges in the headwaters of the Pyrenees. Trimble & Weirich (1976) found in the southwest of the United States that small increases in the forest cover reduced the runoff between 4% and 21%. Zhang et al. (2001) demonstrated that as tree cover expands, losses from evapotranspiration are a significant cause of the decrease in water resources, following the analysis of 250 catchments across the world.

Our results suggest that any future expansion of the forest cover in the catchments where the threshold has not been reached would cause a decrease in water resources additional to that caused by climatic factors. Furthermore, the high variability detected in the hydrological behaviour of the catchments studied confirms the necessity to continue investigating the water resources in the Duero Basin. Further studies would in turn allow the development of a suitable management plan for the control of water resources in this region.

References

Andréassian, M. 2004. Waters and forests: from historical controversy to scientific debate. Journal of hydrology, 291.1-27.

Arnell, N.W. 1999. The effect of climate change on hydrological regimes in Europe: a continental perspective. Global Environmental Change. 9(1): 5-23

Ayala-Carcedo, F.J. & Iglesias, A. 2000. mpactos del posible cambio climático sobre los recursos hídricos, el diseño y la planificación hidrológica en la España peninsular. El Campo de las Ciencias y las artes, 137: 201.222.

Beguería, S.; López-Moreno, L.I.; Lorente, A.; Seeger, M & García-Ruiz, J.M. 2003. Assessing the effect of climate oscillations and land-use changes on streamflow in the Central Spanish Pyrenees. Ambio, 32(4): 283-286

Bosch, J.M. y Hewlett, J.D. 1982. A review of catchment experiments to determine the effect of vegetation changes on water yield and evapotranspiration. J. Hydrology, 55: 3-23.

Calder, I. R. 1998. Water-resource and land-use issues, SWIMP pap. 3, Int. Water Manage Inst., Colombo, Sri Lanka.

Ceballos, A., Martínez-Fernández J., Santos, F., Alonso, P. 2002. Soil-water behaviour of sandy soils under semi-arid conditions in the Duero Basin (Spain). Journal of Arid Environments. 51 (4): 501-519.

De Luis, M., García-Cano, M.F., Cortina, J., Reventós, J., González-Hidalgo, J.C. & Sánchez, J.R. 2001. Climatic trends, disturbances and short-term vegetation dynamics in a Mediterranean shrubland. Forest Ecology and Management, 147:25-37

Gallart, F., Llorens, P. 2002. Water resources and environmental change in Spain. A key issue for sustainable catchment management. En García Ruiz, J.M; Jones, J.A.A. & Arnaez, J. (Eds): Environmental Change and Water Sustainability. IPE-CSIC, Zaragoza. pp 11-20.

Gallart, F. Llorens, P. 2004. Observations on land cover changes and water resources in the headwaters of the Ebro catchment, Iberian Peninsula. Physics and Chemistry of the Earth, 29: 769-773.

García-Ruiz, J.M; Beguería Portugués, S.; López Moreno, J.I.; Lorente Grima, A. & Seeger, M. 2001. Los recursos hídricos superficiales del Pirineo aragonés y su evolución reciente. Geoforma Ediciones, Logroño. 192 pp.

González-Hidalgo, J.C.; Raventós, J. & Sánchez J.R. 2001. Spatial distribution of seasonal rainfall trends in a western Mediterranean area. Int. Journal of Climatology, 22: 1593-1615.

IPCC. 2001. Climate Change 2001: The scientific bases. Contribution of Working Group I to the Third Assessment Report of the Inter-governmental Panel on Climate Change. Houghton, J.T.; Ding, Y.; Griggs, D.J.; Noguer, M.; van der Linden, P.J.; Dai, X.; Maskell, K. & Johnson, C.A. (eds) Cambridge University Press, United Kingdom. 881 pp

Loaciaga, H. A.; Valdés, J. B.; Vogel, R.; Garvey, J. & Schwarz, H. 1996. Global warning and the hydrological cycle. Journal of Hydrology, 174: 83-127.

Milly, P.C.D.; Dunne, K.A. & Vecchia, A.V. 2005. Global pattern of trends in streamflow and water availability in a changing climate. Nature, 438/17: 347-350

Mimkou, M.A.; Baltas, E.; Varanou, E. & Pantazis, K. 2000. Regional Impacts of climate change on water resources quantity and quality indicators. Journal of Hydrology, 234: 95-109.

Morales, C.; Ortega, M.T.; Labajo, J.L & Piorno, A. 2005. Recent trends and temporal behaviour of thermal variables in the region of Castilla-León (Spain). Atmósfera, 18(2): 71-90.

Paredes, D., Trigo, R.M., García-Herrera, R., Franco, I. 2006. Understanding precipitation changes in Iberia in early spring: Weather typing and storm-tracking approaches. Journal of Hydrometeorology. vol 7: 101-113

Parry, M.; Parry, C. & Livermore, M. (Eds). 2002. Valoración de los efectos potenciales del cambio climático en Europa (Informe ACACIA de la Comisión Europea, resumen y conclusiones). Universidad de Castilla-La Mancha e Iberdrola. Toledo, 29 pp.

Scian, B & Donnari, M. 1997. Retrospective analysis of the palmer Drought Severity Index in the semi-arid pampas region, Argentina. Inter.J. of Climatology,17:313-322

Sneyers, R. 1992. Use and misuse of statistical methods for detection of climate change. In: Climate Change Detection Project, Report on the Informal Planning Meeting on Statistical Procedures for Climate Change Detection. WCDMP, (20); J76-J81

Trigo, R.M; Pozo-Vázquez, D.; Osborn, T.J.; Castro-Díez, Y.; Gámiz-Fortis, S. & Esteban-Parra, M.J. 2004. North Atlantic Oscillation influence on precipitation, river flow and water resources in the Iberian Peninsula. Intern. Journal of Climatology.24: 925-944

Trimble. F.W. & Weirich, S.H. 1987. Reforestation reduces streamflow in the south-eastern United States. Journal of Soil and Water Conservation. Vol. 42, no. 4, pp. 274-276.

Zhang, L., Dawes, W.R., Walker, G.R. 1999. Predicting the effect of vegetation changes on catchment average water balance, Technical Report 99/12. Cooperative Research Centre For Catchment Hydrology. Australia, 35 pp.

Zhang, L., Dawes, W.R., Walker, G.R. 2001. Response of a mean annual evapotranspiration to vegetation changes at a catchment scale. Water Resour. Res. 37: 701-708

CHAPTER THREE

HEALTH CARE PROFILES OF THE MEDITERRANEAN BASIN COUNTRIES

MEHMET T. ARITAN

Introduction

As a simple approach for determining healthcare levels in various countries of the Mediterranean Basin, we will examine life expectancy figures. Life expectancy figures, to a large extent, reflect comprehensive health factors and are thus composite health statistics. It is a fact that there are several other factors that play an important part in the longevity of a population. These factors are studied by medical geographers. The factors are as follows: a) source or location of disease; b) the diffusion of disease; and c) the delivery of health care.

Study Area

Although the research and discussion encompass a global perspective, the focus of this empirical study are the healthcare profiles of the Mediterranean Basin countries, including Albania, Algeria, Bosnia, Bulgaria, Croatia, Cyprus, Egypt, France, Georgia, Greece, Israel, Italy, Lebanon, Libya, Malta, Moldova, Monaco, Morocco, Portugal, Romania, Russia, Serbia-Montenegro, Slovenia, Spain, Syria, Tunisia, Turkey, and Ukraine. All of these countries have a coastline on the Mediterranean or its extensions (e.g., the Adriatic, the Aegean, or Black Seas).

Methods

Determining the source and location of disease involves several environmental factors which are related to the global location of a country. These factors include climate, soils, water, and illness-causing organisms (*pathogens*). In terms of diffusion of disease, diseases are passed on from pathogens to hosts (*carriers)*, which are other living organisms such as humans and animals. Once diseases spread and start affecting large numbers of people, they command attention. Therefore, delivery of healthcare is the next important area of study in medical geography.

Results and Discussion

As stated above, determining the source or location of disease involves several environmental factors which are related to the global location of the country. Again, these factors include climate, soils, water, and illness causing organisms (pathogens). The study of the interaction between pathogens and the environment is known as *disease ecology*. Illness-causing organisms thrive in warmer and more humid climatic regions; specifically in the tropics. Unlike the mid- and higher-latitude regions of the world, where there are winters with cold temperatures and hard frost, the tropics have no seasonal change. Warm and humid conditions dominate the whole year. This creates an incubator effect, where the organisms continue to grow, spread, and mutate if they encounter a path of resistance. Such resistance may be natural, or may be the result of human intervention in the form of prevention and or eradication of the pathogens. Therefore, the role of the environment and the investigation of it is a vital and integral part of disease ecology.

Pathogens are omni-present in water, soil, and vegetation. Some of these are visible, such as certain animals that carry enzymes that are poisonous to humans and other animals. Two good examples of this are the mosquito and the tse tse fly. Of course, within certain climatic conditions, it is impossible to eradicate the mosquito population. Similarly, the tse tse fly has not been eradicated. In fact, over the last fifty or so years the tse tse fly has expanded its territory from the el-D'jouf basin near the Fouta d'jallon Mountains to include all of the savanna of West Africa into the adjacent tropical rainforest areas. Even today, the fly is harassing people and livestock with the same results as it did centuries ago. Nomadic tribes such as the Fulani of the West African savanna (also known as the Sahel) lose cattle to the deadly sting of the tse tse fly. In order to avoid such an economic loss, grazing land in the Sahel has been abandoned for this purpose. The effect of the fly on human beings is not so immediate. Human beings contract "sleeping sickness" from the sting of the fly. This is an endemic disease. Endemic diseases have no known cure and they can make a person's life miserable (e.g., sapped of energy where even some menial physical tasks are a major chore for the body). The condition of illness or morbidity of the disease depends on the disease contracted. The disease attacks the immune system and weakens it, making the person vulnerable to some other unrelated simpler diseases (e.g., pneumonia).

Eventually, the disease shortens one's life span (i.e., kills or helps kill the person before his/her natural time). Therefore, the rate of mortality increases, not only by numbers, but also by age. As a result of the high incidence of endemic diseases, tropical regions of the world have lower life expectancies.

Endemic diseases are found more often in the tropics mainly because of the environment, as indicated above. Hot and humid environments nurture illness-causing organisms, and in turn, these become the source of tropics-borne endemic diseases. Simultaneously, the same environment is a contributing factor for weaker immune systems in humans. Therefore, the populations of the tropical regions of the world face a double jeopardy. The rain and the constant humidity, especially in the rainforest areas, wash the soil of its nutrients on a regular basis. Much of the chemical and dissolved mineral matter, salt, and electrolytes that is vital to the blood, heart, and proper functioning of the human body are eventually washed away into nearby bodies of water. From the nutrition deprived soil, plants do not absorb enough nutrients to provide good-nutritious diets for the people. Plants also lose what they take from the soil by way of *evapotranspiration*. The tropical soils lacking the necessary nutrients do not support a wide variety of crops. Limited choices such as yams, cassava, plantain, and bananas do not provide a complete and nutritious diet. These crops are rich in one other major chemical element; potassium (K). Potassium is another vital substance for the functioning of the human body and the strengthening of its immune system. Potassium makes up most of the clayey soils. Clay comes from mud stone or shale. Therefore, the said crops growing in clayey soils carry this necessary chemical. It is a fact that there are locations that are so short of potassium in the normal diet where people resort to eating clay to take in this necessary chemical element. Being mainly of African origin, large number of poor people in Haiti ate mud pies (in other words clay) for sustenance during the last socio-political upheaval in that country. However, in Africa for example, where tropical environments abound, the presence of shale is limited to a few certain locations. Most of African soil comes from igneous rocks such as granite and gabbro, not from sedimentary rocks such as shale and limestone. The igneous classes of rocks such as 'red' granite, which are volcanic in origin, give the soil its reddish color. These are rich in nutrients. The reddish color in the tropical soils however, comes from the oxidation of iron content (Fe) due to continuous rainfall in and around the rainforest areas. Again, these are limited in areas of distribution. The soil and the climate must coincide for the right crops and nutrition of the people. Given the tropical rainforest and the deserts in Africa, for example, the population densities are high only in transition zones (e.g., the savanna). In the savanna of southern Africa, where Zimbabwe and Zambia are located, the rich reddish soils have provided means for agriculture that produces a large variety of crops that have more nutritional value. Although the attempt of this paper is not political, it must be said that such soils have been reserved for plantation agriculture brought in by European colonizers for the benefit of those in Europe.

Resultantly, the poor nutrition that the environment offers eventually results in weaker immune systems or does not help to improve already weakened immune systems. These are cumulative effects of centuries of malnutrition that did not improve or perhaps worsened from one generation to the next. Such conditions render people vulnerable to many diseases that the body cannot fight or naturally cure. Many are endemic; including, but not limited to sickle cell anemia, AIDS, ebola, malaria, sleeping sickness, West Nile encephalitis, hepatitis, dysentery, river blindness, yellow fever, and trachoma. Many of the above listed diseases are still rampant, killing millions of people or shortening their life spans. The fact of the matter is, as global population increases, people push farther into wild and uninhabited areas where they encounter diseases that they had not contacted before. Some of these may have been there for thousands of years lying dormant and affecting only the wildlife in the vicinity, occasionally killing some. Those that escaped death and extinction had found a natural antidote that the same environment had provided. So, until recently, theses diseases did not affect many people.

The second important area of study in medical geography is the diffusion of disease. Diseases are passed on from pathogens to hosts (carriers), which are other living organisms such as humans and animals. Any living organism, in order to escape extinction or eradication, attaches itself to a host where it escapes extinction as well as gets a chance to multiply, mutate, and spread. Once a disease is contracted by a host, although it is not the first priority of the disease, it kills that body (human or animal) in a short period of time (e.g., ebola in Zaire), or it clings to the host for longer periods of time for a chance to spread to a larger population. The diffusion is made possible by people's lifestyles, including philosophy of life, customs, and traditions; in other words their culture. Also, movement of populations and their density are very crucial to the diffusion process.

Cultures change from one region of the world to another. Therefore, there are many forms of diffusion. The availability of educational resources affects how people interact with their environment. Thinking nothing of drinking from the river where wildlife has contaminated it is one example of this. The other, in other cultures, may be sharing of needles for injections. In other parts of the world people may not worry about unprotected sex. Some perceptions and myths also contribute to the diffusion of disease. In some parts of Africa having sex with a virgin is believed to cure AIDS, which in fact contributes to the spread of AIDS to the younger population. Many of these have been built into the cultures over many centuries and are very difficult to change. Traditions and customs change imperceptively over a very long period of time. Also, the stress caused by 21st century lifestyles, economics, and politics added to the basic environmental concerns, problems and perceptions, increase the desire for an escape; perhaps a temporary relief. These are manifested in people's behavior, many of whom live life with reckless abandon.

Movement of people spreads disease to others. Historically, people traveled on horseback, camel, or similar modes of travel. In many locations around the world, a lot of people relied on their own feet. These modes of transportation resulted in slow movements of people. An infected person did not reach others or a larger crowd of people in a short while to effectively infect others. For example, in Africa, the presence of the Sahara desert prevented many African diseases, endemic or otherwise, from reaching places like Europe, the Middle East, or South Asia. The desert acted as a major barrier. Many voids as such existed around the world. Incidence of disease was limited and confined to areas close to their sources. Any infected person traveling across the Sahara in such a slow mode died of that disease and perished in the hot Saharan sand before reaching the destination in Europe or elsewhere. Today, however, these great voids and distances have been closed by ease and speed of travel. People carrying any type of disease may reach any part of their own local region or another continent within hours. Therefore, the diffusion processes have been accelerated.

Diffusion is more accelerated today because of a larger global population (i.e., the more the people, the faster the transmittal of a disease to others). Large numbers of people mean that a disease will spread exponentially, especially in clusters of population. Therefore, there are many diseases that become an epidemic in a very short while. Examples of this phenomenon include the avian or "bird" flu, and SAARS. These were first observed in Southeastern China where the world's largest population cluster resides. Diffusion of these to the immediate regions in Asia was almost instantaneous for the reasons stated above. AIDS (HIV) in Africa is another example. The simple influenza that people contract in the winter in North America or Europe acts the same way because of large numbers of people living close to each other. Needless to say, population densities in certain regions of the United States and in Europe exceed that of many regions of the globe, contributing to the speed and ease of diffusion. Once a disease has spread to all parts of the globe it becomes a "pandemic".

As stated in methodology, once diseases spread and start affecting large numbers of people they command attention. Therefore, delivery of healthcare is the next important area of study in medical geography. This is a little more complicated matter to study. Factors that must be highlighted as critical to the effectiveness of healthcare delivery include:

- **Environment**. Important environmental factors include climatic conditions (e.g., wet impassable roads) and land forms because both determine the accessibility for a population to healthcare.
- **Beliefs and values.** People's beliefs and values relating to health care is important.
- **Government's role.** Governments' involvement in providing healthcare or making it available to its populations is critical.
- **Health care infrastructure and availability.** The availability of medical facilities and personnel is the most crucial element in the study of healthcare in any country.

As economic development in a country depends on economic infrastructure, socio-political infrastructure is necessary to reducing physical and mental stress and strain on the public. Countries that have achieved higher levels of socio-economic and political stability and infrastructure have lesser levels of stress and strain on their populations. These, along with a number of other criteria, compose the formula that shapes life expectancy figures. One way to examine the healthcare profiles of countries is by studying the composite of selected healthcare facilities and personnel in addition to the environment, peoples' culture and lifestyles, and governmental involvement. This may be accomplished by calculating the amount of hospitals, hospital beds, doctors, nurses, dentists, and pharmacists per population. The outcome will show the level of healthcare in a country compared to others. As stated earlier, life expectancy is a reflection of the composite of these elements of healthcare in a country along with other factors, such as the environment and culture of the population. If we take the simplest approach, life expectancy may become a measuring stick for the level and quality of healthcare in a given country (i.e., better healthcare equals higher life expectancy. See Table 3-

1). An examination of Table 3-1, displaying the combined life expectancy numbers for the Mediterranean Basin countries, shows a positive correlation between the "rich" countries (i.e., countries that have high GNPs) and higher life expectancies. Here, high quality of health care is available to the population and also the socio-economic and political environments and infrastructures are relatively stable. In contrast, socio-economic and political turmoil and distress are often major reasons for the lower life expectancies in others. Lower life expectancies also are generally common in those countries that have lower GNPs. Lower GNPs reflect the individual government's financial inability to assist the ordinary citizen with affordable medicines, drugs, or healthcare. These differences, coupled with the differences in cultural backgrounds, the lifestyles of the people, and their given physical environments are displayed numerically in Table 3-1 and spatially in Figure 3-1.

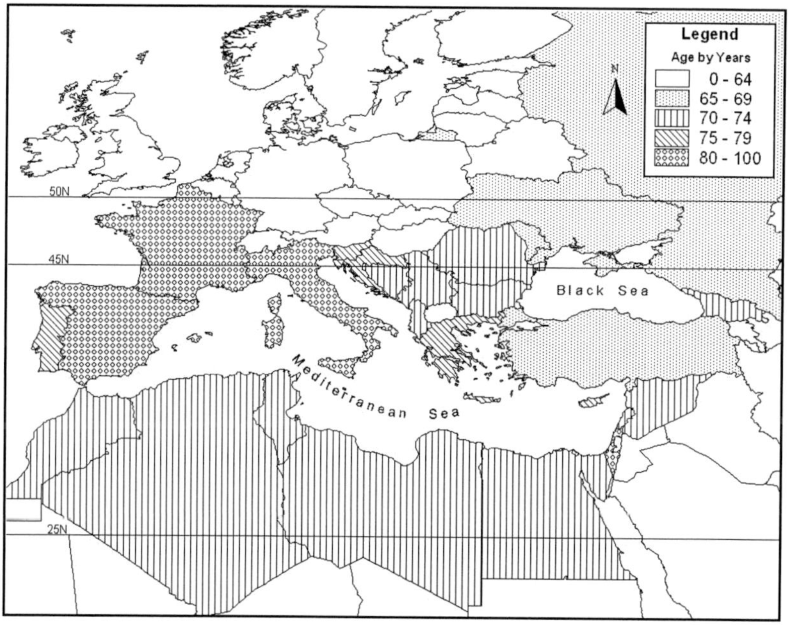

Fig. 3-1: Life Expectancy of Mediterranean Basin Countries

The other correlation is the environment. Latitudinal location, such as the mid- and high-latitudes, make it such that the "cooler" climatic regions display higher life expectancies and "warmer", tropical, and near-tropical regions display lower life expectancies; exceptions notwithstanding (Table 3-2 and 3-3).

Table 3-1: Average Life Expectancies of Mediterranean Basin Countries

COUNTRY	LIFE EXPECTANCY (IN YEARS)
Albania	74
Algeria	71
Bosnia	74
Bulgaria	73
Croatia	75
Cyprus	79
Egypt	70
France	80
Georgia	71
Greece	79
Israel	80
Italy	80
Lebanon	72
Libya	74
Malta	79
Moldova	69
Monaco	80
Morocco	70
Portugal	78
Romania	72
Russia	66
Serbia	74
Slovenia	77
Spain	80
Syria	73
Tunisia	73
Turkey	69
Ukraine	66

* http://www.who.int/whosis/whostat2006.pdf

Conclusion

In conclusion, all Mediterranean Basin countries are not at the same life expectancy level. The range is from an average of 66 years for Russia and Ukraine to 80 for France, Spain, and Italy. The pattern displayed on the map is a telling sign of the above discussed reasons, including, in combination, the economic levels of development reflected in the specific numbers of medical personnel and facilities per person, availability of medicines and drugs, of necessary diagnostic and treatment equipment, of social programs (e.g., free medical care), and socio-economic and political development and infrastructure. The modernized, well-educated, and informed societies, where technology is available for personal and national health care, display higher life expectancy figures.

Table 3-2: Average Life Expectancies of European Countries*

COUNTRY	LIFE EXPECTANCY (IN YEARS)
Austria	79
Belgium	79
Czech Republic	76
Denmark	77
Finland	79
Germany	79
Hungary	73
Iceland	81
Ireland	78
Liechtenstein	80
Luxembourg	78
Netherlands	79
Norway	80
Sweden	80
Switzerland	81
United Kingdom	79

*http://www.who.int/whosis/whostat2006.pdf

Table 3-3: Average Life Expectancies of Middle Eastern Countries*

COUNTRY	LIFE EXPECTANCY
Bahrain	75
Iran	71
Iraq	59
Jordan	72
Kuwait	77
Oman	75
Qatar	74
Saudi Arabia	72
United Arab Emirates	79
Yemen	61

* http://www.who.int/whosis/whostat2006.pdf

Most of the items that are reflective of higher life expectancies are the responsibility of governments. It is expected that health care will be provided by governments for their citizens. However, many obstacles exist that contribute to the lower level of healthcare in a country, some of which have been delineated above, as for the "rich" and well-endowed countries.

References

Cowley, Geoffrey. 2006. "Medicine without Doctors" Global Studies: Africa. Dubuque, Iowa: McGraw Hill/Dushkin, pp. 273:275.

De Blij, Harm, J., and Peter O. Muller. 1994, 1997. "Medical Geography" in Geography: Realms, Regions and Concepts. New York: Wiley, pp. 412:413 (1994) and pp. 328:329.

De Blij, Harm, J., and Peter O. Muller. 2004. "Aids in Sub-Saharan Africa." Geography: Realms, Regions and Concepts. New York: Wiley, p.272.

Meade, Melinda, and John C. Florin, and Wilbert M. Gesler. 1998. Medical Geography. New York: The Guilford Press.

Preston, Richard. October 1992. "Crisis in the Hot Zone" The New Yorker.

Shannon Gary W., Gerald F. Pyle, Rashid L. Bashshur. 1991. Geography of Aids: Origins and Course of an Epidemic. New York: The Guilford Press.

Shannon, Gary W., Dover, GEA. 1974. Health Care Delivery: Spatial Perspective New York: McGraw-Hill.

Schneider, Dona. Vital Statistics: Rates, Ratios, and Proportions March 29, 1994. "Teaching Medical Geography" (Workshop). Annual Meeting of the Association of American Geographers, San Francisco, CA.

United Nations-World Health Organization. 2006. "World Health Statistics 2006". http://www.who.int/whosis/whostat2006.pdf.

CHAPTER FOUR

THE ATTITUDES OF TOURISM ESTABLISHMENTS TOWARD THE DEVELOPMENT OF TOURISM AND TOWARD TOURISM'S IMPACT ON THE ENVIRONMENT IN ÖREN (MILAS), TÜRKIYE

SULTAN BAYSAN

Introduction

In the past several years, there has been an increased awareness of the role of tourism and the impact of tourism on the economy and on employment, along with its social and environmental implications. This awareness which is related to global consciousness about sustainability issues, has led to more research on tourism development (e.g. Pearce 1995, Cooper & Wanhill, 1997), its economic, socio-cultural and environmental impacts (e.g. Mathieson & Wall, 1982; Butler 2000, Diagne, 2004), and sustainable tourism practices and marketing strategies (e.g. Hall & Lew, 1998; Butler, 1999; Okumuş & Topaloğlu, 2004; Hoballah, 2006).

Neto (2003: 216) argues that "it is now widely recognized not only that uncontrolled tourism expansion is likely to lead to environmental degradation, but also that environmental degradation, in turn, poses a serious threat to tourism activities." Degradation caused by tourism activities has been intensively reported in case studies (e.g. Garcia & Servera 2003), some of which were based on researching attitudes towards the environmental impact of tourism. Attitudes towards the environmental impact of tourism vary by different interest or stakeholder groups. For example, Akış (2001), Baysan (Külçür) (2001), Baysan (2002, 2005), Baysan & Teberler (2004; 2006 & n.d.) studied the attitudes of various interest groups (namely tourists, accommodation establishments and local people) and found that there are similar perceptions between these different interest groups as well as different attitudes towards tourism development and its environmental impacts. For example, almost all groups believe that tourism businesses should be developed particularly for its economic income and positive environmental impacts but, their perception may differ on who is responsible for creating these impacts and who will pay for the protection of the environment. Thus, it is important to know about attitudes of different interest groups and to plan the direction of tourism development and to prevent detrimental effects of the practices of distinct perceptions.

As it is commonly known that "prevention is better than cure," the protection of the environment is always easier with the support of governmental bodies and tourism establishments. Knowing local authorities and tourism establishments' perspective towards tourism development and its impacts is essential because, while the former is responsible for decision-making, the latter is at the operational level. Furthermore, these bodies have the power to control and to prevent unacceptable practices of tourism activities. If they are aware of various impacts of tourism development and concerned about the perceptions and expectations of other stakeholders, it is possible to minimize environmental impacts and to maximize socio-economic benefits. This will allow local administrators to facilitate the management of resources in such a way that the economic, social and aesthetic needs of the environment and businesses can be fulfilled while maintaining cultural integrity and essential ecological processes (UN 2001 in Neto, 2003).

The environment in this paper refers to both natural and built environments. It discusses the environmental impacts of tourism as both negative and positive. It should however be noted that this study, along with many others, has concentrated more on the negative environmental effects of tourism than on positive ones because, it can be argued that "there is a wider range of potential negative than potential positive environmental impacts associated with tourism" (Bramwell *et al.*, 1996: 38). Perceptions about environmental issues related to the tourism business arc particularly within the scope of this paper. Attitudes alone may not provide enough information about people's thoughts and behavior. However, they still "involve what people think about, feel, and how they would like to behave toward an attitude object" (Triandis, 1972: 14 in Gibson; Lynch & Morrison, 2005: 88). Therefore, this study focuses on the environmental impacts of tourism by examining the perceptions rather than investigating the actual consequences of tourism or researching the gap between the practices and implementation of environmental law. Thus, it is limited to the "perceived impacts". The study is still important in order to find out if there are complementarities or conflicting interests between parties involved. As part of a complex study, this paper investigates what the tourism businesses consider about tourism development and its environmental impacts. In this respect, it would be possible to compare differences or similarities with the attitudes of tourists and local authorities of Ören, Milas (Muğla) (the results of an earlier study). This study also provides a reference to bodies that might be responsible for paying for protection of the environment.

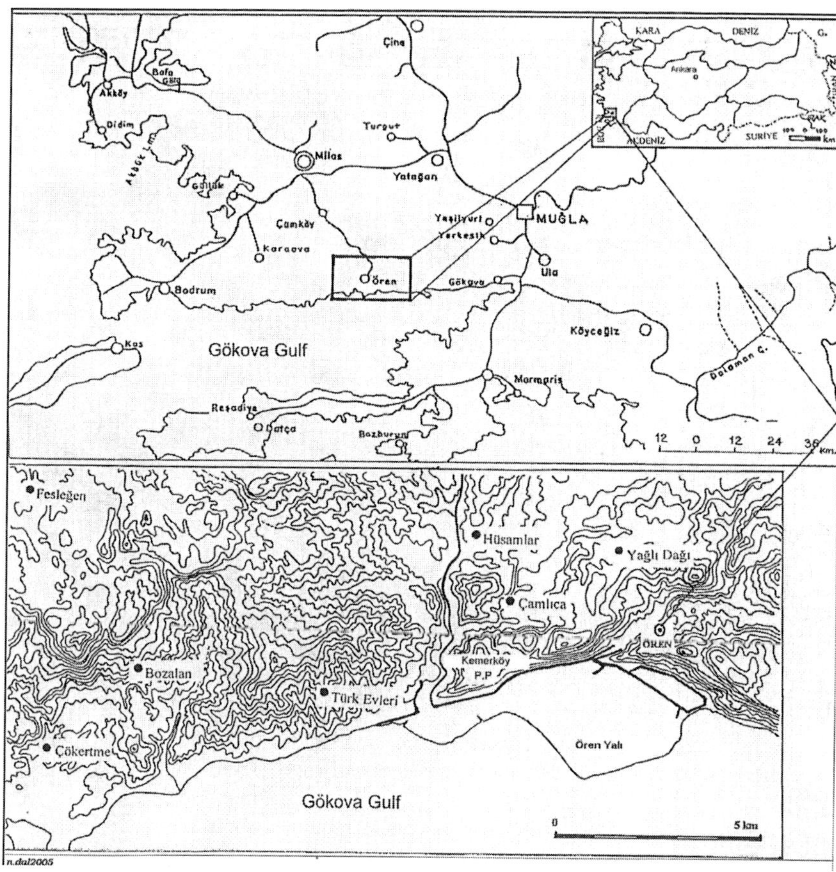

Fig. 4-1: Location map of the study area

Ören (Milas), is a municipality within the boundaries of Muğla Province in Türkiye situated on the North of Gökova Gulf and is subject to this study (Fig. 4-1). Its central town population is 4,500 (9,000 with its villages). Ören's old name is Gereme or Keramos, dating back to the Archaic period. Gökova Gulf is one of the specially protected areas in the coastal zone of Turkey (PAP/RAC 2005). The north of the town is surrounded by around 600 meter hills while it is situated on the alluvial terrain deposited by the Kemerdere stream. Local farmers cultivate citrus fruit, sesame, cotton, tobacco, cereals, olive and various fruits on this terrain. Beside, agricultural and livestock activities, traditional carpet weaving brings extra income to this region.

Among the attractions of Ören, there is not only an 8 kilometre long pebble beach in the centre, but also Keramos ancient city (with its city walls, two temples, aqueducts, necropolis and remnants of a bath), canyons, caves, nature sport opportunities such as paragliding, trekking, mild Mediterranean climate and its pleasant mountain-plain views.

The Ören Tourism Information, Education and Health Foundation (Ören Turizmi Tanıtma Eğitim ve Sağlık Vakfı, 2002: 49) states that the place reaches 30,000 people in summer, particularly with domestic visitors. The Ören Gendarme Commandership reports that 9,100 domestic and 320 foreign tourists visited Ören in 2002 (Ören Jandarma Komutanlığı, 21.08.2003). There is only one three-star hotel in Ören and none with higher ratings, and holiday village type accommodation establishments. There are *pansiyons* (pension or B&B), apart hotels and small motels which are mostly family run establishments. In total, the number of municipality-licensed beds in Ören is around 400 (one fourth of which is owned by the mentioned the owners of the three-star hotel). In this respect, Ören is at the early stages of Butler's (1980) tourism development life cycle. According to the Mayor of Ören Municipality (interview, 21.08.2003), Ören is seeking tourism investments from outside the region but has been unable to attract sufficient investments. When considering the life-cycle of a tourist destination concept (i.e. "evolution from its discovery, to development and eventual decline because of over-exploitation and subsequent deterioration of its key attractions"), it is seen that the destination in the later stages reaches the overdevelopment point (such as Kuşadası) that the environmental and socio-cultural damage are inevitable. Furthermore, tourism revenues loss is due to collapse in not only tourist numbers but also in terms of the quality of tourists. This reality needs to be kept in mind, for in the case of Ören and other newly developing resorts, governmental and local administrational bodies should carefully plan the way in order to ensure sustainable tourism practices.

Methodology

The field study was based on a questionnaire survey. The questionnaire was extracted from several sources, including earlier research applied in other tourist areas, but mainly from Baysan (2002, 2005). The very first aim for the design of the questionnaire was that the questionnaire should engage respondents' interests, encourage their co-operation and elicit answers that were as reliable as possible while trying to extract the maximum amount of information. Every effort was made to avoid response set bias. It should be pointed out that, as Veal (1992) states, the validity of the results is dependent upon the respondents' honesty and what they perceive to be the outcome of their responses.

The questionnaire format included several types of questions about attitudes, activities and the perceptions of respondents and consisted of 36 items, including 10 close-ended, 6 five point Likert-scale and 20 open-ended questions. Some examples of close-ended questions are related to tourism type and the type of establishment desired in Ören, the direction of tourism's impacts, and the bodies responsible to protect the environment. Some other examples for Likert-scale questions are whether tourism should be developed (Very much (5) – None (1)), tourists use more water than local residents (Completely agree (5) – Completely disagree (1)). Some open-ended questions included were meant to address what negative and positive factors in Ören affect the development of tourism, what the municipality should do in order to attract tourists, and, in the future to prevent detrimental effects of tourism on the environment, what are the perceived effects of the thermal power plant of Kemerköy and the planned yacht and hauling area.

The sample population consisted of the 64 respondents who worked in tourism establishments as managers, operators or owners. More than Fifty-four percent (54.8%) of these worked in the catering sector (i.e. restaurant, café, market type of establishments), 11.3% in bed & breakfasts (pansiyon), 11.3% in motels, and 8.1% in the three star hotel. The rest were from the other types of establishments, namely 3 from yachts and one from a small yacht and hauling area. Since there are around 20 accommodation establishments and over 50 other types of tourism businesses in Ören, the sample population represents the perspectives of all tourism establishments. It gives ideas with the help of descriptive statistical tools about what the tourism establishments' attitude towards tourism development and its environmental impacts. While doing this, this paper also compares the results of the tourism business with the findings of local authorities and tourist surveys in Ören. Thus, it investigates whether tourism businesses, tourists and local authority's attitudes differ.

Table 4-1: Perceptions of Positive Factors Affecting Tourism Development in Ören

	Percent
Calm & unspoilt environment	47.2
Clean sea & beaches	44.4
Little construction with few floors	5.6
Historical assets	2.8
Total (n)	64

Tourism establishments in Ören are mainly in the Yalı district (75.0%), 12.5% of each is at the back of Yalı and Çökertme district towards Bodrum. Half of the respondents (50.0%) are owners, almost one third (29.7%) are working at the managerial positions and the rest (20.3%) are the workers at these tourism establishments. The origins of the respondents also show that the tourism business in Ören is at the beginning period of Butler's (1980) life cycle. That is over half (50.9%) of the respondents is from the region (Ören, Milas and Muğla). Thus, this situation represents "insiders" involvement in the tourism business. The rest are from other places in the country (only 2.0% originated outside the country but are still Turkish citizens).

The time period which the establishments started to operate is mostly more than one year (90.3%). Therefore, most respondents are likely to know the past and present of tourism's development in the area. The establishments are small in size, i.e., 84.1% work in a 1 to 5 people working establishment.

Findings and Discussion

Tourism product: Table 4-1 shows the perceptions of the respondents about the positive factors affecting tourism development in Ören. Most of the respondents see that Ören's unspoiled environment, and its clean sea and beaches, are the primary factors for the development of tourism.

Table 4-2 shows perceptions of negative factors affecting tourism development in Ören. While the local authority (38.5%) and tourists (20.0%) see the thermal power plant negatively influencing tourism development, more respondents from the tourism establishments have foreseen unfamiliarity of Ören and lack of publicity (25.5%). Then, difficulty in transportation was perceived by more than 20 percent of the respondents. The latter was perceived by the 15.4% of the local authority and 25.9% of tourist respondents (see Baysan 2005, Baysan & Teberler n.d.).

Table 4-2: Perceptions of Negative Factors Affecting Tourism Development in Ören

	Percent
Unfamiliarity, lack of publicity	25.5
Difficult access (transportation)	23.5
Lack of investment & tourism establishments	9.8
Inadequacy in infrastructure	9.8
Inadequate quay	7.8
Uncontrolled music noise	5.9
Unskilled labour & lack of speaking foreign languages	5.9
Thermal power plant	3.9
Collecting waste late (weekly) in Çökertme	3.9
Over construction	2.0
Having not a sewer system	2.0
Total (n)	64

Despite small percentages, the other perceptions give additional information about negative factors influencing tourism's development. These are lack of investments flowing to the area, and the small size and quantity of tourism establishments, inadequacy of infrastructure, an inadequate quay and its facilities, uncontrolled music noise, unskilled labor, few people speaking foreign languages, over-construction, having no sewer system and inadequate waste collection intervals for Çökertme (Table 4-2). Of this uncontrolled music noise comes from entertainment units and they are usually side by side with accommodation establishments. Thus, loud music makes both tourists and tourism operators uncomfortable after midnight. Hence, conflict between those parties arises. The second perception is about the quay being inadequate. There is some discussion about building a new modern quay. This will therefore allow more yachts to anchor and bring tourists to Ören. A complaint about having no sewer system is the third perception. Instead, every establishment has their own and must treat its sewage directly. Then, they may leave their treated water to the sea. Some establishments (few in number) want a central sewer system to be built, which might be due to the fact that sewage treatment is expensive, forcing their operational costs to increase. Fourth, some respondents from Çökertme reported that collection of waste was once a week. This situation causes bad smell and flies in summer months.

Table 4-3: Reasons Why Tourism is Insufficient in Ören

	Percent
Lack of publicity	41.7
Few tourism establishments	19.4
Lack of investment, difficulty in transportation	13.9
Irregular cleaning/maintenance services	11.1
Short tourism season	5.6
Lack of socio-cultural activities	2.8
Tourism being at the beginning stage	2.8
Lack of marina, boats cannot anchor	2.8
Total (n)	36

Seventy-nine (79.0%) of the respondents view tourism as insufficient in Ören (60.9 insufficient, 28.1% definitely insufficient). This percentage is higher than the perceptions of both tourists and local authority (see Baysan 2005, Baysan & Teberler n.d.). Only 6.3% reported that tourism was insufficient. Among insufficient factors indicated, there are lack of publicity (41.7%), few number of establishments (19.4%), being unable to attract tourism investments and difficulty in transportation to Ören (13.9 %). See Table 4-3 for some other factors perceived as hindering the development of tourism in Ören.

Most respondents think that tourism should be developed (23.4% much, 75.0% very much). Only 1.6% was unsure or did not know whether tourism should to be promoted in Ören. 40.7% preferred to invest on 4 and 5 star hotels. 18.6% each holiday village and 2 and 3 star hotels. 22.0% reported other kinds of establishments (such as camping area, *pansiyon* and time-share accommodation) to be developed in Ören. Considering tourism type which are to be developed in the municipality, respondents preferred a combination of types. For example, 28.1% preferred the sun, sea, nature and yacht tourism, 18.8% sea, nature sports, culture and yacht, 12.5% the sun, sea and nature, 10.9% the sea, while the rest reported other combinations of tourism types such as history, congress, nature and sea sports. 71.1% stated the reason why they prefer these types of tourism as the appropriateness of nature, 21.1% as attracting interest, 5.3% as competing seasonality, 2.6% as lacking these types of tourism.

Table 4-4 shows opinions about how to promote tourism in Ören. Almost half of the respondents think that a web page, leaflets and advertisement are the best choices. Almost one fifth suggests sport and social activities or festivals to promote Ören. Some asked for opening up a new seashore road connecting Ören to Bodrum. They think that Ören is situated behind Bodrum and unless a new seashore road is built, people knowing about the area will be few (see Table 4-4 for the rest of the suggestions).

Table 4-4: Suggestions for Promoting Ören

	Percent
Web page, leaflet, advertisement	48.4
Sport and social activities	17.7
Seashore road connecting Ören to Bodrum	9.7
Quality establishments & investment	8.1
Tours from other resorts	8.1
Yacht harbour services	8.1
Total (n)	62

Respondents also wanted to have yacht harbor services (19.6%), environmental maintenance and inspection (16.1%), sport, camping, festival kinds of organizations (14.3%), controlled development (10.8%) and environmental protection (5.4%) from the local government. Some 1.8% complained about entertainment establishments and suggested these be moved somewhere else and separated from accommodation establishments.

Perceptions about Environmental Impacts

Tourism establishments were also asked if tourism affected the environment and in what direction. 16.1% stated "no effect" of tourism on the environment. Half (50.0%) thought only positive, one fourth (25.0%) stated both positive and negative effects. 8.9% complained about the "only negative" effects. While tourists and local government seemed to perceive more negative effects of the industry, tourism establishments are prone to be more positive on this issue. This perception is "normal" due to the fact that they are gaining the direct benefit from tourism activities. Thus, they want their businesses to be carried on. For example, 43.6% of tourists complained about the negative results of tourism on the environment compare to 8.9% of the respondents from tourism establishments. Among the positive environmental effects of tourism, most respondents (67.7%) stated protection of environmental values. Some 10% (9.7%) stated its effect on increasing environmental consciousness. As the other studies such as Baysan 2005, although economic effects of the industry in Ören was unintended to search, this study once more also revealed tourism as an income generator (6.5%).

As reported earlier, only a small percentage stated the negative effects of the industry. Among these 33.3% stated waste and environmental pollution, 26.7% noise, 20.0% too many buildings, 13.3% environmental degradation and 6.7% forest fires. Other interest groups showed similar attitude, i.e. 43.6% of tourists stated environmental pollution, 23.1% of tourists and 50.0% of local administration indicated too many buildings.

As argued before (Fox-Penner 1997, Creech and Brown 2000, Baysan 2005, Baysan & Teberler n.d.) thermal power plants and yacht and hauling areas have adverse effects on the environment. Ören, having the Kemerköy power plant and the planned yacht and hauling area, some questions were also asked to find out the perceptions of the interest groups. As reported earlier, local administration (67.0%) and tourists (60.0%) seemed to complain about the thermal power plant. 14.3% of the establishments believed it has no effect. The majority of the tourism establishments (57.1%) felt the "negative" and 17.9% "both positive and negative" effects. 10.7% of those who stated "positive" thought that is due to employment opportunities, publicizing Ören and increased services as in better transport. Among the negative effects of the power plant 67.6% saw the visual and sea pollution, 32.4% psychological effect on tourist or investors who could escape from coming or investing in Ören.

Baysan (2005), Baysan and Teberler (2007) reported that almost half of the tourists (48.0%) specified that the yacht and hauling area will have impact on tourism, both positive and negative. 24.0% was only negative and 23.0% positive. Some negative effects were found that tourist respondents were worried about "Ören will develop as Bodrum (the sea and the environment will be polluted)" (65.0%) and, "its quietness and naturality would be spoiled" (19.0%). More than two-third of the local administrators (68.0%) specified that the area would have only a positive impact, while 26.0% believed that both positive and negative effects would occur. Local administrators who perceive both negative and positive effects think that "the sea and environment would be polluted" (50.0%), "overcrowding would occur and public order would be in danger" (25.0%) and "environmental structure would be spoiled and the amount of waste would increase" (50.0%). This study revealed that 67.7% of tourism establishments think the same as the local authority i.e., the area, will only have positive impact. 21.0% believe both positive and negative effects. 93.3% of respondents who stated negative effects of the yacht and hauling area thought that there will be too much construction and environmental pollution, while 6.7% indicated Ören's calmness will be spoiled.

Among the positive effects of the planned yacht and hauling area tourists, local administration and tourism establishments specifically underlined an "increase in income and employment" (44.0%, 72.0% and 70.0% respectively). Besides, the "quality establishments and transport infrastructure and increase in visitors and investors" (23.0% by tourists and 22.0% by local administration and 30.0% respectively) were also indicated (Baysan 2005, Baysan & Teberler n.d.).

When the question of if the environment would be in danger with the effects of planned yacht and hauling area was asked, percentages were close, i.e. while 36.9% of the establishments indicated that the environment would be unaffected, 38.6% believed it is to be affected. 24.6% of which are unsure or don't know.

Environmental Responsibility

This part of the study deals with the issue of environmental responsibility. 31.0% of the establishments indicated that the local administration takes "little" or "no" preventive measures to protect the environment. 22.4% were unsure or don't know, while 46.5% believed that "much" and "very much" measures are taken. Table 4-5 shows what the local authority should do in order to protect the environment in the future. More than half the respondents ask for encouragement to protect the environment. That shows that if some incentives such as low taxes are given to the establishments, they are likely to take more preventative measures. Some also ask for a central sewer and water purification systems and also demand controls over the existing units. The last result takes us if all the establishments do not make their systems run therefore the place where the untreated water goes is a dilemma (see Table 4-5 for the other results).

The literature suggests that tourists use more water than local people (Pattullo 1996; Baysan 2005; Baysan & Teberler n.d.). Slightly less than a half of the tourists in Ören (43.0%) indicated that tourists' use of water was more than locals (Baysan, 2005). This question was important to find out about the level of consciousness of the interested parties towards scarce resource of water and its use. It was found that there was inconsistency among the opinions of the local authority i.e. 12.0% agreed, 24.0% slightly agreed, 24.0% slightly disagreed and 12.0% disagreed on and the rest was unsure/didn't know about (Baysan & Teberler n.d.). It is interesting to see while tourists themselves and the local administration perceived "tourist use of water is more than the local residents' use", almost half (48.4%) of the tourism establishments thought the opposite. 23.4% was "unsure/don't know" and 28.1% indicated "agree" responses.

Table 4-5: What the Local Authority Should do in Order to Protect the Environment in the Future

	Percent
Preventative measures and encouragement	54.0
Sewer, water purification, and fire systems	20.0
Permits to tourist establishments instead of second homes	10.0
Attention to over construction (planned development)	8.0
Control over purification systems of tourism establishments	4.0
Protection of all interest groups' rights	2.0
Control over insects	2.0
Total (n)	50

Three questions were asked in order to find out the perceptions of environmental responsibility. Most of the local administrators thought themselves responsible for protecting the environment and maintaining the balance (83.0%) (Baysan & Teberler n.d.). This result was parallel with the results of tourist survey and other studies, i.e. the majority of tourists believed that the municipality was responsible for protecting the environment (Baysan 2005, Baysan 2002). This study revealed also the local authority's responsibility to protect and maintain the environment (58.1%). The other parties were seen as the local people (12.9%), tourism establishments (11.3%), all the parties (11.3%) and the government (6.5%).

Table 4-6 summarizes the second question on the perceptions of who should pay for maintaining the environment. It is seen that the local administration and tourism establishments firstly perceived that the municipality itself and the government should pay for maintaining the environment. However, tourists indicated tourism establishments and the government as primarily responsible. The municipality was the third party to pay.

Table 4-6: Perception About Who Should Pay for Maintaining the Environment (%)

	*Tourists** *(n=104)*	*Local administration** *(n=19)*	**Tourism establishments (n=59)**
The Municipality	19.5	33.3	30.5
The Government	23.4	26.7	30.5
Tourism establishments	25.0	20.0	22.0
Local people	6.4	13.3	10.2
All	8.7	6.7	1.7
The Tourists	7.0	-	5.1

* (Baysan, 2005; Baysan & Teberler n.d.).

The third question about environmental responsibility was if the establishment pays extra money for environmental protection and maintenance. 46.7% said "yes", 6.5% "definitely yes". Unsure or don't know responses were 25.8%. Only 14.5% and 6.5% responded negatively ("no" and "definitely no" respectively). Having positive attitude on this issue may mean the establishments are aware that the environment is important for their businesses therefore, they should protect and give some extra money for environmental protection. This attitude may help to ease environmental programs to be practiced which are to be enforced by the local authorities or the businesses can take initiatives for the environmental maintenance.

Tourism establishments were asked if they had preventive actions in order to protect the environment and also save money from running costs. 44.7% indicated "no", 31.9% "yes" and 23.4% "unsure or don't know". Another question towards those who indicated the "yes" answer was about what kind of actions they take. 23.1% each indicated sewage purification and environmental maintenance of the establishment and preventing yachts from their bilge drainage to the sea. Among the other actions, putting waste in garbage disposal units, preventing noise, protecting the mature trees, separation of waste, informing tourists about the environment were also reported (7.7% each).

The respondents were also asked if any plan existed to protect the environment and to save money from running costs in the future. It was found that 35.6% were "unsure or don't know", 33.3% had no plans and 31.1% reported having plans for the future. A small percentage of whom reported having future plans, indicated separation of waste (if the municipality takes the waste) (25.0%), planting trees (25.0%), renovating their establishment (increase bed numbers and doing some maintenance) (16.7%), supporting the local authority for building new roads (16.7%), investment in solar energy and informing tourists) (8.3% each).

The establishments indicated that they had some problems when they are running their businesses. These were summarized in Table 4-7. For example, some problems are seen as Ören being less developed and having a small number of facilities, and lack of qualified labor.

Table 4-8 presents other thoughts and suggestions of the establishments. Among these there were beneficent suggestions particularly for the local authority to count. For example, it was reported that harbour services should be provided, investments to be encouraged, the nature be protected and its quietness not to be spoiled, ancient and cultural places to be publicised, handicrafts and carpet weaving to be encouraged, treatment units should be run in full capacity and the local municipality should carefully inspect these processes.

Table 4-7: Problems Faced When the Establishments Run Their Business

	Percent
Ören being less developed with few opportunities	23.9
Lack of qualified labour	10.9
Short tourism season	8.7
Few number of tourists	8.7
Competition with developed resorts e.g. Bodrum, Marmaris	6.5
Bureaucracy	6.5
No problem	6.5
No standardization in price	4.3
Noise	4.3
Ören being preferred by low income tourists	4.3
Reservation cancellation	4.3
Water and power failure, inadequacy in PTT services	4.3
Little publicising, thus, Ören bcing unknown	2.2
High prices in rent	2.2
Having not holidays	2.2
Total (n)	46

Conclusion

This study, along with others, revealed much complementarities and little conflicts between the perceptions of local administration, tourists and tourism establishments. For example, the thoughts of the local administrators, tourists and tourism establishments seemed to be parallel to each other that the sea and nature add to the development of tourism in Ören. Thus, all parties agree about the main itineraries for the tourism business. The majority of local authorities, tourists and tourism businesses believe that tourism should be developed. Besides the sun, sea and sand (3S) type of tourism; all parties suggest a combination of tourism types such as nature, culture and yacht due to the suitable environment of Ören and tourists' interest. The main type of accommodation establishments preferred by these parties is seen as the 4-5 star hotels and holiday villages that may be due to nonexistence of these types.

Table 4-8: General Suggestions by Respondents

	Percent
Harbour services should be provided, investments to be encouraged	11.5
Nature be protected and its quietness not to be spoiled	9.6
Publicise ancient and cultural places	9.6
Transport and roads should be developed	7.7
Increase the number of culture, sport, entertainment places	7.7
Prevent second home development	5.8
Purification units should be run in full capacity and the local municipality should carefully inspect	5.8
Ören should be promoted to attract tourists	5.8
Handicrafts and carpet weaving to be encouraged	3.8
Exhibiting ancient historical artefacts and protect them	3.8
Prevent unfair competition and illicit apart type of accommodation	3.8
Closing down the thermal power plant	3.8
Honest attitude towards tourists by tradesmen	3.8
Location of second homes is wrong (they should be towards the mountain and the hotels should be in front)	3.8
Opening some courses in foreign languages, handicrafts making	3.8
Opening banks	3.8
Water for Çökertme district and collection of waste twice weekly	3.8
Tours from other places as daily excursions	1.9
Total (n)	52

Moreover, all groups perceive that there is a lack of publicity of Ören, and tourists are unfamiliar with this place. Therefore, the majority firstly suggests an advertisement program to publicize Ören. Furthermore, all parties mainly think that tourism has an impact on the environment and the municipality is responsible for protecting environmental resources.

Conflicting areas seem to be small in number; only some little differences appear to occur between parties. A conflicting issue occurs about the perception of what should be developed in Ören. While local administration primarily sees the tourism establishments (particularly accommodation, marina and yacht area), tourists prefer infrastructural improvement (particularly beach and road maintenance) in the first place. Tourism establishments firstly ask for harbor services, adequate environmental maintenance and inspection and organizations of sport and cultural events. Although there is a need for attracting investments to build up new tourism establishments and harbor services, the local authority should carefully take into consideration tourists' interests and needs, because excessive development will lead dissatisfaction of tourists as will bring deterioration and commercialization of environmental values.

It is important to have a perspective to provide the tourism industry with a political platform and improving cooperation and coordination between these interest groups. This is why this paper dealt with the perceptions of the all parties. If the local authority's attitude towards tourism development and environmental issues is known, then it will be possible to compare the attitudes of other interest groups, namely tourists, tourist establishments and local people. Thus, complementarities or conflicting interests become known. If conflicts occur, then these parties can come together to solve the problems and find a way for optimization of tourism development, environmental protection and locals' socio-cultural and economic amenities. Therefore, multiple stakeholder participation and a more strategic and long-term orientated tourism planning can be developed. In this planning approach community can be included therefore, a "bottom up" planning (Ruhanen, 2004) can be accepted in order to develop tourism in a more sustainable concept.

A concerted effort by the municipality, government and private enterprises is needed to reach a viable program of sustainable development and to overcome difficulties. Thus, a private–public partnership is required to protect the environment. Examples of yacht and hauling areas in the world should be investigated, their positive and negative impacts carefully to be considered. As discussed earlier in Baysan and Teberler (2007) potential negative impacts should be carefully evaluated bearing in mind that prevention is always cheaper than the maintenance. The abuse of the coastal environment by uncontrolled development is dangerous. The effects of the thermal power plant also require to be regularly checked out. The regional social, economic and physical impacts of tourism should be reassessed with the aim of preserving the integrity of the local economy, thus ensuring the future generations' needs both of human settlement and of sites for tourist and locals' recreation. Moreover, the area should be utilized as an income generator for local people and the problem of lack of tourism establishments should be somehow dealt with. Difficulties in transportation should be eased and Keramos ancient city to be excavated by finding sponsors for financial support.

This paper has mainly investigated tourism establishments' attitudes towards the development of tourism and current and potential environmental impacts created by tourism activities in Ören. It also discussed the attitudes of two other interest groups, namely local authorities and tourists. It compared the perceptions of local authorities with the perceptions of tourists. It however, remains limited with the perceptions of three parties. The attitudes of local people in Ören are still under investigation. However, this study still provides a considerable amount of information about the environment and sustainability issues. Further research required to be carried out include the attitudes of local authorities, local people, tourism establishments and tourists of other tourism destination areas in order to examine the results of other places and compare the attitudes of different interest groups for rather generalized implications.

References

Akış, S. 2001. Sürdürülebilir Turizm: Bir Alan Araştırmasının Sonuçları. Anatolia, Volume 12, pp. 17-25.

Baysan (Külçür), S. 2001. Perceptions of environmental impacts of tourism: a comparative study of the attitudes of German, Russian and Turkish tourists in Kemer, Antalya. Tourism Geographies. Volume 3, Issue 2, pp. 218-235.

Baysan, S. 2002. Didim Kıyı Alanı Örneğinde Turizmin Çevresel Etkileri Konusunda Yerli Halkın Tutumları: Bir Ege Tatil Merkezi Örneği. In: E. Özhan and N. Alpaslan (eds.) Türkiye Kıyıları 02: Türkiye'nin Kıyı ve Deniz Alanları IV. Ulusal Konferansı Bildiriler Kitabı. Kıyı Alanları Yönetimi Türk Milli Komitesi. Ankara. Vol. 1, pp. 599-608.

—. 2004. Ekolojik Etkiler: Turistler, Konaklama Tesisleri ve Yerel Halkın Tutumları. In: A. Yüksel and M. Hançer (eds.) Turizm, İlkeler ve Yönetim, Turhan Kitabevi, Ankara, pp. 49-72.

—. 2005. Turizmin Geliştirilmesi ve Çevreye Etkileri Konusunda Yerli Turistlerin Tutumları: Ören, Milas Örneği. SOİD Seyahat ve Otel İşletmeciliği Dergisi. Volume 2, Issue 4, pp. 22-30.

Baysan, S. and M. Teberler. n.d. Attitudes of Local Authority towards Environmental Impacts of Tourism in Ören (Milas). to be published.

Baysan, S. and M. Teberler 2004. Environmental Impacts of Tourism with Respect to Attitudes of Turkish Travel Agencies. In: İ. Atalay, M. Ielenicz, D. Balteanu and Y. Arı (eds.) Proceedings of the Third Turkish-Romanian Geographical Academic Seminar, Natural Environment and Civilization. Zeytinli, Balıkesir, 15-24 September: İnkılap, İstanbul, pp. 155-166.

Baysan, S. and M. Teberler 2006. Local Attitudes and Sustainable Tourism Development: The Case of Didyma. Proceedings of Cutting Edge Research in Tourism: New Directions, Challenges and Applications. 6-9 June 2006, University of Surrey, Guildford.

Bramwell, B., I. Henry, G. Jackson, and J.V. der Straaten, (eds.) 1996. Sustainable Tourism Management: Principles and Practice. University Press, Tilburg.

Butler, R.W. 1999. Sustainable Tourism: the State of the Art. Tourism Geographies Volume 1, Issue 1, pp. 2–14.

—. 1980. The concept of a tourist area cycle of evolution: implications for management of resources. Canadian Geographer, Volume 24, pp. 5-12.

—. 2000. Tourism and the Environment: A Geographical Perspective. Tourism Geographies, Volume 2, Issue 3: 337–358.

Cooper, C. and S. Wanhill 1996. Tourism Development: Environmental and Community Issues, London, John Wiley & Sons, Inc.

Creech, D. and N. Brown 2000. Energy Use and the Environment. In: R.D. Bullard, G.S. Johnson and A.O. Torres (eds.) Sprawl City: Race, Politics, and Planning in Atlanta. Island Press, Washington DC, pp. 187–207.

Diagne, A.K. 2004. Tourism Development and its Impacts in the Senegalese Petite Côte: A Geographical Case Study in Centre–Periphery Relations. Tourism Geographies, Volume 6, Issue 4, pp. 472–492.

Fox-Penner, P. 1997. Electric Utility Restructuring: A Guide to the Competitive Era. Public Utility Reports, Vienna.

Garcia, C. and J. Servera 2003. Impacts of Tourism Development on Water Demand and Beach Degradation on the Island of Mallorca (Spain). Geografiska Annaler, 85 A, 3-4, pp. 287-300.

Gibson, L., P.A. Lynch and A. Morrison 2005. The Local Destination Tourism Network: Development Issues. Tourism and Hospitality Planning & Development, Volume 2, Issue 2, pp. 87-99.

Hall, C.M. and A. Lew 1998. Sustainable Tourism: A Geographical Perspective. Longman, Harlow.

Harrison, L.C. and W. Husbands (eds.) 1996. Practising Responsible Tourism: International Case Studies in Tourism Planning: Policy, and Development, John Wiley & Sons, Inc, London.

Hoballah, A. 2006. Sustainable Development in the Mediterranean Region. Natural Resources Forum, Volume 30, pp. 157-167.

Mathieson, A. and G. Wall. 1982. Tourism: Economic, Physical and Social Impacts. Longman, Harlow.

Neto, F. 2003. A new approach to sustainable tourism development: Moving beyond environmental protection. Natural Resources Forum, Volume 27, 212-222.

Okumuş, F. and C. Topaloğlu 2004. Turistik Destinasyonlarda Ürün Geliştirme ve Farklılaştırma: Dalyan Örneği. Seyahat ve Turizm Araştırmaları Dergisi, Volume 4, Issue 1, pp. 54-67.

Ören Tourism Information, Education and Health Foundation (Ören Turizmi Tanıtma Eğitim ve Sağlık Vakfı) 2002. Ören (Keramos). Es Ajans, İzmir.

PAP/RAC 2005. Coastal Area Management in Turkey. Priority Actions Program Regional Activity Centre (PAP/RAC) of the Mediterranean Action Plan (MAP-UNEP), Split.

Pearce, D. G. 1995. Tourism Today. A Geographical Analysis. Longman, Harlow.

Ruhanen, L. 2004. Strategic Planning for Local Tourism Destinations: an Analysis of Tourism Plans. Tourism and Hospitality Planning & Development, Volume 1, Issue 3, pp. 239-253.

Triandis, H.C. 1972. Attitude and Attitude Change. Wiley, New York.

Veal, A.J. 1992. Research Methods for Leisure and Tourism- A Practical Guide. Longman, Harlow.

CHAPTER FIVE

GEOGRAPHICAL ANALYSIS OF NATURAL ENVIRONMENT - HUMAN RELATIONSHIPS AND AGRICULTURAL ACTIVITY ON THE COAST OF THE GULF OF EDREMIT (NW TURKEY)

AYLIN YAMAN

Introduction

In the interaction between humans and the natural environment, coastal zones are one of the most significant areas that are negatively impacted. The reason is that littoral zones and their natural resources are strongly attractive to humans. Coastal zone in most countries have been subjected to severe and increasing pressure as a result of conflicting uses, such as rapid urbanization, environmental pollution, tourism development, and other coastal activities. One example of this is Edremit Bay in Turkey (Irtem, 2005).

The natural environmental features of the coastal zones of the Gulf of Edremit developed as a result of the interaction of physical geographical forces over a very long period of time. The natural environmental features of the area have largely determined the character of its human use, including many of the past and present human characteristics of the area. Due to its natural environmental characteristics the area has been intensvely used from early settlement to the present time. However, the pressures caused by internal and external forces have invariably required changes in the use of this environment over time. With changes in the macro-economic and social environment favoring a rise in global tourism as a industry, tourism has intensified in virtually all coastal zone accessible to European travelers over the past several decades, and this includes the region of the Gulf of Edremit (Ertin, 1992). The gradual population increase in the area has led to many problems in the natural environment.

Since the regeneration of natural environments cannot keep up with population increases and modern human activities in the area, the loss of natural resources has accelerated. The intense pressure placed on the area by human activity has exceeded the carrying capacity of the area and the losses of natural resources have been experienced.

Today, Turkey's coastal zones constitute approximately 30% of the total land, whereas the coastal and related hinterland populations account for about 51% of the total population of 65 million (Burak et al., 2004). Since the early 1960s, the coastal zones of Turkey have gained great importance to the country's economy because of the increase in economic development (mostly tourism) activities. The diversifications of activities and increase in their impacts have resulted in land use conflicts on one hand, and degradation of resource values on the other (Kaya, 2006).

This study focuses on the natural environmental conditions of the coastal region of the Gulf of Edremit and assesses the effects of of human actvities on the area. The aim of the study is to geographically analyze the most intense agricultural activities, particularly olive growing, which is the agricultural activity which is most common to the area.

Study Area

The study area covers the coastal zone stretching from Bababurnu in the westernmost section of the Anatolian peninsula, on the north of the Gulf of Edremit to the east of Ayvalık on the south. In this study, the shores of the Gulf of Edremit that lie below the contour line of about 200 meters are located in the Edremit vicinity on the north of the Aegean region in western Turkey (Fig. 5-1). The study area, which covers a rather narrow area on the north of the gulf includes quite a large area on the east and the south of the gulf. The area measures 818 km², including the islands. In terms of administration, the area on the east of Mıhlı stream is part of Balıkesir Province, while the area on the west of the stream is part of Ayvacık town in Çanakkale Province.

Materials and Methods

In this study, Geographic Information Systems (GIS) was used in order to determine the capacity of the study area and ArcGIS 9.1 was used to create the database and data structure. Topography, geology and soil maps of 1/100 000 scale were used to determine the supporting capacity of the land, and Landsat 2000 ETM+ and 1987 TM satellite images were used as a base to demonstrate land use change. The data obtained from meteorology stations in the area were used to define and explain climatic conditions.

In determining the demographic and agricultural conditions of the area, the data obtained from the Turkish Statistical Institute was used and human layers were defined. Land observations and the interviews with the people living in the area were other important sources of data. After data entry was completed, spatial analyses were performed.

Natural Environment Characteristics

The Gulf of Edremit and the Edremit Plain, which seem like complementary parts of the gulf, are a graben that was created as a result of new tectonic movements. With the epirogenic movements that occurred between the Oligocene and the Miocene, the area between the Kaz Mountains massive and the Madra-Kozak mass fragmented along the fault line in an east-west direction and the graben area of the Gulf of Edremit was created (Bilgin, 1969). The area experienced various corrosion and accumulation periods until the Quaternary. With the Post-Alpine movements that continued during the Quaternary, especially in the west, part of the graben re-collapsed considerably. Thus, the Aegean Sea came close to the graben area and the gulf expanded. The graben area started to be filled with alluvia carried by stream, draining the area and taking its present shape. In the area, the erosion in high zones and accumulation in the river basins are still going on (Buldan and Çukur, 2003).

Morphological conditions of the area are highly important in determining natural environmental conditions. Topographic conditions pave the way for differences in slope, elevation, climate, soil, vegetation, the selection of settlement, the products grown and agricultural systems practiced (Göney, 1979). A volcanic plateau that is 250-300 meters high is located on the north of the gulf, and extends from Bababurnu to the west of Küçükkuyu. In area where the volcanic plateau consisting of andesite, tufa, agglomerate and breccia is located, the high coastal zone draws attention. In the coastal zone, which demonstrates a steady extension with small coves and cliffs, there are small capes and beaches that protrude towards the sea. The area stretching from the west of Küçükkuyu to Akçay has a low coastal zone with several beaches. Edremit Plain is located on the east of the gulf. The plain, consisting mainly of alluvial materials, covers an area of 150 km². There are alluvial cones in sections surroundings the plain, from the north and the south. The Edremit Plain is separated from the Gömeç Plain on the south by the mound of Neogene.

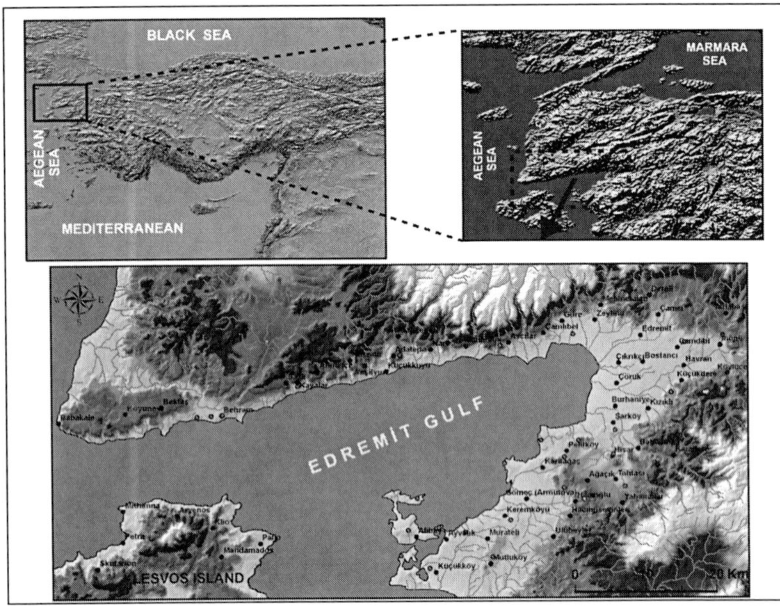

Fig. 5-1: Location of the Study Area.

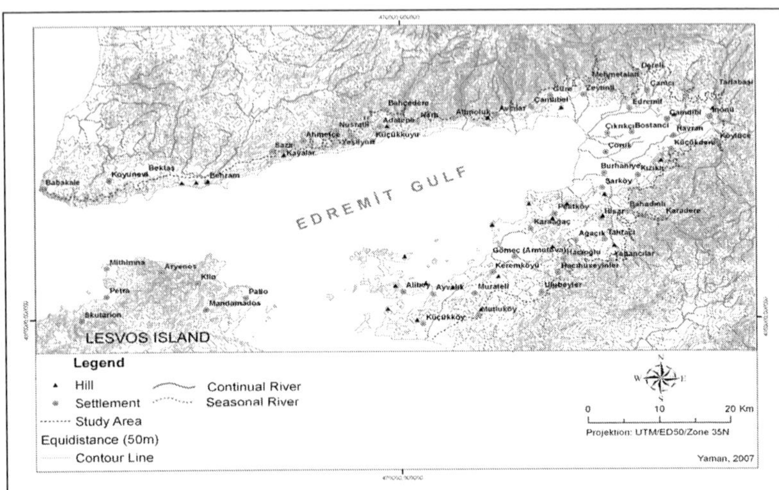

Fig. 5-2: Topographic Map of the Gulf of Edremit and Surroundings

The Gömeç Plain covers an area of 30 km². It reaches the coastal zone of Ayvalık on the south. The coastal zone of Ayvalık is indented with islands, peninsulas, bays and gulfs (Fig. 5-2).

The different environments created by topographic conditions around the gulf have also affected human life. Although the areas convenient for settlement and agriculture on the north of the gulf are restricted owing to the increase in height and slope conditions, such areas are relatively wide on the east and south (Fig. 5-3 and 5-4).

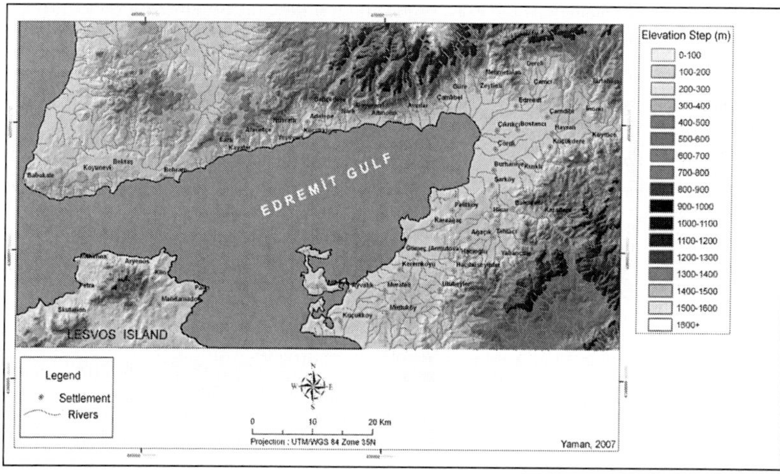

Fig. 5-3: Digital Elevation Model (DEM) of the Gulf of Edremit and Surroundings

One of the factors that determine natural environmental characteristics and capacity is climatic conditions. The coastal zone of the Gulf of Edremit has a Mediterranean climate. The Mediterranean climate is characterized by hot and dry summers, and mild and wet winters. According to the data obtained from the Edremit, Burhaniye and Ayvalık meteorology stations, the annual average temperature is 16.3 °C. The average temperature of each month does not go below 0 °C. The average high temperature 32.4 °C is observed in July and the average low temperature 3.8 °C is observed in January. The annual average precipitation in the study area is 634.4 millimeters. Most of the precipitation falls in winter in accordance with the Mediterranean climate. While the average precipitation that falls in winter is 289.9 millimeters, the average precipitation that falls in summer is 22.4 millimeters. Due to very low precipitation in summer, there is a need for irrigation to maintain agricultural activities. These climatic and environmental parameters have generated varied ecosystems that are home to many endemic species and this has generated attractive conditions over the years for a number of economic sectors.

The number of sunny and sunless days is very important in terms of both agricultural and tourist activities. The frequency of the number of sunless days in the coastal zone of the gulf is low (49.5 days). Especially in summer days, the number of sunless days is nearly zero. However, the number of sunny days is 138.1 and is particularly high in summer months when touristic activities intensify. The duration of sunshine and the temperature of sea water are also important in the area where touristic activities are common. In the coastal zone where the average sunshine duration is 07:43* hour/minute, the sunshine duration between May and September is above 9 hours per day. The average sunshine duration is 12:24 hour/minute in July. The temperature of the sea water is above 19 °C during 6 months from May to October.

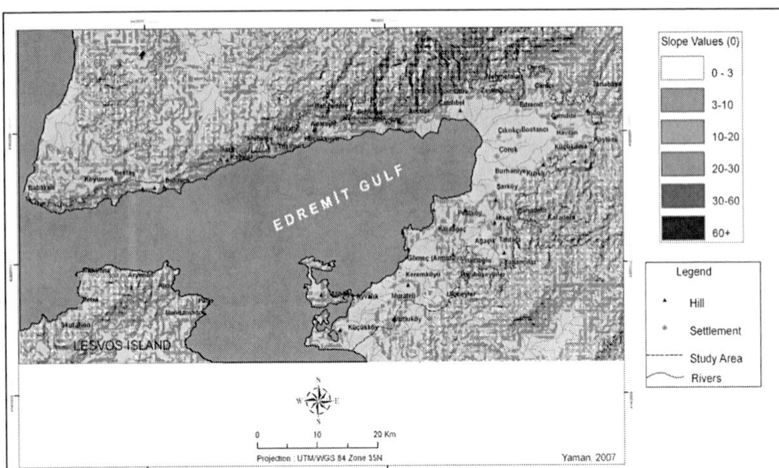

Fig. 5-4: Slope Map of the Gulf of Edremit and Surroundings

Another important factor that determines natural environmental capacity is the soil characteristics of the region. The study area includes zonal, azonal and intrazonal soil groups (Fig. 5-5). In the study area, the soil type with the greatest expansion among the other soil types is non-

* Data obtained from Ayvalık Meteorology Station

calcic brown forest soil with an expansion rate of 33.7%. The land use ability class* of this soil is generally between VI and VII. In the study area, alluvial (12,759.4 ha-16.6%) and colluvial (9,997.3 ha-13%) soils have an important role in local agriculture. These soil types, whose land use ability class is between I and III, are especially found on the Edremit and Gömeç Plains and their surroundings.

Various kinds of vegetables and fruits are grown on these types of soil. Different soil groups affect people's selection of settlement and economic activities, and the variety of the products grown. Most of the settlements around the gulf are located on the lowpitched areas (I-IV class agricultural lands) comprising alluvial and colluvial soil among plain basins and hillsides of hilly masses.

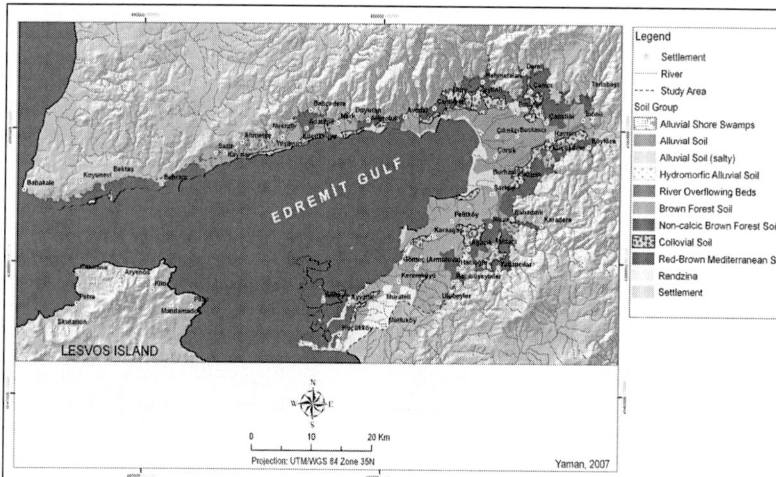

Fig. 5-5: Soil Map of the Study Area.

The study area has an important potential in terms of soil conditions. Among all soil groups, the rate of the soil with a land use ability class between I and III and with considerable significance in terms of agriculture is 29.6%. The potential of the area should be used consciously and properly. The appropriate use of soil in terms of its potential will avoid the loss of natural resources. Another component of the natural environment is vegetation. Anthropogenic factors play an important role in

* There are eight land use ability classes, and soil damages and limitations ascend from I to VIII.

the degradation of natural vegetation. Since the first settlements of the coastal zone of the Gulf of Edremit in ancient times, the vegetation of this area has changed considerably. In recent decades, new settlements have been established on land that was once used for agriculture. Natural vegetation has been cleared to make room for the construction of second homes and vegetative cover is under heavy urbanization and touristic pressure throughout the Gulf of Edremit. Locally dominant species such as *Olea oleaster* and *Pinus brutia* are the major plants that have been affected by on-going human interventions in the area (Efe 2000). In the area, natural vegetation has become attuned to the conditions of the Mediterranean climate and needs high temperatures and high amounts of light.

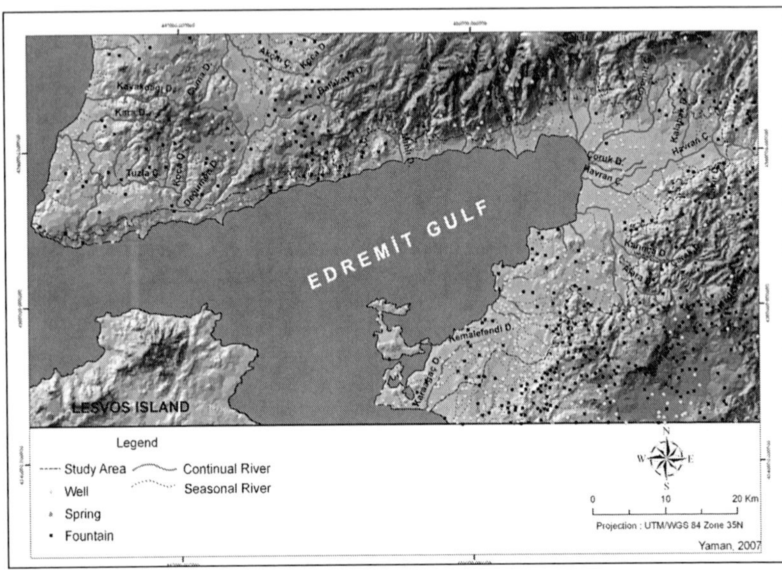

Fig. 5-6: Hydrological Map of the Gulf of Edremit and Surroundings

The fundamental species of the Mediterranean vegetation formation is red pine (*Pinus brutia*). Red pines around the Gulf of Edremit find natural conditions to form forests from sea level to about 700 meters in elevation. However, most of the red pines have been destroyed, and instead, *maquis* and *garrigue* formations are dominant (Güngördü, 1993-1996). The primary cause of the loss of natural vegetation is that it has been destroyed in order to form olive groves. Olive groves extending around the gulf in almost a line are intense between the altitudes of approximately 40-50 to 250 meters. There are red pine forests above the line of olive groves. The end point of thc linc of olive groves and the starting point of red pine forests are very clear. The top border of the olive groves is moved up with forest openings. The higher slopes in the area are being used for olive cultivation, which is not a suitable location for this activity. The change from *maquis* and red pine to olive trees and clearing the red pine on the slopes increases erosion and induces land degradation.

Another factor determining the natural environment capacity is water resources. There are several streams in the study area. While in winter the flow of streams increases due to the high amount of precipitation, in summer it decreases due to the low amount of precipitation. The different sections of the study area have different potentials in terms of groundwater. The plain basins are especially rich in groundwater. In this area there are several wells that supply water for drinking and irrigation (Fig. 5-6). Excessive urbanization and steep population increases apply serious pressure on these groundwater resources. Because of the significant population increase and the intense use of groundwater for irrigation of land in summer, the level of groundwater is gradually decreasing.

Results and Discussions

In the coastal zone of the Gulf of Edremit, the natural environment has provided an abundance of resources appropriate for human life, and human communities have formed settlements appropriate for the conditions of the area (Soykan 2001). The existence of natural environmental conditions that could meet the fundamental needs of humans in the study area has led to human settlement since very ancient times (Stewig 1968, Sevin 2001). The Mediterranean is only 0.7% of the total surface of the world's seas. However, this basin has been the cradle of some of most remarkable civilizations in history. Thousands of years of land use and environmental conditions have shaped a mosaic of diverse patterns (Berberoğlu 2003).

The study area has remained as an area for settlement throughout history. The reason for the continuity of the settlement history in this area is the convenient conditions offered by the natural environment. The study area is one of the most populous parts of Turkey. In the area, the population has generally become dense in settlement centers on plain bases, flat lands around hilly areas and on coastal plains. A total of

201,991* people live in the area and the population density is 247 people. However, the population in 1935 was 70,109 in the area and the population density was 86 people. These data demonstrate the rapid population increase (188%). Due to this rapid population increase, pressure on the natural environment has also increased.

The population is not settled equally in every part of the area. It is clearly seen that the conditions offered by the natural environment to humans are extremely effective. Generally, population density decreases in areas where the slope and the elevation of land forms increase and the lands appropriate for agricultural activities diminish, while it increases in the areas appropriate for settlement and with high agricultural potential.

The intensification of population in certain areas increases the pressure on those areas. In the study area, one of the distinctive features of the population is its mobility. Besides permanent migrations, seasonal and daily migrations are also very common. In summer when touristic activities intensify, the population of the area increases immensely (Fig. 5-7). The natural environmental landscape has started to change with the development of tourism in the area. Land has been used for tourism and this has led to significant differences in the past and present appearance of the land, especially in the last 20-25 years (Fig. 5-8). The demand for already limited natural resources is higher during the summer season.

Human communities settling in the coastal zone of the Gulf of Edremit have been dependent on the conditions offered by the natural environment as in all areas. They have found oppurtunities to develop to the extent the natural environmental conditions permt. However, humans have sought new ways of using the area and developed various economic activities. The natural environmental conditions of the study area have influenced the economic activities and formed the specific structure of the area. When the economic activities in the area are examined, it is apparent that agriculture has a big share. It must be observed that industrial and commercial activities in the area have also developed in relation to agriculture.

Tourism is another kind of economic activity that has developed in the area in recent years. Tourism development may have a negative impact on natural resources, owing to overuse and mismanagement. However, tourism may also play a positive role in natural resource conservation. It is possible to use tourism as a way to balance natural resource conservation and economic development under the preconditions of making effective policies to encourage and help local people participate in, and benefit

* General population census of 2000

from, the tourism business (Li et al. 2006). In the study area, there is a development.

The most important economic sector of the Mediterranean basin is agriculture. It currently provides employment for nearly half of the labour force and it uses most of the land and water resources, leading to a competition between urbanization and environmental conservation (Makhzoumi and Pungetti 1999).

Agriculture is the most important economic activity in the study area as well. As has been noted previously, the natural environmental conditions of the study area are appropriate for agriculture. Generally, tree agriculture is common in the area. The conditions specific to the Mediterranean manifest themselves in extensive olive groves. The area surroundings the Gulf of Edremit is one of the most important olive growing and olive oil producing areas of Turkey. On the north of the gulf, in the undiminished parts there are several olive groves that stretch from very close to the sea to 200-250 meters in elevation.

The alluvial cones and foots around the Edremit plain are also intensively covered with olive groves. The mound that separates the Edremit Plain from the Gömeç Plain, which is on the south, is also covered with olive groves. As has been seen, olive groves surround the gulf in a line (Fig.5-9). In the area, a tract of land 50,957.7 hectares in size is covered with olive groves. This is a very significant value and shows the dominance of olive groves on land use in the coastal zone of the gulf.

Olive is a typical tree of the Mediterranean climate. The Mediterranean is referred to as the olive climate because of the wide distribution of the cultivated olive tree (Makhzoumi and Pungetti 1999). Olive tree cultivation in the Mediterranean goes back to ancient times. This longevous tree integrates and identifies economically, socially, and culturally the inhabitants of this basin and determines its rural landscape. For the residents of the Mediterranean, olive oil constituted the main source of nutritional fats, their most valuable export product, and was identified with their culture. Even now, olive cultivation has a multiple importance for the Mediterranean (Loumou and Giourga 2003). Olive is very much affected by climatic elements, especially temperature and precipitation. It grows well in the areas where the annual average temperature reaches 15-20 °C. In this regard, the study area provides appropriate conditions for olive agriculture with an annual average temperature value of 16.3 °C.

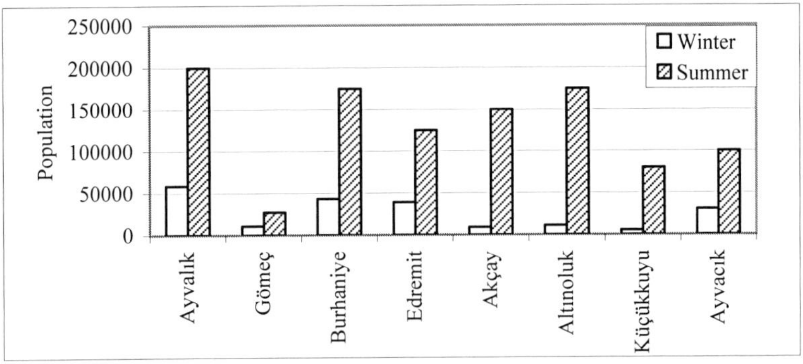

Fig. 5-7: Population distribution of residential districts in winter and summer

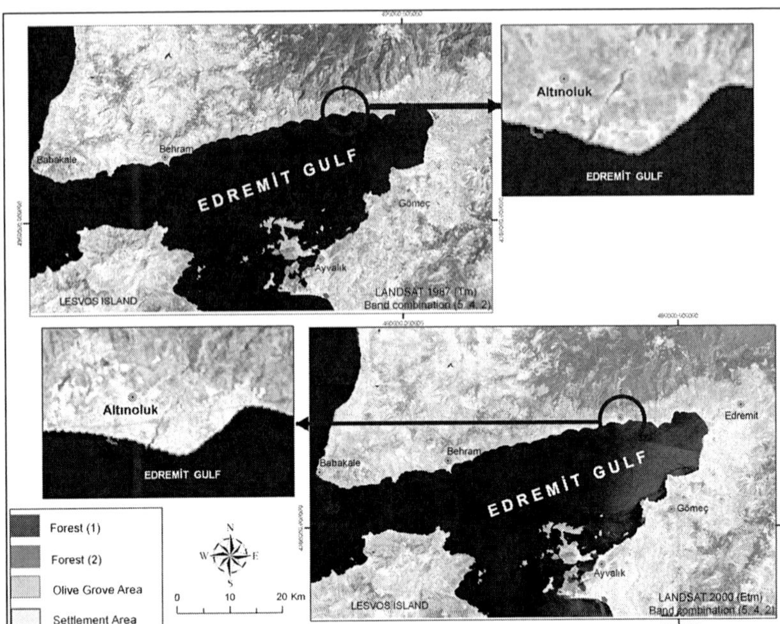

Fig. 5-8: The satellite view of the Gulf of Edremit and its surroundings in 1987 and 2000, and the example of Altınoluk. The increase in the number of secondary housing and touristic facilities due to the intensification of touristic activities is clearly seen in the example of Altınoluk.

In olive growing, another important factor is precipitation. Olive consumes water all through the year, since it maintains its green colour continuously. The annual precipitation need of olive is between 400 and 600 millimeters. There is a need for precipitation of 600-800 millimeters for high efficiency. In the study area, the annual average precipitation of 634.4 millimeters is enough for efficiency in growing olive. In olive agriculture, besides the annual average precipitation value, the distribution of precipitation within months is also important. Olive generally needs precipitation in winter and in the months before and after winter. However, it needs drought during blooming and maturing periods (Aykas, 2004). In blooming and maturing periods, there is almost no precipitation in the area. On the other hand, in other months the precipitation values in the area are enough for olive.

A large number of the world's olive groves are located in the Mediterranean basin. Besides the Mediterranean countries such as Italy, Spain, Greece and Tunisia, Turkey is also one of the world's leading olive growers.

In Turkey, the Aegean region is the leading region in terms of olive agriculture. Our study area is an important area for olive agriculture. In the study area, there are approximately 6,480,221 olive trees, covering an area of 509,577.7 decares. The olives specific to the Edremit vicinity could not be stored for a long time due to their peel that wastes away and splits quickly. Therefore, they are generally used for producing olive oil. There are many olive and olive oil factories in the area.

There are also various fruit trees besides olive trees in the area. The most expansive species among orchards around the gulf in recent years is tangerine. As a result of research on a product alternative to olive, the production of tangerine has increased.

Tangerine orchards have started to become common in the areas that are protected against cold winds and stretch from the vicinity of Avcılar village on the north of the gulf to İnönü village on the east (Mutluer, 1990). In this section, tangerine orchards and olive groves are jumbled. The tangerine species produced is Satsuma tangerine, which has been well adapted to the natural environmental conditions of the area. In the study area, 10,744 dacares is covered with tangerine orchards. There are also other types of orchards, such as fig, plum, apple, pear, quince and pomegranate.

In the study area, various vegetables are also grown on the alluvial plains, thanks to irrigated agriculture. Since climatic conditions are conducive to particular vegetables, and various types of vegetables are grown in summer and winter alike. This includes tomatos, peppers, eggplant, okra, cucumber and beans.

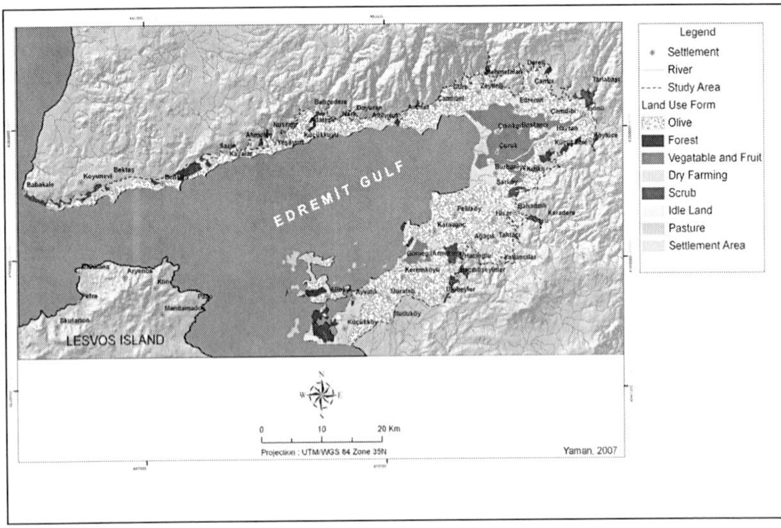

Fig 5-9: Land Use of the Study Area

In recent years tourist activities in the region have intensified. The coastal zone of the Gulf of Edremit has highly appropriate conditions for tourist activities with its natural and historical features, its wide bay and beaches, and its swimming season of 5 months. In addition, the ease of transportation from big cities to the area has increased the popularity of the area. The vicinities of Edremit-Akçay, Küçükkuyu, Altınoluk, Burhaniye-Ören and Ayvalık are areas with intense touristic activities. There are tourist facilities, secondary housing and camping facilities in the coastal zone stretching form the east of Küçükkuyu on the north of the gulf to the coast of Ayvalık on the south. In the coastal zone extending from the west of Küçükkuyu to Babakale, tourism has not developed very much. In this area, Assos-Behram is touristically attractive.

The effect of topography and coastal forms on the distribution of tourist activities in the area is obvious. Besides sea tourism, the area is also popular with thermal spring tourism. Güre and Derman thermal springs are important ones in the area. The Kaz Mountains, stretching on the north of the gulf are touristically attractive places, thanks to its fresh air, its natural beauties, its location near historical sites, and the richness of local social and cultural values. The intensification of touristic activities in the area increases the pressure on the natural environment. Instead of the necessary land use, the land is used for touristic activities. Inappropriate land use causes the losses of natural resources and upsets the natural balance between humans and the natural environment.

According to the observations of the Edremit Olive Enterprise, more than 3 million olive trees have been wasted since 1990 in the coastal areas of the Gulf of Edremit. More than 1.5 million of these trees have been sacrificed to urbanization. Secondary housing or tourism activities and others have only served to destroy olive trees. There are about 1.5 million olive trees which are not cut and are reserved for decorative purposes among buildings. These trees have no economic value and they either dry up with time or are cut by the owners of the buildings. The economic loss resulting from destruction of olive trees at the expense of real estate profits is signifcant, as income from such activities is three times less than olive oil income.

Conclusion

The coastal zone of the Gulf of Edremit is an area where the interaction between the human and natural environment is intense. The natural environmental conditions around the gulf have an effect on human life and direct it. The geomorphological conditions around the gulf manifest themselves mostly in expanded settlement and in transportation and tourism, while climatic conditions and natural vegetation have effects on economic activities.

The convenience of natural environmental conditions has led to the intense use of the land from the past to the present. However, in recent years the increase in the wrong land uses in the area has led to the losses of natural resources. The fundamental economic activity around the gulf is an agriculture which is shaped according to natural environmental conditions. In the area, the natural environmental conditions are appropriate for growing olive which is the most important product around the gulf. Olives have the biggest share in land use. However, the natural environmental landscape has started to change with the development of tourism in the area. Partcularly over the last 20-25 years, the land has been used for tourism and this has led to significant differences in the past and present appearances of the land. The coastal zone is gradually being covered by concrete, due to secondary housing and touristic activities. The existing coastal law is violated in the area and this upsets the natural balance of the coasts. The significant population increase in the area during summer due to tourism boosts the pressure on the area. Fertile agricultural lands are used for purposes other than agriculture, and land is wasted. In order to prevent wrong land use, not only should necessary regulations be passed, but the necessary controls should also be implemented. Coastal Law No. 3621 for the protection of coastal zones in Turkey was promulgated in 1990. Article 5 of this law states that the ‘‘proximity of any building to the shoreline may not be closer than 50 m’’. In the study area, it seems that none of the landowners comply with this act. Most of settlements nearly touch the sea, especially in Ayvalık, Edremit and Altınoluk. In this respect, local administrations and the public have an important role. It is emphasized that there is a need for viable economic and environmental strategies to be designated in the Gulf of Edremit, in order to provide sustainable resource usage.

The agricultural potential of the study area is very high. However, it is hard to say that agricultural potential is properly used. The problems in agricultural activities (wrong land use, insufficient irrigation, erosion, pollution, etc.) should be minimized and necessary support should be given to agricultural activities. The existing balance between natural environment and human use should be maintained.

At present, the study area is an important tourism center and it is obvious that the area will continue to develop in the field of tourism. Therefore, the development should be planned and integrated with other economic activities. Tourism should be developed, but it should also protect natural resources and ensure economic development. It is crucial to designate both environmentally sound and economically viable coastal management strategies that ensure minimum environmental deterioration and maximum attainable protection while benefiting from the blessings of nature.

To maintain healthy relationships between the natural and human environment, and sustainable productive land use in the research area, lands must be used according to their capabilities. Otherwise, loss of natural resources will gradually occur. For sustainable land use, the existing resources must be used effectively within the principles of “protection and use”.

References

Aykas, B. (2004) Zeytinin Yetişme Koşulları, Tesisi ve Modern Yetiştiricilik, Zeytin Yetiştiriciliği Kursu T.C. Tarım ve Köyişleri Bakanlığı Zeytincilik Araştırma Enstitüsü Müdürlüğü, İzmir, pp. 39-56.

Berberoğlu, S. (2003) Sustainable Management for the Eastern Mediterranean Coast of Turkey, Environmental Management Vol. 31, No. 3, pp. 442–451.

Bilgin, T. (1969) Biga Yarımadası Güneybatı Kısmının Jeomorfolojisi, İstanbul Üniversitesi Yayın No:1433, İstanbul, pp. 171-220.

Buldan, İ., and H. Çukur. (2003) Edremit Körfezinde Zeytincilik Doğal Ortam-İnsan, Dokuz Eylül Yayıncılık, İzmir, pp. 55-56.

Burak, S., E. Doğan, and C. Gazioğlu. (2004) Impact of urbanization and tourism on coastal environment, Ocean & Coastal Management pp. 47:515-527.

Efe, R. (2000) The impact of land use changes on natural vegetatiton along the coastal zone between Edremit and Küçükkuyu, International Symposium on Desertification, Konya-Turkey, pp. 222-227.

Ertin, G. (1992) Edremit Körfezi Kuzey Kıyılarının Coğrafi Yönden İncelenmesi, Türk Coğrafya Dergisi, İstanbul, pp. 27:187-209.

Göney, S. (1979) Türkiye Ziraatinin Coğrafi Esasları I, İstanbul Üniversitesi Yayın No: 2600, İstanbul, pp. 8.

Güngördü, M. (1993-1996) Güney Marmara Bölümü (batı kesimi) Bitki Örtüsünün Coğrafi Dağılışı, İstanbul Üniversitesi Coğrafya Dergisi, İstanbul, pp. 4:55-112.

İrtem, E., S. Kabdaşlı, and N. Azbar. (2005) Coastal Zone Problems and Environmental Strategies to be Implemented at Edremit Bay, Turkey, Environmental Management Vol. 36, No. 1, pp. 37-47.

Kaya, G. L. (2006) Critical Barriers to Rational Planning Processes for the Coastal ZoneManagement: the Case Study of Antalya, Turkey, Ph.D. Thesis, State University of New York College of Environmental Science and Forestry Syracuse, New York, USA, pp. 45.

Li, W., Q. Zhang., C. Liu, and Q. Xue. (2006) Tourism's Impacts on Natural Resources: A Positive Case from China, Enviromental Management, pp. 38:572-579.

Loumou, A., and C.Giourga. (2003) Olive groves: "The life and identity of the Mediterranean", Agriculture and Human Values pp. 20: 87–95.

Makhzoumi, J., and G. Pungetti.(1999) Ecological Landscape Design and Planning: The Mediterranean Context, GBR: Spon Press, London, pp. 15-27.

Mutluer, M. (1990) Edremit-Burhaniye Ovası ve Yakın Çevresinin Beşeri ve İktisadi Coğrafyası Üzerine Bir Araştırma, Ege Üniversitesi Sosyal Bilimler Enstitüsü, Yüksek Lisans Tezi, İzmir, pp. 145-146.

Sevin, V. (2001) Anadolu'nun Tarihi Coğrafyası I, Türk Tarih Kurumu Basımevi, Ankara, pp. 44.

Soykan, A. (2001) Kazdağı Milli Parkında Doğal Ortam - İnsan İlişkileri ve Zeytincilik, Kazdağları I. Ulusal Sempozyumu, Edremit, Balıkesir, pp. 251-280.

Stewig, R. (1968) Batı Anadolu Bölgesinin Kültürel Gelişmesini Gösteren Kartografik Bilgiler (Çeviren Ruhi Turfan), İstanbul Matbaa Sanat Enstitüsü, İstanbul, pp. 64.

CHAPTER SIX

RESEARCH ON CONTRIBUTIONS OF WINTER TOURISM TO LOCAL SOCIETIES (IN TERMS OF ECOTOURISM): ÇOBANISA VILLAGE (ISPARTA), TURKEY

SALIH CEYLAN AND HILMI DEMÍRKAYA

Introduction

Although the preservation of natural resource, especially at the end of 18th century, was emphasized considerably by naturalists in many countries of the world, the need to preserve and protect the environment was not at that time viewed as urgent as it is viewed today. In the period from the 1970s to late to the early Twenty-first Century, efforts to protect the environment on global scale have significantly increased (Yalınkılıç and Yenilmez Arpa, 2005). Indeed, the first serious and systematic steps to promote sustainable development and the conservation of natural resources on a global scale were taken in the 1970s. In more recent years, efforts to prevent pollution and to set specific consumption limits and benchmarks to preserve the environment have increased. These developments are represented by international summits like the European Union Summit in 1972, held in Paris. Additionally, the conference at Stockholm, held on 5 June 1972, with 113 countries attending, began to deal with broader environmental issues. These conferences had an important impact on international discussions of environmental problems from the end of the 1980s onwards (Cavuş and Tanrısevdi, 2000:151).

In the Environment and Development Report with the title "Our Common Future", which was prepared by the United Nations World Environment and Development Commission in 1987 (also known as Brundtlant Report), it was emphasized that the environmental issues had risen to such a level as to threaten the lives of all living things in global terms. Also, in the report, traditional approaches to development were criticized, and a "sustainable development" approach was proposed. The term "sustainable development", which means making economic and social development continuous, without damaging the ecological structures of the natural surroundings, was considered, together with actions which could take place at a global level. This was the topic of the United Nations Environment and Development Conference, known also as the World Summit, held with the participation of the leaders of 116 countries, in Rio de Janeiro in June 1992. In addition, the issue of sustainable development became the main theme of the Sustainable Development Conference of UNEDC, also named "Habitat II", and held in İstanbul in1996.

The term "sustainable development" is closely related to tourism, as it bases economic development on protecting and sustaining environmental values. Tourism is an economic sector which serves humans, but which, at the same time, most relies on the human work force. It is perhaps for this reason that, traditionally, the primary focus of tourism policy has been mainly the economic effects of tourism. However, locations which have natural landscapes and cultural environments which are conducive to the rise and development of tourism have the potential to promote an increase in incomes as a result. Consequently, it is important to realize that the effective use of such places is essential to the future of the tourism industry.

Globally, substantial changes in the understanding of tourism and sustainable development have begun to be observed in recent years, parallel with economic, social, political and technological advances. Since the 1980s, the increasingly destructive and environment-damaging developments in the tourism sector and their negative effects have been the main concern of some international conferences and symposia. Therefore, enterprises have been formed to apply new tourism policies and strategies that conform to the principles of sustainable tourism and that don't threaten the very future of the industry. In this context, the UNSDC declared the year 2002 as "International Eco-tourism Year." Between 19 and 22 May 2002, under the leadership of UN Environment Programme, the World Eco-Tourism Summit was held in Quebec. The aim of the summit was to provide and support an atmosphere where eco-tourism activities could develop under the concept of sustainable tourism. Eco-tourism was viewed as a part of sustainable development and was proposed as one possible development strategy, and a possible solution to poverty in some less-developed areas (Lee et al., 1998: 40-47).

The type of tourism which is remembered first when we consider sustainable tourism is eco-tourism. Eco-tourism is a small-scale type of tourism which is based on natural and cultural resources and which doesn't damage them. The philosophy upon which ecotourism is based also rejects mass-tourism (Soykan, 2002:158). In this context, protection of the environment and the opening of natural and cultural resources to tourism through an understanding that such tourism must be sustainable have begun to develop as a concept and is at the top of the policy agenda. Turkey, in terms of its mountains, forests, plateaus, lakes, rivers, floras and faunas, caves, canyons, amazing geological formations, cultural heritage and authentic cultures has richness which can not be compared with other countries having coast on the Mediterranean. Because many Mediterranean European countries have been considerably industrialized and urbanized, they have substantially lost their authentic cultures. Turkey is still a developing country and maintains cultures which are traditional and rural. Its culture and its heritage offer great potential for sustainable eco-tourism in many of its regions and local areas (Cengiz, 2002). This richness makes Turkey an attractive and privileged one with respect to eco-tourism activities. And Turkey's mountainous areas constitute the most important part of this potential.

For example, mountan-climbing has become a sports activity that is mostly engaged in by tourist. Relatively-speaking, mountain climbing has an adventurous side, and requires preparation, and special knowledge and talent (Doğanay, 1989:165).

Until the 1992 Rio Summit, there were very few international studies of mountains and of mountainous areas. The 13th part of Agenda 21 of the Rio Summit was an important outcome of the 1992 Rio Environment Summit. This part of the report held the title "Continuous and Stable Mountain Management". The summit was thus a turning point for the world's mountains (Gönençgil and Güngör, 2002:38).

Sports activities, particularly climbing and skiing, were mainly responsible for the opening of mountains to tourism. The provision of guidance services to tourists, and accommodations and transportation to climbers led to the start of a tourism type known as "alpinism". This has led to the transformation of parts of mountains, which are suitable for skiing, and to the development of skiing centers and the introduction of winter tourism with the help of accommodation and entertainment facilities (Doğaner, 2001:176).

Winter tourism is a kind of tourism in which sports activities are concentrated at a altitude able to get snow, depending on the available snowfall in any given season (Doğaner, 2001:178). In the Mediterranean, winter tourism, specifically skiing tourism, relies on a number of factors, such as altitude, slope, exposure, snow and its duration on the ground. The season generally runs from the end of autumn until the beginning of spring (Gönençgil and Güngör, 2002: 40). Mountains have continued to be the main focus on interest in winter tourism..

"Winter tourism" and "winter sports" constitute the most important aspects of mountain tourism and mountain sports. Winter tourism and sports centers have features peculiar to themselves that make tourism and sports activities possible, besides their sub-activities and relations (Ülker, 1987:144). Today, there are 20 close mountainous areas that are winter sports centers, and one of them is the Davraz Mount Winter Sports Tourism Center, which is newly developing. Because it is located on the route to the Davraz Mount Winter Sports Tourism Center, our study area. "Çobanisa Village" has recently been introduced to tourism. It is a settlement where village tourism has lately started to develop.

The aim of this study is to examine the eco-tourism characteristics of Çobanisa Village, which have developed around the idea of using natural surrounding.

Location and Transportation

Çobanisa Village is in the east of Isparta and is 18 kilometers from the city center (Figure 1). In Turkey, villages are traditional settlements where municipal administrations have not yet been founded. Villages are a small settlements usually formed by fewer than 2,000 inhabitants. A "village administration" has a jurisdiction within the boundaries of its village. In general, small settlements, together with their properties, fields etc. constitute a village. A village administration has three organs: 1) a village society; 2) a council of elders; and 3) a village headman, or Muhtar (Turkish Ministry of Interior; 1999:89-95). The village is located north-west of Davraz Mount and the west side between the altitudes of 1250 and 1300 meters. Transportation to Davraz from Isparta, as well as to Eğirdir (Büyükgökçeli Village) is provided through the highway passing through Çobanisa Village. Mount Davraz is a very suitable site. It is 26 kilometers to the Isparta city center, 154 kilometers to Antalya, 452 kilometers to Ankara, 420 kilometers to Izmir, and 650 kilometers to İstanbul. It sis also conveniently located in terms of air transportation. The distance to Süleyman Demirel Airport is 55 kilometers. The distance to Antalya Atatürk Airport is 130 kilometers. A railway line extends to Isparta and Eğirdir.

Çobanisa Village and Mount Davraz, whose road standards are in a good condition, are linked to both Eğirdir town center and Isparta city center by asphalt roads. Although, private cars are used, during the skiing season (December- May) it's possible to catch municipality buses. Çobanisa Village, because of its convenient location regarding transportation, fulfills the required condition necessary for rural tourism to develop.

Mount Davraz- Winter Sports Tourism Center

Mount Davraz situated 26 km far from Isparta city center and declared as a tourism center by the decision of the Government on 17th February 1995. Infrastructure subsequently began with the construction of a ski-house (Isparta Special Provincial Administration Management) and this establishment was completed and opened in 1997. Today, there is a privately-run hotel (with a restaurant) which has a capacity of 62 beds and 150 people. Additionally, the four-star Davraz Sirene Hotel -- with the capacity of 280 beds and built by a private company -- was approved by the Culture and Tourism Ministry, opened in December 2005.

In Davraz Winter Sports Center, which has two chair lifts, a first length of 1211-1674-1961 meters and it has a carrying capacity of 1000 people per hour ; and second line lengths of 936.6-1965-2150 meters with also a carrying capacity of 1000 people per hour. There are also "telexes" (T-bar) with a length of 624 meters (between 1950-2150 meters), and it carries 800 people per hour. There are two baby lifts 300 meters long. At the Sports Center, one can do "North Discipline", "Alp Discipline", "Tour Ski", Snowboard, and "Tele Mark". The skiing season runs between December and April and the thickness of the snow ranges between 50 and 200 centimeters.

Winter Tourism and Cobanisa Village

Each year, winter tourism has begun to gain more importance. Many tourists who traditionally have taken summer holidays have begun to take winter holidays. Thus, interest in "Winter Tourism" and "Winter Sports" centers have increased generally, including in Turkey. In recent years, thanks to a number of factors, including the expansion of automobiles, an increase in individual car ownership, the fact that weekend holidays are two ore more days in length, and a rise in personal incomes, interest in winter tourism has risen, just as summer tourism originally increased (Ülker, 1987: 148)

Winter tourism may be defined as tourism activity which is done in areas with snow cover, and is generally intensified in areas where modern winter sports infrastructures and centers are developed (Ülker; 1987; 145). Çobanisa village, which is 8 kilometers from Davraz Mount Winter Sports Center and 18 kilometers from Isparta, is developing a rural tourism (village tourism) which is largely focuses on winter activities.

In Çobanisa, a mountain village, local incomes traditionally depend mainly on agriculture and stock raising. In Çobanisa in 2000, there were 70 houses and 350 people. Local people raise approximately 7500 goats and 2500 sheep and about 150 cattle. The literacy rate is 99%, and 27 students from the village commute daily to Gökçeli Village Primary School. Twenty (20) people from Cobanisa work at the Davraz Skiing Center Hotel and the center's cable car.

Davraz Mountain cottages were built by the village mukhtar and the special provincial administration, in accordance with local architectural features, and began service at the beginning of 2006. The cottages, which were rented to a private company for 10 years, have 10 rooms and a total capacity of 85 people. In addition, the village mukhtar and his family run their own seven-room boarding house, also catering to tourists.

To encourage the development and operation of home-pensions supported by the Ministry of Tourism, courses were organized in Çobanisa Village. After these courses, it was determined through the interviews with local people that many were planning on operating boarding-houses.

In the boarding-house in Çobanisa, lower and middle class skiers are usually served. Both for-a-day and for-a-night organizations connected to Antalya are involved. In addition, people staying in Sirene and Selene Hotels come for dinner to have experience a village atmosphere. Additionally, for-a-day visitors come to the area from Isparta and Eğirdir.

Foods produced in the village, such as eggs, walnuts, molasses, and almonds, are served to visitors. Additionally, some of villagers market their goods to visitors along the road passing through the village.

The Davraz Çobanisa Village Skiing Training Club was established so that children and the young of the village would become interested in the sport and learn to ski. The club has now has 14 licensed athletes, all of whom are from the village.

The Ministry of Tourism included part of road crossing through the village into the tourism zone, and its development plan and project have been completed by the General Directory of Tourism Ministry Investment and Management. The project aims to improve the institutions, as hotels and motels, and also to develop village pensions under the Regulation of Tourism Institutions.

Discussion and Conclusion

Rural tourism is a type of tourism made up of natural surroundings, rural settlements, and local economic activities. For this reason, in many countries of the world, rural tourism serves as a means to promote rural development. However, in Turkey, it does not have such a function, as it has not yet been efficiently developed. In fact, the concept of rural tourism is based on an understanding of eco-tourism, and ecotourism may be seen as promoting small-scale activities and as a rejection of the massive scale of conventional tourism. Rural tourism also gives much importance to the *authentic* (Soykan, 2002:154). In this context, the tourism type beginning to develop in Çobanisa Village has a feature which gives importance to local culture and does not encourage distorted and fast building, for example. Additionally, this type of tourism focuses the importance of the survival of nature, with minimum destruction.

Rural tourism discourages migration of rural populations to urban areas by integrating local people in income-generating activities, and by promoting locally-based socio-economic wealth. Even many developed countries in Europe seek solutions to the problem of shrinking populations in mountainous areas through rural tourism. In some countries (e.g.; Italy, France, Spain and Austria) weak agriculture, stock-raising, and forestry in rural areas are enlivened by the tourism industry. Thus, population is kept in its place (Soykan, 2000:21-33).

It is well-known that Çobanisa has traditionally lost population because its income sources have been limited. It is expected that tourism activities will – to a great degree – help to stop migration from the village.

The fact that rural tourism is relatively cheap provides liveliness to domestic tourism, while spreading the benefits to the masses. Other aspects of its effects include its provision of employment to rural women, its combining of agriculture and tourist-related activities, and also its tendency to slow out-migration. These are the most important benefits derived from this type of tourism. Furthermore, rural tourism is a type of tourism which will have important functions of enlivening Turkish tourism, whereby options offered by the tourism industry will be enriched and extended through 12 months, and also it will be brought to inner parts of the country, rather than being limited to coastal areas. There will also be principles that local potential will be used; that cultural heritage will be preserved; and that local cultural riches will be introduced to the rest of world (Soykan, 2002:158).

Mountains provide many opportunities to tourists who are sensitive to the environment and who believe in the principle of preserving nature and have chosen to engage in eco-tourism. Mountain villages are particularly suitable for tourism where local architectural characteristics are preserved (Doğaner, 2001:177). Tourists are able to witness and experience local living styles and situations. Çobanisa village preserves its authentic features, with its traditional lifestyle and present these rich local features to tourists and to the tourist industry. In Cobanisa, this has been done with a minimum of changes, under the leadership of the village mukhtar and other local authorities.

People living in mountains have life experiences which are often unique, with their own life styles peculiar to them. Their traditional culture is admiringly amazing. Despite this, relative to people in many other areas, most of the people the mountains are in desperate poverty. Generally, these people have less food input, and less overall production capacity. Factors such as the lower fertility of the land, shorter growing seasons, smaller cultivatable parcels, difficult climatic variables, and individual higher calorie need due to lower oxygen level, are important factors in the mountain survival situations (Karadeniz and Güneş, 2002:24).

The traditional income sources of Çobanisa Village are very limited. In this area, where only the cultivation of grains, and raising of sheep or goats are available, young people constantly migrate to cities like Isparta and Antalya in order to find jobs. The fate of the village is gradually changing, with the opening of Davraz Mount Winter Sports Center.

The aim of mountain tourism is to the contribute to the development of mountainous and forest areas which haven't been developed socially or economically, and to spread the tourism season to the whole year; to use natural values and surroundings, protecting them; to enable the expansion of the variety of choices offered by the tourism industry (Ülker, 1987:150). Through interviews done by the present researchers, it was determined that local people planned to extend local tourism activities to 12 months.

The alternative tourism represented by this study make people conscious of the variety of local things that have value, and also supports cultural awareness. Additionally, it causes people to come closer and improves their life-styles (Soykan, 2002:154). Thc beginning of tourism activities and services has made local people in the study region conscious of preserving the natural environment and their own cultural values.

In the development of rural tourism, dense populations and the difficulties caused by city life play important roles. Therefore, it is important to provide people with the facilities that they can not find in big cities. Through the concept of rural tourism, it is possible to promote conservation of nature, to enliven local traditions, and to promote simplicity, peace and friendship. However, it is most important is that we maintain the continuity of constructive concepts and practices after tourism has been established in rural villages. Even if some European tourists visit the same Turkish village for years, and can find the same old house, the same markets and clean nature, and thus, may continue to view the experience as enriching and positive. Such is the common understanding of the concept of sustainable tourism (Soykan, 2000: 32). In this context, in the establishments being built around Çobanisa for tourism, the primary goal is to reflect local culture, not to promote distorted construction typical of conventional tourism, and mostly to restore village houses and structures, being loyal to original local patterns.

The continuation of eco-tourism plays an important role in bringing local people, authorities, tourism companies and tourists together and in supporting the development of local culture (Björk, 2000;189-200). Individuals engaged in tourism in Çobanisa are conscious of maintaining the local culture, as they know that tourists coming to the area are primarily interested in the local life-styles and natural resources.

The negative effects of tourism can be seen in the areas where the participation of local people is little. Actually, as local governments were kept out of the development process of tourism in the South-West Antalya Tourism Development Project, developments there caused social-economic alienation (Eke and Özdemir, 2000:177). In Çobanisa village, local governments lead tourism development activities. In this context, the village mukhtar and his coworkers established a skiing training center on mount Davraz and helped plan and implement tourism strategy. This shows the important role of local governments in tourism development activities.

References

Björk, P., (2000) Eco-tourism from a Conceptual Perspective, an Extended Definition of a Unique Tourism Form. The International Journal of Tourism Research, 2, pp.189-200.

Cengiz, Y., (2002) The Risen Value in Tourism: Eco-tourism.TÜRSAB Ar-Ge Departmant, April 2002.

Çavuş, Ş. & Tanrısevdi, A., (2000) Sustainable Tourism Development in Globalization Progress. Anatolia: Tourism Investigation Review, Year:11, Septmber-December, p.151.

Doğanay, H., (1989) Ski Sports Tourism in Erzurum and Major Problems. Science and Literature Fculty Publications, No:55, Erzurum, pp.165-182.

Doğaner, S., (2001) Türkiye Tourism Geography. Çantay Bookstor, İstanbul, p.176

Eke, F.,-Özdemir, N., 2000, The Effect of Tourism to Local Residents. Commercials and Tourism Review, Year:2000, No: 3, p.177.

Gönençgil, B. & Güngör,Y., (2002) Mountains Tourism in Turkeyi, Potantials, Problems.Turkeys Mountains I. National Symposium, 25-27 June, 2002, Ilgaz, Kastamonu, p.40.

Karadeniz, N. & Güneş, G., (2002) Mountains Ecosystems and Sustainable Approaches. Turkeys Mountains I. National Symposium, 25-27 June, 2002, Ilgaz,Kastamonu, p.24.

Lee, C. H., Lee, J. H. & Han, S.Y. (1998) Measuring the economic value of ecotourism resources: The Case of South Korea. Journal of Travel Research, 36 (4): 40-47.

Mansuroğlu, S., (2006) Determination of Local People's Approaches of Tourism Developments: The Case of Akseki/Antalya. Akdeniz University Agricultural Faculty Review, 19(1), p.36.

Özkan, H., (2000)Touristic Service Potantials and Capaciy of Turkey. I. National Turkey Tourism Symposium 2-3 November 2000, İzmir, p.62-84.

Soykan, F., (2000) Rural Tourism and Gained Experientials Skills in Europa. Anatolia Tourism Invesigations Review, Year:11, september-december, p.21-33.

Soykan, F., (1999) Natural Envionment and A Type of Tourism Integrated to Rural Culture; Rural Tourism. Anatolia:Tourism Investigations Review, Year:10, March-June.

—. (2002) Socio-economic Effects of The Rurl Tourism and Turkey. Turkey Mountains I.National Symposium, 25-27 June 2002, Ilgaz,Kastamonu, p.154.

The Mountain Institute, 1998,Community-Based Mauntain Tourism. 13 April-18 May 1998, Synthesis of an Electronic Conference of the Mountain Forum.

Turkish Ministry of Interior, General Directorate of Local Authorities, 1999, Local Authorities in Turkey, Ankara.

Ülker, İ., (1987) The Mountains Tourism in Turkey and World.Turkey Development Bank Tourism Book 1987, Ankara, p.144.

Yalınkılıç, M. K & Yenilmez Arpa, N., (2005) Ecotourısm And The Protected Areas in Turkey. Protectd Natural Places Symposium, 08-10 Septmber 2005, SDÜ, Isparta.

CHAPTER SEVEN

PROSPECTS FOR RURAL CULTURAL TOURISM IN TURKEY: CASE STUDY OF KARABURUN

GÖZDE EMEKLÍ AND FÜSUN SOYKAN

Introduction

Although tourism industries can flourish in countries with relatively little in the way of geographic variation and geographic richness, there are many instances where physical geography determines not only the variety of options available in the planning and development of tourism industries, but also the quality of offerings and the distribution of viable tourist attractions within a region. Geographical position is an important factor in the emergence and development of tourism. Additionally, there is a very close relationship between climate and tourism. A country possessing physical geographical attributes favorable to large-scale tourism can, in turn, invest in key transportation infrastructures as well as in programs to promote preservation of the physical environment.

Countries possess physical properties that offer opportunities for a viable tourism industry often also possess opportunities to develop their cultural resources. Tourism can help to maintain the viability of small farms and rural communities through additional income, a market for handicrafts and organic farm produce (including traditional foods and drinks), and social interaction for country people. Tourism can also enhance opportunities for women (Hall 1998b; Swarbrooke 1996c).

However, businesses associated with tourism need to be well-managed as well as broadly sustainable if they are to be an effective component of rural development and to justify significant levels of investment made at the outset by the government. In several transition economies, there is the question of environmental quality, requiring careful regulation and monitoring of trade-offs between economic growth and environmental protection.

Swarbrooke (1996b:465–8) reiterates a forward looking approach through a 'rural culture continuum [to] combine the best of traditional cultures with exciting, dynamic new cultural infuences' and he also recognizes the need to 'balance culture with the economy and the environment, as these are all interrelated and interdependent in rural areas'. According to Swarbrooke, economic well-being and environmental quality go together.

Rural cultural tourism covers a range of activities that take place in non-urban areas. Its attraction is based on communication of tourists with local people (and local nature), and includes a broad range of options. Rural cultural tourism may include ecotourism, sport and health tourism (exploring nature, skiing, trekking, biking, horse riding, adventure, hunting, and fishing), ethnic tourism, educational tourism, shopping, and heritage and arts (OECD, 1994). A distinctive feature of the assets of rural cultural tourism is to make visitors go through a personal experience, enjoying the natural and cultural environment, attending residents' activities, and observing traditional practices and their daily life (Ehrlich, 2006). The benefits of rural cultural tourism include creating different revenue sources for rural society, decreasing unemployment rates by generating employment, employing women, creating awareness in local communities that cultures have value, and supporting local entrepreneurship (Cawley, M.E. and others 1999, p: 50-55). Tourism can also support local culture in rural areas by encouraging restoration of local and regional historic sites. Such sites can play an important role in developing regional tourism.

Study area, materials and methods

The main objective of this study is to highlight the importance of rural cultural tourism in Turkey, especially in Karaburun, and to give several examples of rural cultural tourism. Our goal is to help to develop an awareness of rural cultural tourism among people in the private and public sectors, among organizations of social workers, among instructors in tourism, and among local people. The Karaburun Peninsula is located on the Turkish Aegean coast at the entry of the Bay of Izmir. It forms the northern part of the Urla peninsula. The peninsula was introduced to marine tourism in the late 1980s. Hotels and boarding houses simultaneously were built and second houses rapidly increased. Crops and products of the peninsula such as olive and citrus trees, vineyards, and farms and gardens dedicated to artichokes, narcissus, hyacinth, sage, thyme, were gradually related to tourism.

In the peninsula, the old Greek and Turkish villages that are now deserted may be visited for sight-seeing, and trekking tours are available around Akdağ (the White Mountain), which is almost 1200 meters high.

Common consciousness to protect and preserve the natural, cultural and economic structure of the peninsula has been aroused and new projects with many participants have begun to be planned and applied. As a result, the peninsula really deserves to be fully protected. Valuable animal and plant diversityof this area include the Mediterranean seal (*Monachus-monachus*), the endemic island sea-gull, which is gradually becoming extinct, and many pharmacognostic, medicinal herbs with fragrant aroma. In additon, agricultural crops and fishing that support the local culture are also worthy of protection and preservation.

Rural Cultural Tourism Potential in Turkey

In terms of tourism, the geography of Turkey offers a great variety and wealth of natural resources and cultural values. Each of the seven major districts of Turkey not only offer unique human and natural resources, but collectively offer tourist opportunties that are universally valued, such as major archaelogical sites, favorable climate, access to the Mediterranean, as well as to striking mountain settings. Tourism in Turkey has a serious potential to flourish, particulary on the shores of Meditarrenean and in the Aegean and Marmara regions. Other districts are not at such a satisfactory level.

When there is a disproportionate demand for a region, it causes some negative evironmental problems. Turkey has a significant potential supply of rural cultural attractions. However, the demand for rural cultural tourism continues to be inadequate. Tradtionally, Turkey's rural population was relatively high, accounting for about 75% of the total in 1950. Since that time there has been a steady decrease in the rural population, and a steady increase in the urban population. The main motivation the development of rural cultural tourism is to provide support for traditional local cultures, to improve rural incomes, and to increase support to local geographical environments.

In Turkey, there are 36,527 villages and 42,098 sub-village settlements (in total 81,841, DPT, 2006, page 25). Although Turkey's rural population has decreased dramatically over the past several decades (75 per cent in 1950, 61 per cent in 1970, and 35 per cent in 2000), 35 per cent of country's population was employed in agriculture sector in 2006 (DPT, 2006, p.19). Today, rural populations have mainly a sedentary structure. However, in mountainous regions, we still see a semi-sedentary structure due to animal husbandry. Moreover, low productivity, low income levels, segmented agricultural lands, and increasing mechanization in agricultural production in rural areas push people away from rural areas. At the same time, opportunities for employment, education, health, and socio-cultural activities in urban areas attract people towards cities. In addition, in villages and towns which are near main roads, historic sites, and large cities, tourist-related establishments – including restaurants -- have recently developed. Since boarding-houses are not yet sufficiently developed in some areas, one-day tours have been organized. On the other hand, some projects related to reviving old villages like Kayaköy-Fethiye have been on the agenda and the ancient houses in some villages, such as Şirince near İzmir, have been used as mansions (Soykan, 2005a-b).

In Turkey, the use of rural areas for recreational purposes is multidimensional and goes back hundreds of years. Today, daily or longer trekking tours to mountainous areas such as the Taurus, Kackar and Kaz Mountains and to Cappadocia, Izmit-Golcuk, Yalova, Iznik, Sakarya-Akyazi, Istanbul-Sile-Agva, Dilek Peninsula, Bafa Lake, Fethiye have begun to be arranged by experienced and well-known travel agencies (Ekin Yazım Merkezi, 2000, p.877-911). Another way local communites in Turkey can benefit is by making more rural second homes available to tourists. Second homes that are mainly concentrated in coastal areas have shown a tendency to diffuse to other, non-coastal areas in recent years. The number of rural second homes has been increasing in the areas around major cities (Istanbul, Ankara, Izmir), in the mountains where mountain and winter sports have developed (Uludag, Ilgaz Mountain), in hinterlands of tourism centers (Antalya, Alanya, Marmaris, Fethiye, Bodrum, Kusadasi, Cesme, Ayvalik), and in regions which have agricultural value (Aegean, Marmara and Mediterranean Regions). Moreover, due to the traditional practice of 'recreational migration to plateaus' which involves moving from coastal zones to higher places, increasingly there exist 'plateau houses' (http:/www.turizm.gov.tr).

National parks are rural places in Turkey which can be used for ecotourism and rural tourism purposes. In Turkey, there are 37 national parks and many officially declared protected areas (Natural Park, Natural Protected Areas, Special Protection Areas, and Ramsar Sites etc.; Küçük, Arpa Yenilmez, Özmen, 2004, p.129). The natural resources of Turkey's rural areas offer many different alternatives for rural cultural tourism. There are also places with natural and cultural attractiveness on the World's Heritage List or other similar lists. The places included on the World's Heritage List are Goreme and Cappadocia National Park, Hattushash, Nemrut Mountain, Xanthos-Letoon, Pamukkale-Hierapolis,

Troy, and in indicative list are Bursa and Cumalikizik, Seljuk caravanserais, Ishakpasa Palace, Harran and Sanliurfa, Ahlat and Tombstones, and Sumela Monastery (www.kultur.gov.tr).

During the last 10-15 years rural settlements (villages and towns) located along coastal tourism centers, big cities, archaeological sites, and near major highways have increasingly been integrated into the tourism industry. Rural restaurant and stands have been opened in many rural areas, and these areas have increasingly come in contact with tourists, particularly with daily visitors.

In the present study, it is not possible to determine – in a detailed, systematic fashion – the potential for rural cultural tourism in Turkey. Rural cultural tourism in Turkey has the abilty to create external demand. As there are numerous opportunities for such tourism, advantages are also abundant. For example, in some plateau villages on the mountains near the Black Sea and the Mediterranean regions, traditional village houses are carved into the volcanic tufa in Kapadokya. In some vllages in Eastern Anatolia, there are unique buildings which are famous for their handicrafts, and are candidates for the rural cultural tourism. Therefore, it is vital to introduce rural cultural tourism after they are properly planned and programmed. Village projects can be accomplished successfully provided the local public is educated and organized, infrastructure is improved, village houses are changed into boarding houses, and local crops are standardized and transformed into well-known trademarks. With these changes, traditional village culture can be kept alive become increasingly well-known with tourist networks. In this paper, our approach is to evaluate the many rich physical and cultural features which are available in rural areas of Turkey, focusing on Karaburun, and to support turning these resources into viable touristic products. In turn, revenues gained from tourism in rural areas can be used for preservation and protection of unique and fragile cultural and natural heritage sites.

Description of Karaburun

Location and accessibility. The Karaburun Peninsula is located on the Turkish Aegean coast at the entry of the Bay of Izmir. It forms the northern part of the Urla headland (Fig. 7-1). The distance from İzmir center by way of land is 100 kilometers (approximately 60 miles) and there are regular bus services and a three-lane modern highway until the toll near Çeşme. The rest of the road is narrow and curvy in some places and it may take up to two hours to reach Karaburun from İzmir. The road along the west-coast takes you to the popular holiday resort of Çesme (46 kilometers). North of the Karaburun Peninsula lies the Greek island of Lesvos and westwards, the island of Chios (15 miles). The district area covers the peninsula of the same name (Karaburun Peninsula) and the town is situated close to the northern tip of the peninsula. The peninsula is bordered by the districts of Çeşme and Urla in its south and faces the Greek island of Chios. The municipalities of Karaburun and Mordogan are making great efforts to install a regular connection with Izmir via the sea.

Karaburun City is situated on the northern part of the peninsula. Karaburun has two small islands in front of its coast. Most of the peninsula has a difficult terrain. The north-south oriented Bozdag mountain range (1,218) splits the peninsula into two parts.

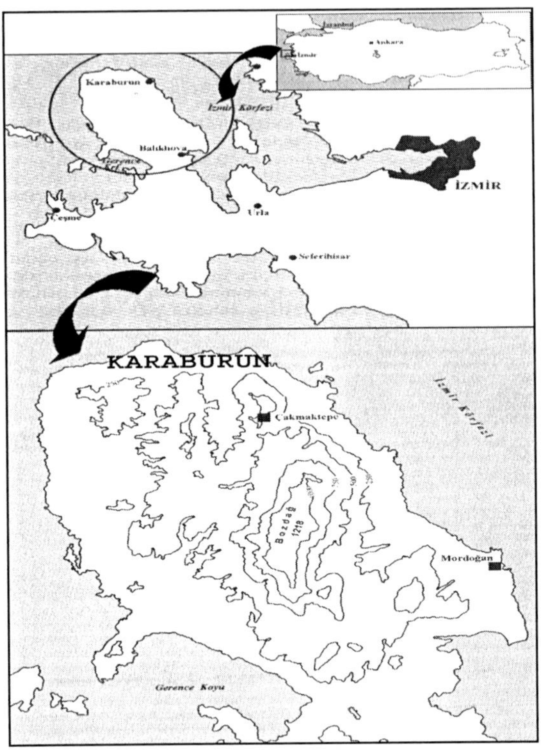

Fig. 7-1: The Geographical Location of the Karaburun Peninsula

Karaburun Peninsula is a rather mountainous region, and large farm tracts are very scarce. In old days agricultural land was mainly used for grape and tobacco agriculture. Today, only a few vineyards have survived, while tobacco has completely disappeared. Nearly all crops are grown in a natural way, which makes them – by definition – organic. Because of the nature of the soil and the climate, the cultivated products of this area, i.e. olives and olive oil, artichokes, narcissi and mandarin oranges, are of very high quality. Some products, like the Hurma olive (an olive which through a microclimate effect ripens directly on the tree and need no further treatment before consumption) grow only on the Karaburun Peninsula.

Society and Economy

The Karaburun Peninsula has 13 villages of which 5 are administrated by the Karaburun Municipality. The number of inhabitants is approximately 3000. The villages are quite different from each other with regards to their culture and natural scenery. Two of them (Ambarseki and Saip) lie on the east coast of the peninsula, while the other three (Bozköy, Tepeboz and Haseki) lie on the west coast. The fishing village of Yeniliman is administrated by the village of Tepeboz. This little port is certainly worth a visit. Small scale agriculture, goat and sheep breeding, flower cultivation and olive growing are some common activities in these 5 villages. Villagers of Saip, Tepeboz and Yeniliman are also involved in some fishing actvities. Migration to İzmir for a long time caused some villages to be deserted, and some agricultural fields were either sold or left uncultivated and converted into wilderness. Over time, the average age of the population changed in favour of older people.

Large plots of land have been very rare in the region, because of the mountainous nature of the peninsula, and a significant number of families have survived off small cattle rising. Certain regional specialties are made with the milk of goat and sheep, including kopanisti cheese, basket cheese, goat skin cheese and a dessert made from cheese which is called "Hösmerim" or "Sündürme". The most important agricultural activities of the Karaburun Peninsula consist of cut-flowers (narcissi and hyacinth), mandarins and artichokes. The yearly narcissi and hyacinth production is about 22000 flowers. One hundred forty (140) hectares are used for artichoke production, resulting in 6,450,000 artichokes. Lemon, mandarins and regular oranges are grown on about 90 hectares. The yearly production amounts to 1060 tons. Since the peninsula is surrounded by the sea, a signifcant number of people earn their livelihood from fishing, and the most important catches include the lesser gray and the red mullet. Some families continue to produce honey. The flavor and taste of the Karaburun honey is quite well known and many people who are familiar with the quality of Karaburun honey go and buy their honey directly from the villagers (www.karaburun.bel.tr/)

The main aim of the Karaburun Women's Agritourism Cooperative is to develop agritourism and to start a local house-boarding business in the Karaburun Peninsula. For this reason, some women in Küçükbahçe, Parlak and Sarpıncık villages on the peninsula took the initiative for the first time in Turkey and founded an agritourism cooperative in 2003. Membetship consisted of 18 people. Six of these have detached houses to be considered as boarding houses. Three are trying to prepare their buildings for use as boarding houses. The remaining nine members are planning to sell food and other local products.

The boarding house business in the area began in the summer of 2005. Reservations can be made by phone. The cooperative's office is in the district of Karaburun and the local service place is in the village of Küçükbahçe. Through the cooperative, local skills and crafts have the possibility of surviving. Additionally, the women who are involved can contribute to family and household incomes. Representatives of the women cooperative have visited Petra on the island of Lesvos in Greece in an attempt to make acquaintances and to learn from the experiences of women engaged in similar activities.

The Prospect for rural cultural tourism in Karaburun

Karaburun was introduced to marine tourism in the late 1980s. At that time, hotels and boarding houses were built and second houses rapidly increased. While crops and other products, such as olive and citrus trees, vineyards, artichoke, narcissus, hyacinth, sage, thyme, and cake-cheese were gradually decreasing, the shore area began to turn into a concrete area through constructions. The overbuilt nature of the shore areas gave rise to a strong movement led by a small group of highly motivated local people called the Karaburun Civil Initiative in 1998. Subsequently, the Civil Initiative was changed into the Local Agenda-21 of the Karaburun Peninsula (www.karaburun.bel.tr/).

The development of a movement to protect and preserve the physical, cultural and economic structure of the peninsula led to the development of new projects, with many participants. As a result, it is commonly acknowledged that the peninsula really deserves to be protected, partcularly due to the existence of native animals, such as the Mediterranean Seal (*Monachus-monachus*) and the endemic, nearly extinct, island sea-gull. In addition, there must also be an aim to preserve local culture, and promote the continued production of medicinal herbs, local agricultural crops and local fish products.

The Karaburun Peninsula has the potential to offer a very wide range of activities. Underwater activities, trekking, village accommodation, eco and agro-tourism programs are all perfectly realizable as major tourist activities.

Historical and Cultural Snapshot. Ever since ancient times the Peninsula has been inhabited by different cultures. From old sources we know that the Peninsula was known as "Mimas". Excavations in the Çakmaktepe area have revealed artifacts, including cutting tools, stone axes and primitive pottery dating back to the Chalcolithicum period (4000 B.C). Around 3000 BC, the Hittites appeared on the Karaburun Peninsula. Following this period the peninsula came under the influence of Ancient Greece, Persian, Rome and Byzantium. During the classical period, the peninsula gained cultural and commercial importance because of the city of Erythrai (www.karaburun.bel.tr/). Under Roman rule, the importanceof the area declined.

Later, during the Byzantine period, the peninsula again gained in importance. Once the peninsula came under Turkish control, the city of Karaburun was called *Ahurlu* or *Ahırlı*, meaning "The stables". It was only in Ottoman times that it took the name of Karaburun. Therefore, it has a rich historical and cultural inheritance.

Coast and Beaches. The peninsula's villages are usually located close to the sea. Therefore, it is still possible to swim right off the shore, for the coastal area is not yet been covered with concrete structures. Sahip Island demarcates the small bay of Karaburun from the center town of the district, with some of the purest beaches within Turkey. The coasts of the peninsula have beautiful bays, as yet very often undiscovered by outsiders, although there is one German vacation village slightly to the north of the district center. In and around the city of Karaburun, one can find many beaches. Some of these beaches are reachable by road, while others can be visited only by boat.

Karaburun City has two beaches which bear the "Blue Flag" label. This includes Bodrum Beach and Aquarium Beach (Incirlikoy). The other beaches are the Kuyucak Beach and the Mimosa Beach. The "Blue Flag" label is an award for the guarantee of the quality of the sea water and of the environment. It is therefore very important that, together with local authorities, we protect our beaches and environment overall. There are two more little beaches located left and right of the little fishing port of Karaburun. Further away from the center you can enjoy the calm of the beaches of Esendere, Saipalti, Igdealti, Büyükkent, Dolungaz, Yildizkent, Akçakilise, Yeniliman and Kumbükü.

Villages. The Karaburun Peninsula has 13 villages of which 5 are administered by the Karaburun Municipality. However, the population of these villages has dropped dramatically over the last 80 years. According to records, all five villages have old histories. In addition, one can still see the ruins of what once were the villages of Çullu, Hisarcik and Manastir (Kalecik). The Karaburun Municipality has developed a project for the restoration of the old mosque of Çullu (Fig. 7-2).

The villages are quite different from each other with respect to their culture and natural scenery. Two (Ambarseki and Saip) lie on the east coast, while the other three (Bozköy, Tepeboz and Haseki) lie on the west coast. The fishing village of Yeniliman is administratively dependant on the village of Tepeboz. Small scale agriculture, goat and sheep breeding, flower cultivation and olive growth are activities common to these 5 villages. The villagers of Saip, Tepeboz and Yeniliman are also engaged in fishing. Traditional handicraft skills, such as basket making, are transmitted by the old masters to the younger people. During the last couple of years dwellers from outside Karaburun settled in Ambarseki and some of them introduced the idea of starting pension tourism in order to contribute to the development of the village. Bozköy (or "Boz" as it is commonly called) is known to be one of the oldest villages of the Karaburun Peninsula. Typical of the village are the many original stone houses, the old fountain and mosque, as well as the old cemetery. The Valley of Uzundere and its fertile soil are a wonderful place for agricultural activity. The area of Akçakilise is renowned for its fertile land and for beautiful bays and beaches, where one of the Mediterranean Monk seal caves is located. Narcissi and hyacinths are the most important crops. Mandarins, olives, and goat and sheep breeding are other sources of income of Bozköy. Until the 1950s, Bozköy was one of the major grape producing villages of the peninsula.

Local products. The peninsula offers a couple of unique products, such as the "Hurma" olive (an olive which through a microclimate effect ripens directly on the tree and need no further treatment before consumption), the narcissus flower with its unique scent, and artichoke. Since ancient times, the Karaburun Peninsula was known for its vineyards and olive groves. But plant diseases and emigration made the villagers abandon grape growing.

The Karaburun Peninsula has a lot of local products which may be sold in the agritourism sector and most of them are ecological. They are partly sold as fresh and just-picked (artichoke, grape, citrus, narcissus, etc.), and some are processed (olives, olive oil, dried medicinal herbs like sage and thyme, and cake-cheese made from goat milk). Home-made bread and macaroni, regional pastries, honey, special jams and marmalades are also for sale. Additionally, there are other products made by the women such as lace, hand-woven things and embroidery. They are either sold in the places where they are made, or taken to market places nearby, into the town center of Karaburun, even to festivals and fairs.

Healing Plants. The Karaburun Peninsula has a rich natural collection of healing plants. In the hills and valleys, one can find some 67 medicinal herbs and plants. Spurge, pennyroyal, common poppy, St. John's-wort, caper, thyme, silifa, Aaron's rod, sea squill, sage and wild lavender are only some of them.

Fauna. The Karaburun Peninsula has a rich stock of wildlife. On land as well as in the sea, one can see various species, of which some are quite rare. Wild boar, fox, marten, badger, otter, hare, squirrel, birds of prey (eagle, buzzard, falcon etc.), water-turtles, river-crabs, chameleon, snakes, lizards, many types of butterflies and other insects like beetles, grasshoppers, praying mantis etc. are part of the Karaburun fauna.

Certainly one of the most important species present along the coasts of the Karaburun Peninsula is the extinction-threatened Mediterranean Monk seal (*Monachus monachus*) of which only 500 are presently surviving worldwide, about 100 of which are in Turkey. The shores of the Karaburun Peninsula hide several sea caves, where the Mediterranean Monk seals feed their pups during their first months. Most of the caves have been located and great efforts are made to protect the seals in order to guarantee the survival of this most vulnerable species.

Accomodations. In an effort to understand local tourist activities, it is certainly important to figure out total arrivals, total number of nights spent, and the average length of stays. But overnight stays by foreign visitors are very few in Karaburun. In 2007, there were 11 licenced establishments and 680 beds in Karaburun. The accommodation capacity of the agritourism cooperative in Karaburun villages is about 30 beds. The average length of stay in Karaburun is 2-5 days. The Occupancy rate is 40 per cent. The total catering capacity of cafes, bars, restaurants averages 1800-2000 persons. The capacity of accommodation facilities meets present needs. However, current tourism promotion by the Karaburun Municipality in cooperation with other official bodies and private enterprises will require a certain expansion which can easily be obtained by upgrading the existing accommodation potential and creating some new accommodation facilities.

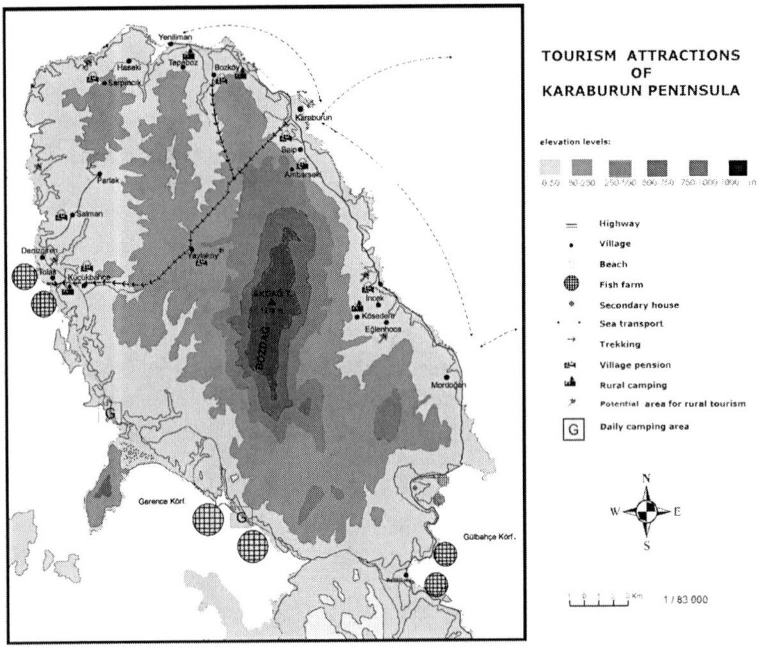

Fig. 7-2: Major Tourism Attractions of Karaburun Peninsula

Conclusions and Recommendations

Karaburun is one of İzmir's most attractive mountain regions where a rural cultural tourism initiative is now well-established on a modest scale with local control of a business which clearly projects a living culture. All of this meets one of the key criteria for sustainability. In spite of the regions unspoiled nature, its historical past, its cultural variety and its climate, the Karaburun Peninsula along with the City of Karaburun has not yet received its rightful place on the tourism scene. However, we are confident that, critical problems, such as transportation, accommodation, and a skilled work force have been solved, tourism development will automatically grow. Like many other places in the world, where mass tourism has been abandoned, we are aware that "alternative tourism" is the right choice and represents the best development model for Karaburun. It is obvious that we will gain a lot more from such a model of tourism, particularly where unique local features of culture and nature are combined with the areas unique sea and climate factors.

In order to contribute to the development of the tourism sector, the Municipality of Karaburun has started a number of services and projects and will continue these efforts. With the right policy and with the support of the local authorities and local people, the Karaburun Peninsula can get its due place on the international tourism scene.

It is important that the region protect its monuments. There is good coordination among NGOs at the county level and planners are cooperating with advisors in several areas to ensure that the character of the region is not adversely affected by the expansion of tourism. The region is not well-adapted for mass tourism. Rather, it is one of the best places in the İzmir area to escape the stress of the modern world and to discover a culture that works to a leisurely pace set in an environment that is rich in terms of both scenery and biodiversity. However, marketing stands as a barrier to further growth and, while the environmental hazards have not had a disastrous impact on the area; they demonstrate the need for much more effective regulation systems if tourism is to play a more central role in the rural economy.

References

Cawley, M., Gaffey, S. and Gillmor, D.1999. The Role of Quality Tourism and Craft SME in Rural Development: Evidence from the Republic of Ireland. Anatolia: An International Journal of Tourism and Hospitality Research, *10 (1), 45-59.*

DPT 2006. Dokuzuncu Kalkınma Planı (2007–2013). Kırsal Kalkınma Politikaları Özel İhtisas Alt Komisyonu, Son Taslak Raporu. (9th Development Plan (2007–2013), Rural Development Politics Private Specialization Commission, The Latest Draft Report), Ankara.

Ehrlich, K. 2006. Rural Tourism: a Promising Segment for a Sustainable Development Strategy. UNWTO European Conference: Impact of European Union Enlargement on Tourism Development in Europe, 1-2 March 2006, Vilnius, Lithuania.

Ekin Yazım Merkezi 2000. Gezi' 2000, Türkiye Tatil Rehberi, (Trip' 2000, Turkey Holiday Guide), İstanbul.

Emekli, G. and Soykan, F.2005. Diversification des Types de Tourisme dans la Région Egéenne (en Turquie): Possibilités du Paysage Naturel et Culturel et sa Répartition Géographique. Xe Séminaire Internationale de Forum UNESCO Université et Patrimoine: Paysages Culturels au XXIe Siècle, 11–16 Avril 2005, New Castle, UK.

Hall, D.R. 1998a. Central & Eastern Europe: tourism development and transformation. In Tourism and Economic Development: European

Experiences, ed. A.M. Williams and G. Shaw, pp. 345–37. Chichester: Wiley.
—. 1998b. Tourism development and sustainability issues in Central & South Eastern Europe. Tourism Management 19: 423–31.
—. 2000. Evaluating the tourism–environment relationship: Central & East European experiences. Environment & Planning B: Planning and Design 27: 411–21.
Jurdant, E. 2003. Le Tourisme Rural : Un Acteur Méconnu du Développement Rural Global. Très Petites Entreprises, PME, Artisans: Les Nouveaux Atouts Du Développement Rural, Une Conférence du Mouvement Européen de la Ruralité, 20 November 2003, Bruxelles. http://www.ruraleurope.org/MER
OECD, 1994. Les Stratégies du Tourisme et le Développement Rural, OECD/GD (94) 49, Paris. http://www.oecd.org
Rural Europe, (1998). Un Atout Pour le Développement Local: La Ressource Patrimoine Source: LEADER Magazine no:17. http://www.rural-europe.aeidl.be/rural-fr/biblio
Soykan, F. 2002. Kırsal Turizmin Sosyo-Ekonomik Etkileri ve Türkiye. (Socio-Economical Effects of Rural Tourism and Turkey) I. Türkiye Dağları Ulusal Sempozyumu, 25-27 Haziran 2002, Ilgaz-Kastamonu. Bildiriler Kitabı, 154-159.
—. 2005a. Le Tourisme Rural et ses Effets à Village de Şirince. Vth Mediterranean Conference: Tourism Beyond the Coastline: New Trends in Tourism and the Social Organisation of Space, 22-24 September 2005, Thessaloniki.
—. 2005b. Un Village où Rencontre de la Culture Turque et Grecque: Le Tourisme Rural et Şirince (Turquie). VIIIe Congrès Culture Européenne, 19-22 Octobre 2005, Pamplune, Espagne.
Swarbrooke, J. 1996a. Towards a sustainable future for cultural tourism: A European perspective. In Tourism and Cultural Change, ed. M. Robinson, N. Evans and P. Callaghan pp. 227–56. Newcastle: University of Northumbria.
—. 1996b. Culture tourism and the sustainability of rural areas in Europe. In Managing Cultural Resources for the Tourist, ed. M. Robinson, N. Evans and P. Callaghan pp. 447–70. Newcastle: University of Northumbria.
—. 1996c. Towards the development of sustainable rural tourism in Eastern Europe. In Tourism in Central and Eastern Europe: Educating for Quality, ed. G. Richards, pp. 137–63. Tilburg: Tilburg University Press.

Tunçdilek, N.1978. Türkiye'nin Kır Potansiyeli ve Sorunları. (Rural Area Potential and Their Problems in Turkey) İstanbul Üniversitesi yayın no: 2364, İstanbul.

Turnock D. 2002. Prospects for sustainable rural cultural tourism in Maramures, Romania Tourism Geographies 4(1), 62–94 Tourism Geographies ISSN 1461–6688 print/ISSN 1470-1340 online © 2002 Taylor & Francis Ltd http://www.tandf.co.uk/journals

WTO (2004). Rural Tourism in Europe: Experiences, Development and Perspectives. World Tourism Organization Publications.

http: //www.karaburun.bel.tr/

http://www.eurogites.com

http://www.gap.gov.tr

http://www.geziturkiye.com

http://www.kultur.gov.tr

http:/www.turizm.gov.tr

CHAPTER EIGHT

AGRICULTURE AND TOURISM CONFLICT: CASE STUDY OF MUDANYA REGION IN THE WEST OF TURKEY

SELMA AKAY ERTURK

Introduction

The study places its main emphasis on the interaction and conflict between rural and urban areas, especially agricultural and tourism areas, and the factors which have brought about substantial change in land use in Mudanya and its environs. The study area has an attractive landscape and location, which have allowed for the development of active tourism and several other economic activities. The most important subsistence activities in Mudanya are fisheries and the cultivation of fruit, olives and vegetables. Industrial activities are not at an advanced level in this region. Mudanya has a great historical tradition and many cultural treasures. Recently, internal tourism has been showing high potential and has been increasing. Lately, the increasing demand for second homes has caused the destruction of olive grove areas. Agricultural land has been in transformed building areas. Change in land use is going on now and is supposed to carry on in the future, but such change causes degradation of the environment. The study focus is therefore on the necessity for developing policies for the successful management of prevailing conditions, in order to make appropriate changes to the area's geographical potential and to provide sustainable development therein.

Study Area

Mudanya is located in the northwest of Turkey, the southern part of the Marmara region, on the southern coast of the Marmara Sea. Mudanya is close to Bursa, which is Turkey's fifth largest city, and provides a connection between Bursa and Istanbul, Turkey's largest city, by seaway.

The site of the study area is between local hills, which are about 250 meters high, and the Marmara Sea coast. The length of the coastal area is 7.6 kilometers. The study area also comprises Güzelyalı, which had been one of Mudanya's villages until 1960, then became a quarter of Mudanya and was later accepted as a municipality in 1994. The increasing demand for second homes has caused changes in land use in Güzelyalı. The transformation of agricultural land into building plots has been dramatically realized in Güzelyalı since 1990.

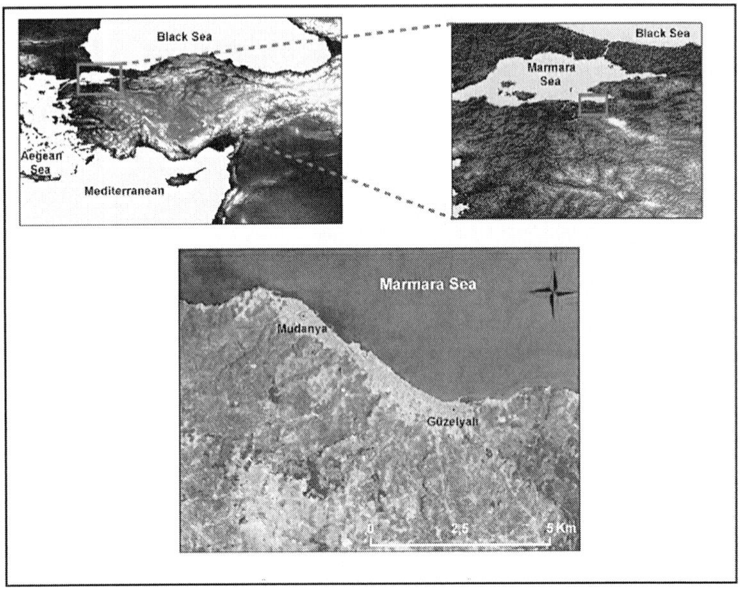

Fig. 8-1: Location of the Study Area

Material and Methods

This study examines the interaction and conflict between agriculture and tourism. A variety of information and analytical methods are employed, including analysis of satellite images and field surveys of changes in land use. The present situation in the study area and the evolution of rapid growth in urbanization were directly observed. In this study, Landsat 2000 Etm and 1987 Tm satellite images with a 30 meter resolution were used. The study is spread over a thirteen-year time span (1987-2000).

Findings and Discussions

Tourism is the temporary movement of people to destinations outside their normal homes and workplaces, the activities undertaken during their stay and the facilities created to cater to their needs (Mathieson and Wall, 1982). Tourism is very often based on the development of residential estates aimed at external stakeholders, who establish their residence in the area for part of the year. Some "outsiders" choose the areas to retire to, whilst others use it to develop their professional aspirations which are connected to the new tourist activity, and hence become permanent residents (Esparcia and Buciega, 2005). In natural areas, where tourism either exists or is proposed, there is the potential for both beneficial and adverse environmental and socio-economic impact to occur (Newsome, Moore and Dowling, 2002). The range of tourism-related impact is similar to that of many other environments. The impact of tourism may be more problematic because tourism is often concentrated on or near a ecologically and geomorphologically dynamic coastal environment (Hall and Page, 2001). The tourism of the study area is based on residential tourism linked to second homes, which are only occupied during certain periods of the year. The other recreational activities are based on day trips and weekend excursions from Bursa.

The study area has some geographical and historic characteristics which make it very attractive to urban consumers. It is formed by the coastal region and the hills, lying between two depression areas in the north and south. Gemlik Gulf is in the northern part of the study area and Bursa Plain, Manyas and Uluabat Lakes in the south. The physical structure of the study area is formed by hills, whose heights are no more than 250 meters. The hills lie behind and parallel to the coast. The altitude of the hills gently decreases from the northeast to the southwest. The coastal zone between Mudanya and Güzelyalı is a low coastline, very suitable for recreational activities. Owing to this fact, the coastal zone of the study area is the most preferred place for secondary housing cooperatives, whose number has increased in the last decade and is where multi-storied second homes are being built.

The climate of the study area is similar to the Marmara region's climate; in general terms a transition climate between the Mediterranean and Black Sea climates. Its main characteristics are that summers are not as hot as those in the Mediterranean, nor is it as rainy as in the Black Sea region. The annual average temperature is 15.3°C. The average temperature in the coldest month is 5°C and 22.4°C in the hottest month. The annual average precipitation value is 622 millimeters in the study area. Because of the effects of the sea, the study area is also more temperate than inland, which is very crucial for land cultivation and tourism activities. The most influential winds are the north winds. As the study area is located in the northern part of Turkey, the seawater temperature is only suitable for bathing for a period of 2-3 months/year. On the other hand, the bathing period is 6-8 months per year on the Mediterranean and Aegean coasts. Accordingly, the vacation season in Mudanya is shorter than that of the Mediterranean and Aegean coasts and tourism is not at an advanced level. Tourism is limited only within the Bursa region and there are not many hotels and motels available for temporary residents, yet there are numerous second homes in the study area.

Mudanya has a great historical background and many cultural treasures. It has been a settled area since ancient times. The study area has some characteristics that make it attractive to people who would like to relocate there. The main draw is that Mudanya is on the coast, which makes it an important harbor. Mudanya also has climatic conditions especially suitable for olive cultivation. Fisheries are also a very significant survival activity for the people inhabiting the coastal zone. For all these reasons, Mudanya and its environs are accepted as an attractive place for human settlement.

When Mudanya was founded is not known. Its original name was Myrlea (Umar 2004), and then Apemeia (Strabon, 12 IV 3). There is no archaeological heritage from ancient times (Umar 2004). The name of Mudanya was Montaneia during the Byzantine period and it has been Mudanya since the Ottoman period, when it was a very important harbor city, and has continued to be ever since for export and import, owing to the easy transportation of goods and people between Bursa and Istanbul. The seaway between Bursa and Istanbul is presently being considered for research. Bursa is highly industrialized, which affects the region as well.

Those buildings surviving from the Ottoman period are being protected today. The protected area includes the four quarters of Mudanya. Recently, the importance of the cultural heritage has been realized. Restoration of the historical buildings has been increasing gradually. The building where the Mudanya Truce was signed at the end of the Independence War is a real historical treasure. There has been an increase lately in the number of people joining cultural excursions, especially by those who live in metropolitan areas. The restoration of old buildings introduces a huge potential for culture tourism in Mudanya. In contrast, there are no such historical and cultural treasures in Güzelyalı.

Fig. 8-2: Agricultural land transformed into a building area

Transportation between Mudanya and Istanbul had only been provided by sea buses up to the end of April 2007. Since then, ferry-boats started to set sail between Güzelyalı and Istanbul, with Mudanya no longer being used as a transfer point. Güzelyalı seems to be the new transfer point. The demand for second homes in Güzelyalı will increase, especially due to the demands of the people who live in Istanbul. Correspondingly, pressure on agricultural land will increase in the study area as well.

Mudanya and its environs have some geographical characteristics which make them attractive to urban consumers for recreation and tourism. The study area is located on the coast, where the climate is very favorable, especially for the elderly. Because of its climatic conditions, the area is recommended to older people by physicians. The study area is very close to Bursa and the topography is well suited for ground transportation. The people who live in Bursa tend to seek time away from the noise and the crowds of the metropolitan area, preferring Mudanya and its environs for day trips, weekend excursions or even for second homes. All of these factors make the study area the recreational site for Bursa.

Table 8-1: Mudanya's Population Growth (1935-2000)

Year	Population Inhabitants	Increase Inhabitants	Growth Rate (%)	Population Density (per square km)
1935	5030	-	-	39
1940	4823	-207	-4.1	39
1945	5624	801	14.2	43
1950	5824	200	3.4	49
1955	5911	87	1.4	71
1960	6026	115	1.9	72
1965	6849	823	12.0	72
1970	7938	1089	13.7	73
1975	8399	461	5.4	78
1980	10606	2207	20.8	80
1985	12152	1546	12.7	83
1990	17196	5044	29.3	112
2000	20682	3486	16.8	162

Demand for second homes keeps growing and prices are increasing accordingly. This situation increases the pressure put upon agricultural land, especially olive groves. As a result of the current situation, the owners of agricultural areas are preparing to sell their land. Olive groves are easily adaptable to the land plots where the multi-storey buildings or villas are to be built. The types of the settlement and the landscape have been changing over time and have caused conflict between tourism and agriculture (Fig. 8-2).

When the increasing population of the study area was analysed, it was observed that the population of Mudanya was 5,030 in 1935 and 20,682 in 2000 (Table 8-1 and Fig. 8-3). The rate of increase quadrupled between these dates, most notably after the year 1975. The population exceeded 10,000 in 1980. The population doubled in the forty-five years between 1935 and 1980, and doubled again in the twenty years between 1980 and 2000. The national annual population growth rate was 18.28‰ per annum between 1990 and 2000, but the rate was 53.35‰ in Mudanya. For this reason, it can be said that urbanization has progressed very quickly.

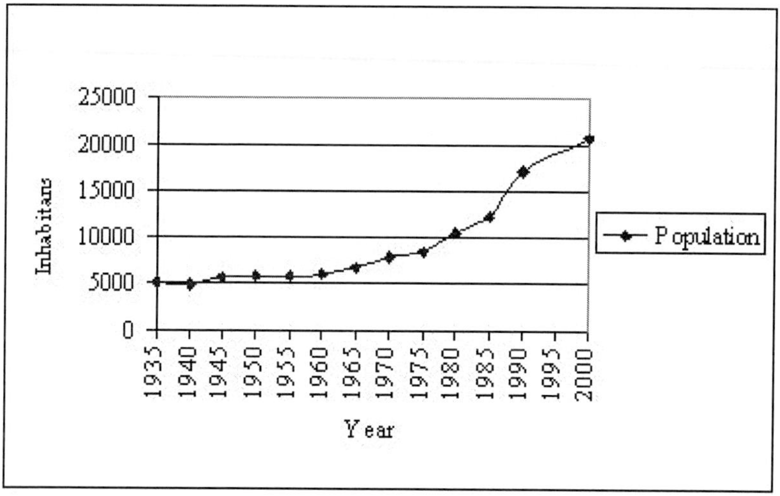

Fig. 8-3: Mudanya's Population Growth (1935-2000).

Table 8-2. The Population of Güzelyalı (1935-2000)

Year	Population (Inhabitants)
1935	252
1940	366
1945	318
1950	566
1955	676
1960	915
2000	8341

Güzelyalı, which was a village before 1960 and was known then as Burgaz, was a quarter of Mudanya until 1994. The population of Güzelyalı was 252 in 1935, 915 in 1960 and 8,341 in 2000 (Table 8-2), and has been a municipality since 1994. The marked population growth has caused changes in land use in Güzelyalı. In the last decade a great number of multi-storied second houses have been built in Güzelyalı. The distance between the sea and the hills lying in parallel is much greater than in Mudanya. For this reason, both first and second homes are built here, but the area near the coast is full of buildings and there aren't any more olive groves. The new buildings are built on the slopes of the hills. The hill's slopes are also covered by olive groves. So the transformation of agricultural land into building areas is still going on.

The population of Mudanya and Güzelyalı increases in the summer months because of the second home owners coming from Bursa. Mudanya's winter population is 20,682 and more than triples to over 70,000 during the summer months. Güzelyalı's winter population of 8,341 increases by more than six times, to over 50,000 during summer periods, due to local tourism. In the study area, the population growth exerts great pressure on natural resources (soil, water, biodiversity) in the summer.
Of the total general budget tax revenues, approximately 10% is transferred to local authorities via the central government, and of this only 6% is paid to the municipalities on the basis of local population figures recorded in the official census of the resident population. In fact, coastal municipalities cannot provide adequate services for the tremendously increased population in their areas during the summer months (Burak et al. 2004). Because of rapid tourism development, direct discharges from rivers, surface run-off and drainage from port areas, local fuel storage terminals, domestic and industrial effluent, all discharge through outfalls, and various contaminants from ships and long-range atmospheric transport, the anthropogenic and environmental activities show complex interactions in the region. Most of the coastal settlements in the Gemlik Gulf use creeks or simple outfalls for their waste water discharge. Mudanya is one of the most polluted areas on the southern coast (Ünlü et al. 2006).

A sewage treatment plant was installed last year in Mudanya. Because of infrastructural and financial problems, not all of the waste water in the study area can be treated. The collected wastewater after preliminary treatment has been discharged by deep-sea outfall to the sea. This situation has caused water pollution and ecological problems.

The population of the study area has continuously increased since 1935. However, in the last two decades the increase has been significant. Natural resources, however, are fixed and do not increase. Thus, degrading of natural resource makes the maintenance situation more difficult. The main reason for the land use change in the study area is the increase of Mudanya's and Bursa's populations which have occurred since 1970 and the rise in living standards of medium-income people after the 1980s. For this reason, the tendency to own a second home on the coast has risen since 1990.

Fig. 8-4: A view of Mudanya in the 1920s.

Fig. 8-5: A view of Mudanya in 2006.

Industrial activities have increased since the 1970s in Bursa. Automotive, spare parts and textile industries have especially developed in the area. Industrialization has caused migration to Bursa from rural areas and other undeveloped urban areas since the 1970s. The urbanization process was quick and unplanned. The population of Bursa was 289,864 in 1970. As a result of internal and external migrations, especially from Bulgaria, the population of Bursa increased to 1,289,699 in 2000. Overpopulation has been the cause of many social, economic, spatial and environmental problems. Owing to this fact, people who live in Bursa and would prefer to get away from a crowded society are in favor of spending their weekends and summer holidays close to nature, especially along the coast. Mudanya is very close to Bursa and accessibility has been a key factor in increasing the number of second homes in the study area. This is why Mudanya and its environs have been the recreational centers of Bursa. The people, who are mainly from the metropolitan area of Bursa, see Mudanya and its environs as a place to relax and the number of second homes is gradually increasing.

There were two different periods when the improvement of local tourism occurred in the study area. The first period was during the 1970s and 1980s. Retired people who lived in Bursa chose to spend the summer season in bungalows and camps located by the sea near Mudanya and Güzelyalı. In this period, the number of local tourists was low and was not the cause of the changes in land use. The increasing demand for second homes had not yet taken off. The impact of recreational activities on natural resources also had not been seen yet. A few of these camps are still active, but they are surrounded by many of the new multi-storied buildings. Increasing demand and the prices of new second homes will entirely destroy these camps in the near future.

In the second period, since the 1980s, the rise in the living standard of medium-income people has changed the types of places where people prefer to stay during the summer holidays. The demand for has picked up in the last decade. The rise of automobile ownership has also strengthened the trend. It has become much easier to get to and from Bursa and Mudanya and its environs. Multi-storied buildings have primarily been built by the coast. The delay in planning systematic zoning destroyed the olive groves on the coast. During that period coastal areas filled up. Nowadays, the multi-storied buildings gravitate toward the slopes of the hills (Fig. 8-4 and 8-5). The slopes are also covered by olive groves. The pressure on the agricultural fields on the slopes has been going on continuously. Different characteristics have appeared during this period, as compared to the first: Demand for second homes is increasing; wealthy people prefer spending their holidays in privacy, so they buy residential houses and the prices of second homes have been continuously rising. The environmental impact of tourism has also increased in this period, for instance, water pollution (sewage, organic and inorganic waste, fuel oil from sea traffic, etc,), soil erosion, destruction of vegetation, direct loss of habitats and noise pollution.

Use of GIS techniques, coupled with remote sensing, are essential elements for the preparation of an integrated development plan. Together, they can provide rapid, accurate and up-to-date databases at all levels for planners. The inherent flexibility of GIS techniques can make an important contribution towards improving the speed and qualitative performance of urban and regional planning (Fazal 2000). Satellite images were used as a working base map and with GIS techniques it provided data about urban expansion in the study area. The surface area of the building area was 1.97 km² in 1987, but this has increased to 5.42 km² in 2000. This shows that urbanization has progressed very fast and the study area is losing its agricultural characteristics (Fig. 8-6).

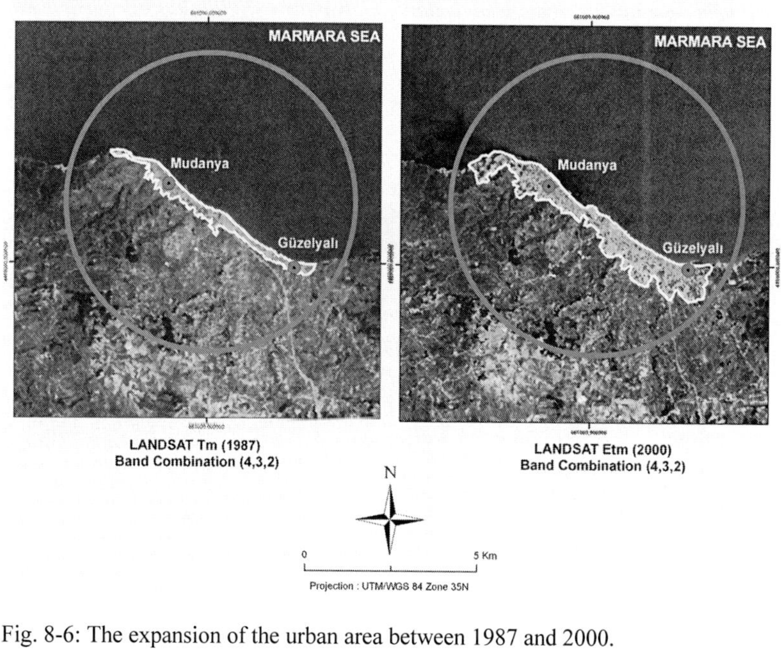

Fig. 8-6: The expansion of the urban area between 1987 and 2000.

Conclusion

In light of the above discussion, it can be pointed out that for the development of sustainable tourism, planning has a key role to play in resolving conflicts. In the absence of planning, tourism development will become unregulated and this will cause negative social, economic and environmental impacts to occur. Therefore, the planning of tourism development should take place at national, regional and local levels. Tourism should be planned according to the principle of sustainability.

Recreational activity began in the 1970s and has been most popular since the 1990s in Mudanya and its environs. Up till now, tourism development has been unplanned and negative impacts have occurred.

The findings of this study on Mudanya and its environs have demonstrated that:

- The population of the study area has increased continuously. Increase of the population has been significant in the last two decades and has caused changes in land use. The number of first and second homes has increased.
- Demand for second homes keeps growing and this increases the pressure on agricultural lands, especially olive groves. Agricultural lands have been transformed into building areas. The loss of agricultural lands seems to be irreversible. Urbanization and tourism have put more pressure on the nearby rural areas.
- The building area has increased on the slopes of the hills, especially around the Bursa-Mudanya road.
- The surface area of the building area was 1.97 km^2 in 1987, but this surface increased to 5.42 km^2 in 2000. This shows that urbanization progressed very quickly and the study area is losing its agricultural characteristics.
- Transport between Mudanya and Istanbul had only been available by sea buses up to the end of April 2007. After that, ferry-boats started to set sail between Güzelyalı and Istanbul. Therefore Mudanya is no longer used as the transfer point and Güzelyalı seems to be the new transfer point.
- The demand for second homes in Güzelyalı will increase, especially due to demands by the people who live in Istanbul. Therefore, the pressure on the agricultural land in the study area will also increase.
- Because of legal, infrastructural and financial problems, not all the required precautionary and preventative measures have been taken so far. The increase in population causes water, noise and air pollution.

- Local people have been participating in the decision-making processes concerning land use planning through local meetings in the City Council, which was established in 2006. It has been a good evolutionary development in civil participation (NGO's) in land use planning.

There are some recommendations;

- Legal regulations for the protection of agricultural land have to be strongly implemented.
- Farmers, especially young farmers, should be supported.
- Local people should be informed about cultural tourism. They should believe in this type of tourism.
- Recently there has been an increase in the number of people who join cultural excursions, especially by those who live in metropolitan areas, in particular Istanbul.
- The restored old buildings have a huge potential for cultural tourism in Mudanya, but this potential is not being properly realized.
- Restoration of old buildings should be part of a regular maintenance programme, as should that of the interesting historic hotels.
- To introduce these places to tourists, there has to be good publicity.
- Local foods, especially seafood, should be widely advertised.
- Local agricultural products, especially olives, olive oil and wine, should be marketed to day trippers and tourists. In this way, "symbiotic" relationships will be established between tourism and agriculture.
- The coasts are places where many human activities are concentrated. For this reason, the sustainability of the coastal zones is important.
- As Mudanya is quite close to Bursa and located on the coast, it might be a suitable station for the ecological education of the younger generation.

In conclusion, the study area has many geographical and historical characteristics which make it attractive to urban consumers. Therefore, effective preventative measures should be implemented for the protection of the environment. As Mudanya has a high potential for cultural tourism, there is also an excellent opportunity to make progress in more sustainable forms of tourism.

References

Atalay, İ. 2004. Açıklamalı Doğa Bilimleri Sözlüğü Coğrafya-Ekoloji-Ekosistem. Meta Basım, İzmir.

Esparcia, J., and A. Buciega. 2005. New Rural – Urban Relationships in Europe: A Comparative Analysis Experiences from The Netherlands, Spain, Hungary, Finland and France, RURBAN Project Rurban Building Rurban Relations. http://www.ruralurban.org/files/5fb1d5fe03b938775f1c535bc705d63b.pdf.

Burak, S., E. Doğan, and C.Gazioğlu. 2004. Impact of urbanization and tourism on coastal environment. *Ocean & Coastal Management* 47: 515-27.

Fazal, S. 2000. Urban expansion and loss of agricultural land-a GIS based study Saharanpur City, India. *Environment and Urbanization* Vol 12 No 2: 133-49.

Hall, C.M. and S. J. Page. 2001. The Geography of Tourism and Recreation Environment, Place and Space. Routledge, London and New York: 151-58.

Mathieson, A. and G. Wall. 1982. Tourism: Economic, Physical and Social Impacts, Harlow, Longman.

Newsome, D., S. A. Moore, and R. K. Dowling. 2002. Natural Area Tourism Ecology, Impacts and Management. Chanell View Publications, Clevedon: 4.

Özgüç, N. 1998. Turizm Coğrafyası Özellikler-Bölgeler. Çantay Kitabevi, İstanbul.

Strabon. 2000. Geographika Antik Anadolu Coğrafyası (Kitap XII-XIII-XIV). Pekman, A. (translated), Arkeoloji ve Sanat Yayınları, İstanbul: 56.

Umar, B. 2004. Bythinia, İnkılap Yayınevi, İstanbul: 261-62.

Ünlü, S., B. Alpar, Ş. Aydın, C. Akbulak, N. Balkıs, İ. Barut, E. Meriç, A. Aksu, and Ç. Kırbaşoğlu. 2006. Anthropogenic Pollution in Sediments From the Gulf of Gemlik (Marmara Sea, Turkey); Cause-Result Relaitionship, *Fresenius Environment Bulletin* 12: 1526-29.

CHAPTER NINE

THE INFLUENCE OF TOURISM ON ALANYA AND ITS CLOSE ENVIRONMENT

AYHAN AKIŞ, BAŞTÜRK KAYA
AND RAHMAN SEFEROV

Introduction

Tourism activities, which began to attract attention with the *"Industrial Revolution"*, became more effective, particularly after the 1950s. As some societies became industrialized, the upper classes within those societies began to travel as a luxury. From the onset, tourism began in Turkey, as in some other parts of the world.

Our research area, Alanya, Turkey has one of the most beautiful coastlines on the Mediterranean, with many beaches, bays, caves, plateaus and forest areas. The Damlatas Cave, which was discovered in 1948 and became a destination for Health Tourism in 1954, was the initial reason for tourism in the area. In a study of the area, Kökten (1957) states that "in the Alanya Kadıini Cave, it was shown that the history of the region goes back to the Paleozoic Era (20.000 BC-17.000 BC)". Most of the well-preserved historical sites in the Alanya area, however, go back to Turkish-Muslim era, however, particularly to the Seljuk and Ottoman Empire periods. Works of art found from these eras which are found in the area of study include: the Alanya Castle, the Red Tower (Kızıl Kule), Seljuk Dockyard, Seljuk Armory, Hıdırellez Dede, Alanya Lighthouse, Castle Ramparts and Adam Atacağı (Man Thrower).

Indeed, the area is rich in natural and historical features that are valuable to the tourism industry. However, the most fundamental source of richness in the research area is the human, as mankind has known how to utilize the natural richness existing in the area and has left lasting cultural elements to this day. In addition to the historical and natural richness of

Alanya, the area is also remarkable for its cuisine culture and its traditional clothing.

Study Area, Materials and Methods

Alanya, which constitutes the research area of this study, is an eastern district of the Province of Antalya (Fig. 9-1). The area of the research field is about 2.085 km^2. Alanya is surrounded by Manavgat (Antalya) to the west, Gündoğmuş (Antalya) to the north, Taşkent (Konya) and Hadim (Konya) to the northeast, Sarıveliler (Karaman) and Gazipaşa (Antalya) to the east and the Mediterranean Sea to the south. The research area is located between the northern latitudes 36°30'07" and 36°36'31" and the eastern longitudes 31°38'40" and 32°32'02".

Fig. 9-1: Location map of the research area

In the present research, while the effects of tourism on Alanya and it's the local environment are presented, previous studies conducted on the topic have also been reviewed. In addition to the field surveys that were conducted in certain parts of the area, especially where tourism has had a greater effect, interviews were also conducted in the field in order to better present the effects of tourism. Some of the maps in this study have been produced using the 1/100.000 (1966) scaled map of the General Command of Mapping-Turkey.

Findings

Not counting the low number of tourists who visited the research area in the years before the 1950s, we can say that tourists began to come to the area in the 1950s. As has happened in other parts of the world, tourism has had various effects on the area. Not only were positive effects observed, but certain negative ones were also experienced.

Some Major Positive Impacts of Tourism

In terms of its positive impact, it is useful to present the annual tourism income of Alanya and the number of tourists that annually visit the area, as a way of showing the effect of tourism on Alanya and its immediate environment. Positive effects include the following:

- an increase in incomes of residents of the Alanya area;
- the modernization of dwellings and other buildings;
- a modernization of infrastructures.

The Effect of Tourism on the Annual Income of Alanya. The most significant positive effect of tourism on Alanya and its close environment is an economic one. Tourism has considerably increased the levels of income of the people living in the area. In 2005, 1,464,686 foreign tourists visited Alanya and those tourists spent $942 per tourist. The total income generated by foreign tourists in Alanya in 2005 was $1,379,734,210 (Table 9-1).

Considering that the amount of income which Turkey received from all sectors in 2005 was approximately 364 billion dollars, it is of high importance that the amount of income Alanya gained from foreign tourists alone was approximately 1,380 billion dollars. The economic income level of the area is high, owing to tourism activities. When agriculture, stockbreeding and the other sectors are added to the list, the economic importance of the area becomes remarkable.

As of 2005, a total of 13,024 foreigners from 33 different countries had bought 9,332 immovable properties in Alanya. Germans, with 3,016 properties, became the largest group among foreigners who bought immovable property in Alanya. The second largest group are the Dutch with 1,539, while the Danes come in third with 1,398, the Irish fourth at 1,116 and the Norwegians fifth with 817 (Data of the General Directorate of Land Registry and Cadastre, 2006).

The Effect of Tourism on the Modernization of Dwellings and other Buildings. Modern hotels, restaurants, home dwellings etc., were built in the area in order to meet the needs of tourists. These new buildings created a modern appearance for the city of Alanya. Today, one component which first attracts attention in the area is the modern dwellings. The demand for modern dwellings by people whose income levels rose with tourism also increased. Not only do Turks living in Alanya have a need for modern dwellings, but foreigners living in Alanya also have such needs. Therefore, the demand for modern dwellings and other closed spaces has been gradually increasing in Alanya.

Table 9-1: Tourism annual income of Alanya (1996-2005)

Years	Average Expenditure Per Tourist ($)	Number of Foreign Tourists	Total Income ($)
1996	812,0 $	592.870	481.410.440 $
1997	758,0 $	698.628	529.560.024 $
1998	727,0 $	617.312	448.785.824 $
1999	743,0 $	418.537	310.972.991 $
2000	823,0 $	677.340	557.450.820 $
2001	932,0 $	866.130	807.233.160 $
2002	934,0 $	1.029.350	961.412.900 $
2003	943,0 $	988.785	932.424.255 $
2004	969,0 $	1.133.616	1.098.473.904 $
2005	942,0 $	1.464.686	1.379.734.210 $

(Alanya Chamber of Commerce and Industry; 2005: 68)

The Positive Effect of Tourism on Alanya's Infrastructure. The need generated by the tourism industry for infrastructure has easily been met by using earnings from tourism. The city of Alanya had a sewerage system and sewage treatment plant earlier than some cities in Turkey. For example, the Tepe neighborhood, which is one of the highest-elevation neighborhoods in Alanya (450 m-700 m), does not have any infrastructure problems, which is also an indicator for the situation in other neighborhoods of Alanya. As the Tepe neighborhood is located in a high area of the city, it is quite difficult to implement the infrastructure needed. That the infrastructure problems of such areas have been solved in the city is definitely an outcome of the prosperity that tourism has provided to the local government and to the region.

A Critical Analysis of Negative Effects

Tourism has had a number of negative effects on Alanya and its close environment. The main reason for the negative effects of tourism was the domination attained by mass tourism over Alanya, after the boom which was experienced in the region in the 1980s. In those years, due to the fact that agriculture and stockbreeding were up to that point the most effective means of making a living, the area was caught unprepared by the emergence of tourism on such a large scale.

1) The Effect of Tourism on the Development of the City of Alanya. The factors which had a great impact on the determination of the morphological characteristics of the city settlement of Alanya were:

- The settlement was founded on the coast;
- the effect of the highway;
- the development of tourism on the coastal area;
- the lack of economic activities which would emerge;
- and, the connection of the area with the interior parts and the sea.

The first factor which has had an effect on the determination of the morphological characteristics of the city of Alanya, is that the place on which the settlement was first established is a coastal plain. As the Alanya coastal plain extends approximately in easterly and westerly directions, the development of the town inevitably occurred in these directions. Before the urbanization of the city, a loose settlement structure of dwellings and buildings was observed along the coast, on the coast in patches and on the inner part of the coast in patches. This is also the main characteristic of the settlement. While Alanya showed a loose settlement structure between the years 1950-1970, it has demonstrated the development of a dense settlement structure from 1970 to the present day (Kocakuşak, 1993:240).

The second factor which has had an impact on the development of the city of Alanya is the Alanya-Antalya Highway (Fig. 9-2).

With the completion of the construction of the Alanya-Antalya Highway D-400 in 1958, the settlement cores started to appear around the highway. As a result of the improvement of the settlement cores, the morphological image of the Alanya settlement in those years showed a *"linear"* characteristic; whereas in the following years, the settlement center started to show an urban characteristic.

The third factor is an obstructive one: that is, the region is surrounded by the Taurus Mountains in the north and by the Mediterranean Sea in the south. As this situation has prevented north-south oriented economic activities, important streets and avenues could not be constructed in a north-south direction in the region.

Fig. 9-2: The view of Alanya city from Hacımehmetli district (North-South).

The forth factor is that, while the plan of the city was being formed during the development stage of the city of Alanya, local governors and experts ignored the geographical features of the region in order to please interest groups and people who wanted to take advantage of the situation. Rushed planning of zoning ordinances, as a result of the demand explosion caused by mass tourism, can also be added to this factor.

Therefore, the settlement of Alanya completed the morphological development stage in an east-west direction on the Alanya coastal plain. The plans of the streets and avenues of this city were determined according to the range of characteristics of the coastal plain and the D-400 Highway.

The city center of Alanya has very hot and humid weather in summer. It is inevitable to construct streets which are parallel to the effective wind direction, so as to eliminate the hot and humid weather in summer. In order to accomplish this, the mountain and sea breezes which prevail in the town from the mountainous region in the north should reach the inner parts of the city and, the streets and avenues should have been planned in the north-south direction in these areas. Indeed, in the research area "the winds are directed towards the Mediterranean from the Central Anatolian anti-cyclone in the winter. By crossing the Taurus Mountains through the River Dim and River Oba valleys, these winds prevail in a northeast-southwest (NE-SW) direction, parallel to the range of these valleys in the area. The effective wind direction of the research area is southeast and the effective winds in the region blow from the eastern sector all year, except for May and June. The winds have been effective from the southern sector in May and June" (Sür; 1977:58-63). The River Oba and Dim Valleys have the characteristics of carrying the fresh air of the Taurus Mountains to the city. However, with the construction of multi-storey buildings around the slopes of the mountainous areas and valleys, and furthermore, the main streets of the town not being constructed in the direction of the effective wind, the air circulation in the town is effected in a negative way (Kocakuşak; 1993:244). Air circulation is necessary to provide comfortable surroundings. Streets and buildings must be oriented in such a way as to allow breezes" (Soyak; 2003:110). "When the planning and structuring of the town is examined, it is clear that there is no main street appropriate to these directions and moreover, effective wind directions are nearly prevented by the high buildings (Kocakuşak; 1993:245).

For this reason, the valley and wind systems are not being utilized in order to cool the city and to eliminate urban air pollution in the city of Alanya.

2) The Effect of Tourism on Plateaus. "Plateaus" are one of the regions which have been most affected by tourism in the research area. *"Yaylacılık"* (plateau activities), which earlier had a significant place in the socio-economic and cultural life of Alanya and its close environment, lost its importance in the years around 1985, when a tourism boom was experienced in the area and again over the following ten years. The reason for this loss is that it was impossible for the people living in the region to earn the same level of income pursuing traditional plateau activities that they could earn in the tourism industry. Thus, a group of people who used to be engaged in plateau activities in the summers started to work in the tourism sector.

As tourism had an effect on a number of sectors, it also affected agricultural activities in the area. Because of the increase in the number of tourists, the population of the city of Alanya rises to 500-600 thousand in summer months. Quite a lot of gardens have been built in the villages and the plateaus of Alanya, in order to especially meet the tourists' needs for vegetables and fruits in the summer months. An important number of gardens are being built in the plateaus of the area in order to meet the summer demand for fruits in Alanya. Cherries are grown in these gardens as monoculture.

The first reason why the plateaus of the area became areas of attraction after 1995 is, as stated above, fruit farming provided remarkable incomes. The second reason for the regaining of this popularity is that these plateaus, especially the ones like Gedevet Plateau near to the city of Alanya, became recreation areas for the people living and working in the city. The people who get bored of the hot and humid weather and the noisy environment of the city have the opportunity to commute daily to the plateaus close to the city in the summer months. For this reason, people who work in the city during the day can go up the plateaus in the evening, and after taking a rest in the fresh and silent environment of the plateaus, return to Alanya city center and work daily. While people can stay permanently on the plateaus, which are midway to the Alanya city center, such as the Çayarası Plateau, they can also get away from the depressive atmosphere of the city by visiting these plateaus on the weekend. Today, because of the two main reasons discussed above, the plateaus in Alanya and its close environment have regained the socio-economic and cultural importance they once had in the past.

The infrastructure has not been completed on any of the plateaus in the research area. The brooks, streams and the other natural settings on the plateaus are being polluted by solid and liquid wastes. Concrete buildings contrast with the natural beauty of the plateaus (Fig. 9-3). As Cimrin noted "While the traditional houses on the plateaus are one or two storey buildings made of wood and stone which are in harmony with the nature, apartment blocks have been rising recently" (Cimrin; 1996:283).

3) **The Effect of Tourism on the Coastline.** In the research area, the most striking negative effect of tourism is that it is causing Alanya to become a concrete jungle. As the city was planned without considering human geography and physical geography, it has turned into a concrete jungle. When the city is viewed from a high location, the picture of a concrete jungle is presented.

In the region, sandy beaches were destroyed by constructing numerous hotels and second homes along the coastal zone. In the research area, it is possible to come across lots of second homes and hotel buildings.

Another factor which has caused the destruction of the coastline is the "Antalya-Mersin Highway". While this highway was being constructed and the highway expropriation border was being determined, the road was passed through an area which is very close to the coastal line along the coast. As a result of this, either during the construction of the road or after the extension of the road within the expropriation limit, a certain amount of serious destruction occurred on the coast. During the construction of the "Antalya-Alanya Divided Highway," the destruction reached its maximum level. In Yeşilköy-Incekum, at the entrance to the city center of Alanya and in Alanya-Mahmutlar, because of road extension work, several hectares of sandy beach disappeared under the highway. Today, this destruction continues. The most essential and important reason for the destruction of the coastline in the area is that the public institutions responsible, in the first case being the General Directorate of Highways, which is responsible for road construction works, did not take into consideration the needs that would arise in the area in the future while they were determining the expropriation border and route in the 1950s.

Fig. 9-3: Concrete high buildings contrast with the natural environment of Çayarası Plateau.

Today, a similar danger is a case in point in the coastal zone between Alanya and Gazipaşa. Due to the divided highway system, the construction of hotels and second homes in the area, some cliffs, beach rocks, small bays and a few forests adorning the coast have been disappearing.

4) The Negative Effect of Tourism on Traffic in Alanya. There are serious traffic problems in the area during the summer months. The population of the city center, which is 88,346 during the winter months, increases to a population of 500-600 thousand in the summer months. Due to the increase in population, it is observed that the roads which are hardly adequate in winter are definitely inadequate in summer. To overcome this problem, the construction of the alternative road being carried out to the north of the current belt highway should be accelerated. Moreover, the traffic problem in the city can partly be solved by rearranging the roads reaching the beaches and the city center. As the Antalya-Mersin Highway carries most of the traffic load of Alanya, it significantly hinders the connection of the establishments with the sea (Kocakuşak; 1993:240). It has been observed that the E-24 Highway which passes through the Alanya city center causes noise pollution, especially during the summer months (Öğüt; 1993:109).

5) The Effect of Tourism on the Disappearance of Terrain in the Area. One of the most important problems of the current mass tourism and unplanned development of tourism areas is the superfluity of the second homes in tourism areas. Because of mass tourism, the costs of houses and domains have increased to excessive levels. Some building contractors who wanted to take advantage of the situation entered the area and built thousands of houses. Most of the houses built in the area are used as second homes. Second homes are common in the city center, and in the towns of Demirtaş, Mahmutlar, Kestel, Konaklı, Türkler and İncekum towns, either on the coast or on the inner parts of the coastal area in Alanya (Fig. 9-4).

Second homes, which are only used in the summer months, do not serve a large number of people. As we have mentioned before, properties bought by foreigners are mostly second homes. As the second homes are only used in the summer months and do not serve many people, they take up too much room. This situation causes both a decline in visual quality and a decrease of tourism areas in the region. Well-planned hotels should be built instead of second homes. The benefit of this would be that many more tourists could be accommodated and far more revenue could be generated.

6) The Negative Effect of Tourism on Vegetation. While forests are not widespread in areas bordering the sea, pine forests have been protected in the areas extending to the shore in certain places between İncekum and Alanya and Alanya and Gazipaşa. Apart from these areas, damaged forests and maquis (shrubbery) areas are seen in the coastal areas (Fig. 9-5).

Throughout history, the forests have been damaged in order to meet the need for timber and firewood in the area. From the 1950s to the present, certain areas have been deforested as a result of the high demands of mass tourism, not only in Alanya, but also in other parts of Turkey's southern and eastern coasts. In the beginning, the reason for this trend of deforestation was to create agricultural land and for constructing buildings. However, forests have been destroyed in Alanya in order to build tourism facilities and to gain previously unearned income from the land, which became valuable with the tourism boom of the 1985s.

Fig. 9-4: Secondary housing and hotel construction in Kargıcak and the area occupied by construction

Fig. 9-5: A hotel view showing the damage done to the coast, tourist complexes, and natural vegetation between the Alanya and Gazipaşa coastal region

7) The Effect of Tourism on Sea Pollution. The sea was considerably cleaner before the start of tourism activities in the region. However, due to the excessive demand of tourism, partial pollution of the sea attracts attention, especially in the summer months. The most important reason for this pollution is that there is a high demand for the sea in the region and the majority of tourism activities have developed in the areas close to the coast.

Although the sea looks partly polluted in the summer months, it can be considered as clean in the area. The primary reason for this is the sensibility of the Alanya Municipality. Some hotels in the area were allowed to drain liquid wastes directly into the sea until the 1990s. In order to prevent this, the Alanya Municipality first built a liquid waste treatment plant in 1997. Furthermore, it took the necessary legal measures to keep the shores and the sea clean. Thus, this problem has been solved by the sensitivity shown by the administrations of the Alanya Municipality. The second reason is that the sea has a deep shelf and a coastal current. In the area, the solid and liquid wastes which reach the sea are carried away from the coast by means of the Antalya Bay current and out to the inner parts of the sea.

8) The Problems of Solid Waste Pollution and its Removal in the Region. Although there are certain problems related to solid waste in Alanya, the Alanya Municipality has been making a serious effort to solve the problem. The most important indicator of this effort is the solid waste treatment plant being established by the municipality that will come into service in 2007. Despite this effort, it can be observed that solid wastes have been disposed of randomly in some places in the research area and this action hasn't been able to be prevented by the authorities. It is striking that a substantial amount of solid waste is randomly thrown 400 meters away from Cleopatra Beach at the end of Kızlar Pınarı Street in Alanya city center (Fig. 9-6). This solid waste has become a breeding-ground for insects, especially in the summer months and they have been threatening the health of the people living in the close environment. The solid waste materials also create visual pollution.

The Kestel Municipality has been using an area close to the Regulator in the River Dim Basin as a waste storage area. The wastes are dumped into this waste storage area without using any solid waste treatment techniques. Consequently, some villages and tourist establishments in the area suffer from the stench coming from this garbage dump.

Fig. 9-6: Bulk waste in Alanya city center in Kızlar Pınarı Street (400 m from the beach)

Results and Suggestions

The most prominent positive effect of tourism on the research area is economic. By the year 2005, the number of foreign tourists who had visited the research area was 1,464,686. Those tourists, by bringing in an income of $942 per tourist, generated a total income of $1,379,734,210. Another positive effect of tourism on the research area is that dwellings and other buildings of the city developed a modern appearance.

The rapid growth of the city of Alanya after the 1980s brought about certain planning problems. Due to the fact that a suitable city plan was not implemented, certain difficulties are experienced in the city's traffic flow, especially during the summer months. In order to solve the traffic problem, the construction of an alternative road project to the north of the current belt highway is being carried out by the Municipality of Alanya. In this project, the roads that will reorganize the traffic flow in a north-south direction should be given high priority, in order to have faster-moving city traffic.

The easy circulation of mountain and sea breezes and the dominant winds have not been taken into consideration in the city plans implemented in the city of Alanya so far. Efforts should be made to plan the streets and the alleys, which will be newly constructed in the north-south and northeast-southwest directions, in order to make the hot and humid weather more tolerable in summer months.

An excessive number of hotels and second homes exist on the coastline in the research area. These buildings cause problems, such as the destruction of the beaches and some forests which extend to the shore, and the deterioration of view. Hotels should be preferred in the area, as they cover a smaller area when compared to second homes and many more tourists can be accommodated in such structures. Tourism activities, which are indispensable sources of income, should certainly be maintained in the area. However, organized and planned tourism activities should be introduced in order to cause minimum damage to the environment.

A certain amount of solid and liquid waste pollution is observed even in the city center in the research area. For this reason, local governments should be more careful about preventing environmental pollution in the area.

"Yaylacılık" (high plateau activities), which lost their importance in the years after 1985, revived again, when after 1995, the plateaus became the fruit gardens of the tourism facilities in the area. However, excessive demand on the plateaus, which regained its importance in the area, has brought about a number of problems such as deforestation and unplanned construction.

References

Cimrin, E. 1996. Traditional "Yayla" Houses in the Vicinity of Alanya, Ankara: The Middle East Technical University in the Department of Architecture, Unpublished Master Thesis.

Öğüt, A. 1993. Alanya İlçesi Doğal Potansiyelinin Kentleşme Sürecindeki Durumu. İstanbul: Yıldız Teknik Üniversitesi, Fen Bilimleri Enstitüsü, Peyzaj Planlama Yüksek Lisans Dalı, Basılmamış Yüksek Lisans Tezi.

Kocakuşak, S. 1993. Alanya'da Kentleşme-Turizm ve Sorunları. Ankara: A. Ü. Sosyal Bilimler Enstitüsü Coğrafya Anabilim Dalı, Basılmamış Doktora Tezi.

Manisa, K. 2002. Turizm ve Mimarlık Penceresinden Alanya Kent Kimliği. İstanbul: Yıldız Teknik Üniversitesi Basım-Yayın Merkezi Matbaası.

Soyak, C. 2003. Alanya'da Turizm ve Kentsel Mekanlar. İstanbul: Yıldız Teknik Üniversitesi Basım-Yayın Merkezi Matbaası.

Sür, A. Alanya'nın İklimi. Ankara: A. Ü. Dil ve Tarih-Coğrafya Fakültesi Yayınları, No: 270.

Yetkin, H. 2001. Alanya (Dünden Bugüne Alanya'da Yaşam). İstanbul: Çali Grafik Matbaacılık A. Ş. Alanya Ekonomik Rapor. 2005. Alanya Ticaret ve Sanayi Odası.

www.alanya.bel.tr

www.altso@org.tr

www.tkgm.gov.tr

CHAPTER TEN

FUNCTIONAL CHANGES AND THEIR REASONS: IN YAYLA (HIGH PLATEAU) SETTLEMENTS IN ANTALYA BEY MOUNTAINS

SERHAT ZAMAN

Introduction

Human beings adapt to their living environments and benefit from the opportunities offered by these environments as much as possible. It is possible to observe many examples of this situation. Almost every community gets familiar with the opportunities and boundaries of their natural environment and develops a life style parallel to these characteristics. In the mountainous regions of southern Turkey, these lifestyles sometimes include *yaylas* (high plateau or upland) and *yayla activity*, a type of transhumance where people traditionally move between environments in order to find adequate fodder and water for this grazing flock. An agreed definition and practice types of *yayla* and *yayla activity* in many parts of the world has not yet been made, but in Turkısh history and settlement culture, yayla and yayla activity practices have a profound role. Yayla activity is both a life style and one of the sources of cultural heritage.

A number of researchers have introduced definitions for yayla settlements and yayla activities. The large distribution and regional characteristics of yaylas make it difficult to list all characteristics in a single definition. Therefore, in the present paper, regional and local characteristics were set aside and common characteristics were emphasized in definitions. Grazing areas in high places, where summers are cool and winter conditions are harsh in mountainous regions, are called *yayla* (İzbırak, 1992:339). When considered as a settlement place and economic activity area, yayla is defined as an area where groups of people living in separate settlements, with different socio-economic status migrate in certain parts of the year and carry out human and economic activities (Emiroğlu, 1977:9).

The facts that animals are taken to high places kept and grazed there in plains and hillsides of mountain ridges for grazing in the arid summer season and are taken to coasts and valleys in winters led to introduction of the terms “yaylak” and “kışlak” (Darkot, 1968:131-132).

It is possible to increase the number of definitions and expand the portfolio with studies carried out by different scientific disciplines. As it is clear from the definitions above, yayla settlements are areas which are higher than their surroundings and thus cool in summers, and where rural production methods including animal husbandry activities and sometimes cultivation activities are carried out. Yayla activities, which have been practiced for hundreds of years in Anatolia for the reasons we mentioned above, and which are substantially identified with our national culture, have been undergoing change in recent years. In addition to the increasingly limited gains from yaylas, the traditional areas used for yayla activities have declined, and some of the same areas are also being used for tourism and recreation.

Today in different geographic regions of Turkey, many yayla activity practice types with very different purposes are observed. One of the regions where yayla grazing activities are intensively practiced is Beydağları, which extends to the western coasts of Antalya bay.

The principal factors that lead us to carry out yayla grazing research can be listed as follows:

Determining the current situation of yayla settlements in Bey Mountains and yayla grazing activities practiced in this region;

Summarizing the historical development of yayla activity practices carried out in Bey Mountains;

Analyzing the reasons for change in yayla activities in the region;

Finding and describing the changes in functions of some yayla settlements in the region;

Discussing the effects on yayla grazing of the economic and social change of permanent settlements that participate in yayla activities;

Classifying yayla grazing activities and yaylas in Beydağları and studying the geographic reasons for their distribution;

Analyzing the interaction between geographical environment characteristics and yayla and yayla activities;

Indicating future trends and development possibilities of yayla and yayla activities and making assumptions.

The Research Area

Bey Mountains, selected as the research area, is located in the Antalya area of the Mediterranean Region. The study area is Teke and Antalya, mostly the Teke area of this region. Beydağları, where the research was conducted, extends along the west coast of Antalya Bay, in a northeast-southwest direction. This mountainous area, which extends as a mountain range get close to Finike bay in the south and the Lakes Region in the North (Fig. 10-1).

This high mountainous area, where yayla settlements and yayla activities have a large distribution, is separated from its surroundings by prominent natural boundaries. These boundaries are Antalya Bay in the east, the Elmalı Polje Plains in the west, Finike Bay in the south, and the Lakes Region in the north (Fig. 10-1).

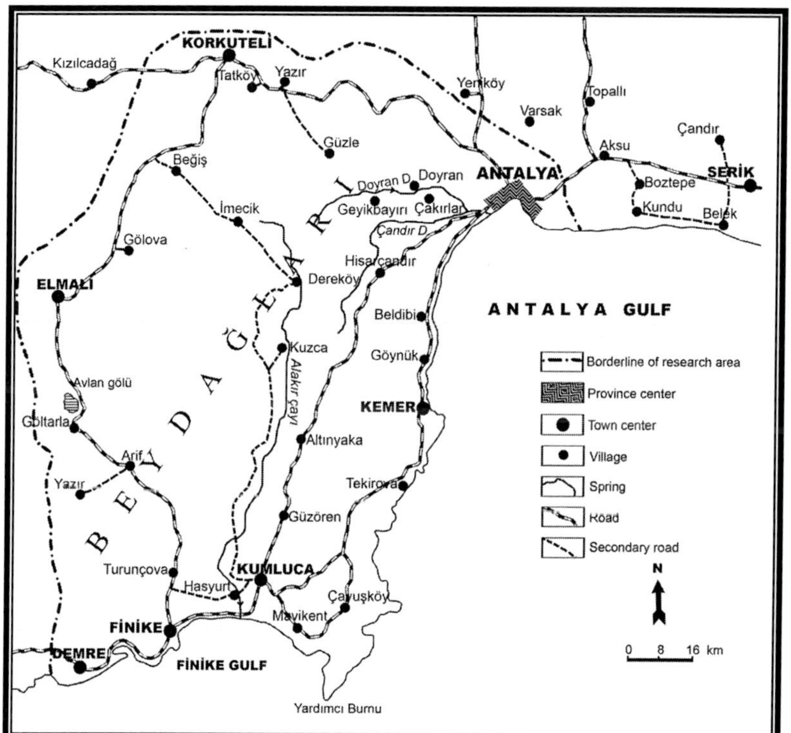

Fig. 10-1: Location map of the research area

The settlements participating in yayla activities, the yaylas and Beydağları as a whole are located inside the administrative boundaries of Antalya Province. This mountainous area is surrounded with Korkuteli in the north, Kumluca, Finike and Kemer in the south, Elmalı in the west and the city center of Antalya in the east (Fig. 10-1).

Beydağları, the study area, is composed of nested elevations. It is possible to consider Beydağları as two mountain ranges extending parallel to each other. These two elevated zones are divided into two distinct parts by the Alakır Creek which flow into sea at Finike Bay. The first one of these is the coastal mountain range that extends between Antalya Bay and Alakır Creek Valley, and the second one is relatively higher Asıl Beydağları system, located between the Elmalı Plain and Alakır Creek.

The yayla settlements in the research area are located between 680 and 2050 meters in elevation. However, while some settlements taking part in this activity are located in the coastal belt, the altitude in the pastures that the settlements use reaches up to 2500 meters. Thus, it can be stated that the vertical boundary of yayla activities in the area is practiced approximately between sea level and the belt at 2500 meters.

In the research area, whose boundaries were determined in general terms, today there are approximately 30 yayla settlements with different functions. The population coming from the center of Antalya, and the Finike, Kemer, Serik, Kumluca, Korkuteli and Elmalı districts participate in yayla activities in the research area. In addition, a population outside Antalya comes to the area and generally takes part in yayla activity mainly for recreational purposes.

The yaylas in the research area are mostly used by settlements in south and east of Beydağları. The fact that the climatic characteristics of these places force the population in the near surroundings to yayla activities can be one of the main reasons for its continued existence. In addition, the dense population of the area is also a factor.

Method

This study, in which change observed over time in yayla and yayla activities in Bey Mountains, is analyzed, required field work. A number of stages were involved in the study from the collection of data to analysis.

As was mentioned earlier, this study is based on field research. Field observation and field research was supported by sophisticated office work. In this process, after determining the research topic, place, limits, reasons, objectives and content of the study, the literature on the topic was reviewed.

The research area, Bey Mountains, has a number of different settlements, due to its large size. Yayla settlements in Beydağları number around 30. While the study was being conducted, all of the yaylas were visited. However, many similarities between the settlements, such as the purpose of settlement, the functional change process, the relationship between settlements and the natural environment, were observed. Despite this, there was the feeling that interesting results could be obtained is many of the yaylas were studied separately. However, since some yayla settlements gave similar results and other yaylas were studied in functional classification, some of them were analyzed by giving general information only. Thinking that similar samples often give similar results (Doğanay, 2002:7), sampling models were used in the research. The sampling group was expanded and 15 yayla settlements were analyzed in terms of the objectives of the research.

Following the completion of the processes summarized thus far, office work was carried out and the data were arranged, according to the objectives of the research. Transfer and interpretation of the obtained data to maps constituted one of the most important stages of office work. By mapping the data, distribution maps in line with the objective were produced.

Fig. 10-2: An example of yayla tent-like houses called "alaçık" from Imecik Bey Mountain yayla.

Discussion

Beydağları is one of the rare yayla settlements, each of which is used for different purposes. One side of Bey Mountains overlooks the Mediterranean Sea and the other side overlooks inner regions. The area has long had yayla settlements with different functions.

Before giving information about functional change in yayla settlements and the reasons of this, it is best to group yayla grazing activities. There are 4 yayla groups. These are;

1. Traditional yaylas, which were kept away from the effects of urban population:

a) based on animal husbandry;

b) based on cultivation (agriculture). There are yaylas where production was affected by urban demands.

2. Vacation yaylas for recreational purposes serving urban population;

3. Yaylas for recreation and cooling needs of the rural population;

4. Multi functional yaylas (transition type. For example, the Geyikbayırı, Uçoluk, Karagöl yaylas)

Fig. 10-3: Goat hair tents are symbols of yayla activities for animal husbandry purposes.

As it is clear from Table 10-1, Bey Mountains yaylas served different functions in line with changing yayla activity demands in time. In these processes, former functions either completely disappeared, or regressed to second a function situation. Among function groups in Beydağları, the group which was most affected by the change process was undoubtedly yayla groups with animal husbandry purposes. In these yaylas, animal husbandry lost its importance and yayla activities for recreation or agricultural purposes became important. Another interesting point which can be made from the Table 10-1 is that some yaylas were formed in totally new areas.

Fig. 10-4: Goat is the dominant animal in Beydağları yaylas for animal husbandry purposes, as it adapts to environmental conditions best.

The Main Function Groups in Beydağları Yaylas: Yayla (Upland) Activities with Animal Husbandry Purposes

The longest known functions of yayla activities practiced in Beydağları and Antalya depend on animal husbandry and cooling. Many of the yaylas in Antalya and in research area are passing through a process of losing former functions and gaining new functions. Through the questionnaire and interviews conducted in the area, as if acted on a consensus, all the participants stated that, "in last twenty years everything has changed."

It is known that in Beydağları, 30 years ago animal husbandry was practiced almost at every yayla. However, today the numbers of these yaylas are gradually decreasing. The most prominent characteristics of yaylas where yayla activity for animal husbandry is the main purpose, is that this is the yayla activity group which was most affected by the functional change process. Many yaylas with an animal husbandry function have lost this characteristic today and began to exist for recreational purposes. Particularly in recreation-based yaylas used by rural settlements, it is possible to see the traces of yayla activity for animal husbandry purposes. Although the main purpose of some yaylas is recreation, it was observed that many families in these settlements kept a few cattle. This practice indicates traces of animal husbandry, which was among the former functions of yaylas.

The yaylas where animal husbandry is the main function are generally located at the upper path of rivers, at plateaus at an altitude of more than 1500 meters.

Fig. 10-5: In yaylas for animal husbandry purposes tent and sheepfolds (ağıl), built for the keeping of animals are the most significant elements of settlement.

Table 10-1. Yayla settlements at Bey Mountains and their principal functions

Name of Yayla	Altitude (meters)	Settlements Yielding the Yayla	Current Function	Former Function
Feslikan	1810-1890	Çakırlar, Doyran, Antalya kenti	Recreation	Animal husbandry and recreation
Karçukuru	1960-2030	Bahtılı	Recreation	Animal husbandry
Sakarpınar	1725-1770	Antalya, Çakırlar	Recreation	Animal husbandry
Narpızlı	1610-1660	Antalya, Çakırlar	Recreation	Animal husbandry
Elmelik	1570-1630	Akdamlar, Antalya	Recreation and animal husbandry	Animal husbandry
Moryer	1630	Antalya kenti	Recreation	-
Geyikbayırı	570-680	Antalya kenti	Recreation	Had kışlak function
Doyrankozu		Antalya kenti	Recreation	Has recently become yayla; no former function
Saklıkent	1790-1850	Doyran, Antalya kenti, Manavgat, Ankara, İstanbul, Almanya	Recreation yayla for urban population	Pasture for Doyran village
Üçoluk	1100	Göynük, Kemer	Recreation	Animal husbandry
Söğütcuması	1350	Kemer, Göynük, Beldibi	Recreation and animal husbandry	Animal husbandry
Ördübek	1430-1500	Finike, Turunçova	Recreation	Recreation and animal husbandry
Avlan	1010	Finike, Mavikent, Beykonak, Hasköy	Recreation	Animal husbandry
Karagöl	1360-1430	Kumluca	Recreation	Animal husbandry
Göllü	1220	Kumluca (Hızırkahya, Toptaş, Beşikçi)	Recreation	Animal husbandry
Beydağı	1620-1680	Dereköy, Kumluca	Animal husbandry	Animal husbandry
Tufa	1700-1780	Körler köyü	Animal husbandry, recreation and agricultural yayla grazing	Animal husbandry
Kepiralanı	1150	Yazır, Geyikbayırı	Animal husbandry	Animal husbandry

Kartal	2000-2080	Geyikbayırı	Animal husbandry	Animal husbandry
Elmayanığı	1430-1530	Hisarçandırı	Animal husbandry and recreation	Animal husbandry
İmecik	1900-1950	İmecik	Animal husbandry	Animal husbandry
Yazır-Beydağı	1710-1750	Yazır	Animal husbandry	Animal husbandry
Sinandeğirmeni	870-1020	Geyikbayırı, Akdamlar	Agricultural yayla grazing and recreation	Animal husbandry
Tekirmınarı	1550	Bahtılı	Agricultural yayla grazing and recreation	Animal husbandry
Yazır	940-1010	Antalya	Recreation yayla for urban population	-
Yavşanlı	1620-1650	Dereköy	Animal husbandry	Animal husbandry
Altınyaka (Gödene)	1050-1100	Kumluca-Finike	Recreation and animal husbandry	Animal husbandry and recreation
Ardıçgediği	1050-1100	Geyikbayırı	Animal husbandry	Animal husbandry
Yazırgüzle	1250-1300	Yazır	Animal husbandry	Animal husbandry
Çakmak	1600-1650	İmecik	Animal husbandry	Animal husbandry

These yaylas were able to keep their traditional yayla activity styles due to two prominent reasons. The first one is that, these yaylas were located in areas away from the recreational demands of the urban population. The more accessible the yaylas, the quicker functional change occurs. The yaylas with animal husbandry purposes are located in areas far from cities, when compared to yaylas of other function groups. The second reason that Beydağları yaylas keep their animal husbandry purpose is that permanent settlements by people have a close relation to location. If the population in yayla belongs to a settlement located in a coastal belt, functional change occurs fast. If the population belongs to a settlement located in an inner area, functional change occurs slowly. For example, yayla activity carried out near the Kortkuteli and Elmalı villages still have an economic purpose (animal husbandry) but in coastal settlements, rather than an economic activity, yayla activity is practiced for recreational purposes such as vacation.

The economic structure of yaylas with animal husbandry purposes in Beydağları generally depends on livestock marketing. Wool, hair and dairy products of animals do not have an important economic value, and livestock–raising for sacrificial animal is prominent (Fig. 10-4).

The most prominent characteristic of yayla activity practiced for animal husbandry purposes is yayla houses. Since this kind of yayla activity is mainly based on leaving dry pasture lands and moving to green pastures, the portable houses are built with simple construction materials supplied from the surroundings. The characteristic house of people in yayla is typically a "tent-type" structure. Tent-like houses known as an "alaçık" has almost disappeared in the area, due to an overall reduction in number of yaylas with husbandry purposes (Fig. 10-2).

Yayla (Upland) Activities for Agricultural Purposes

Beydağları yayla activities are based on two historical functions. The first one, which has been practiced longer, is yayla activity for animal husbandry purposes. The second one is yayla activities for recreation and leisure. A very new and rare yayla activity was added to these two historical functions in recent years. This new yayla activity type is based on agricultural activities.

In Beydağları and in the majority of yaylas in Turkey, there are small cultivated pieces of land around the houses. However, the yaylas and yayla activities for agricultural purposes are primarily based on agricultural activities, and on economic gains from these activities.

Yayla grazing based on agricultural activities appeared in Beydağları after 1990. The former functions of these yayla settlements were based on animal husbandry. However, because of the increase in urban population in Antalya, a new function appeared for yaylas. Agricultural production areas in coastal plains and open areas are sometimes interrupted in certain periods of the year (June-August). The reason is that that are high temperatures prevailing in the coastal belt between June and August, excessive evaporation, and thus, water problems. During this period, the unfavorable conditions in the coastal belt forces people to move to mid-altitude areas which have favorable conditions for agriculture. Agricultural activity carried out in this period is called "yayla production" in Antalya. Vegetable and fruit products supplied by some settlements, including Bucak, Çeltikçi, Çavdır, Gölhisar in Korkuteli, Elmalı and Burdur Province were also brought to Beydağları yaylas due to the increasing demand from tourism facilities and an increase in the Antalya urban population. In summer months, the majority of vegetables and fruits in the Antalya marketplace have the label of yayla production.

It is widely accepted that yayla products are healthier than the ones which were produced in greenhouses. Therefore, the demand for yayla products is quite high, and parallel to this, the commercial value of yayla products is higher as well.

In addition, a new population with higher purchasing power has appeared in recent years, due to changes in Antalya Province. This population tends to buy yayla products from yayla markets and has caused the products to gain value which is higher than the normal price. Additionally, high demand for yayla products in settlements such as Korkuteli, Geyikbayırı, Çağlarca, Üçoluk and Söğütcuması can explain the increase in agricultural lands in yaylas. This situation has led to the introduction of a yayla activity, which depends on agriculture, in Beydağları.

Fig. 10-6: Agricultural yaylas were initially based on cereal agriculture. The most significant proof of this is cereal storages where wheat and barley were stored. These storages are used as a warehouse today.

This yayla grazing type is practiced in area, where water sources and agricultural lands are abundant. However, it should not be forgotten that this yayla grazing is practiced in certain elevation belts. According to our information, these activities are practiced between 950–1200 meters.

Yayla production starts during the last week of June and the succeeding weeks, when agricultural production in coastal belt is interrupted. However, it should be stated that yayla production can never supply the production deficit in winter. Starting from the first week of July, vegetables and fruits began to be sold in yaylas and this period continues almost until mid-september.

In the study area, the number of yaylas whose main function is agriculture is 2. These are Sinandeğirmeni and Tekirpınarı. Sinandeğirmeni is the most typical example of this kind of yayla activity. This yayla is specialized in agricultural yayla activity.

Today, agricultural yayla activity is principally focused on vegetable and fruit cultivation (Fig. 10-5). Cereal production, which was the economic activity carried out by these yaylas when the settlements were first formed, has almost disappeared. Vegetable and fruit production is not an economic, but a commercial activity. The economy of all of the families who continued these activities in yaylas depend on fruit and vegetable agriculture in villages as well. However, the cultivated products vary according to elevation.

Yayla Activities for Recreational Purposes

The most important yayla group, both in terms of participating populations and number of yaylas, is yayla activity for recreational purposes. It is important to note that many of yayla settlements used by urban population today for recreation and vacation were vacation yaylas serving rural populations in the past.

This kind of yayla activity can be studied under two groups. The first one is recreational yaylas used by rural populations. The second one is recreational yaylas serving urban populations. In both groups, the purpose is living at a cooler, more pleasant altitude, and vacation.

In recreational yaylas formed by urban populations, the population does not have a long yayla activity past. In the research area, these kinds of yaylas have a history of a maximum 30 years. This period is parallel to the functional change process of yaylas in Beydağları. However, the yayla activities carried out by rural settlements have a long history.

Recreational yayla activities in Beydağları have – at the same time – some general similar characteristics and differences from other types of Yaylas. Some of these include:

- Recreational yaylas are generally located at places over 1000 meters in elevation;

- Recreational yaylas are frequently located in woodland areas (especially among cedar and juniper forests);
- Recreational yayla settlements are distributed along the coastal belt of Beydağları and in areas near to urban settlements.

Many of the recreational yayla settlements are located near the center of the city of Antalya. A significant part of the urban population which led in the formation of this kind of yayla grazing lives in the Province of Antalya.

In recreational yaylas, yayla houses are generally neat houses, built with expensive materials (Fig. 10-8). However, there are significant architectural differences among these houses.

Unlike rural yayla settlements, populations of recreational yaylas come from different settlements. In normal, more traditional yaylas, grazing population get acquainted with each. In recreational yaylas, the yayla activity period is generally determined by temperature conditions in coastal belt. Yayla activity populations in recreational yaylas used by urban populations generally belong to professional groups, including to high-income physicians, academicians, lawyers, artisans, tourism agents, pilots, etc). While rural populations in yayla settlements do not engage in activities such as gaieties and festivals, these activities are more important in recreational yaylas.

Although the principle recreational yayla activity purpose is cooler environments and vacation, many families raise small cattle. A part of the settlements in recreational yaylas have either exceeded their physical capacities or have reached to their limits. Water cuts are the most significant indication of this in these kinds of yaylas. Today many recreational yaylas have reached growth limit due to water shortages.

Functional Change in the Beydağları Yaylas and Primary Reasons for Change

In addition to yayla activities practiced in Beydağları, change in these activities through time is an inevitable reality. Yaylas and yayla activity characteristics are passing through a process of structural change. Obviously, there are various reasons of this change. All settlements and economic activities are certainly affected by changing external processes, either slowly or rapidly. From these aspects, it can be stated that they look like human beings or at least reflect human beings.

Yayla activity does not simply mean the reciprocal interaction between yaylaks and kışlaks. It is possible to state that this activity branch is easily affected by regional, national and international developments. Possible economic, social and cultural changes of any settlement from which people move to yaylas can influence yayla activity and its practice.

Almost 30 years ago, mainly two kinds of yayla activity were practiced in Beydağları. Today the number has increased to 4. In the past, yayla activity was practiced mainly as an economic activity. However, today economic activity has lost its importance, and yaylas have turned into vacation and cooling places. To agricultural yayla activity and yayla activity practiced by rural settlements have been added yayla activity practiced for animal husbandry purposes and yayla activity practiced by rural settlements for cooling purposes. In this process, despite the formation of new yayla areas, generally former yayla areas gained new functions. Undoubtedly, yayla activity for animal husbandry purposes was most affected by this process. As has been noted, the most important factor forcing yaylas to functionally change has been the demands of urban populations.

Fig 10-7: Yayla houses in recreational yaylas are built in modern architectural design

There is a close interaction between socio-economic status and yayla activity purposes. The locational characteristics of yaylas influence the functions of yaylas. In yaylas near to coast and where transport is favorable, functional the change process was faster. However, the yaylas in higher elevation, to where transport is hard and which are located in the Alpine belt, generally continued to have traditional functions.

The reasons that forced yaylas to functionally change are the following:

1. The population increase in the Province of Antalya, and the plains in its vicinity, and changes in the social and economic characteristics of the population.
2. The addition of economic activities which increased dependence on local farming, such as citrus fruit agriculture.
3. The dominant role of the tourism sector in coastal parts of Antalya.
4. The coincidence of the summer tourism season and the yayla season.
5. A decrease in the will of young people to move to yaylas.
6. The appearance of a new population groups who imitate tourist attitudes, and the imitation of tourists as the income levels of the yayla people increases.
7. Pensioners moving to Antalya.
8. An increase in populations who demand recreational activities in the Province of Antalya and in neighboring provinces.
9. Putting mountainous areas into the service of tourism as a result of efforts to add new horizons to tourism activities, which otherwise had a limited zone in a narrow zone along the coast.
10. The transition to intensive-stock production methods in animal husbandry activities and the negative effects of this has had on yayla animal husbandry.

Conclusion

Beydağları yaylas lost their traditional functions in time and gained new functions. The new functions generally appeared parallel to the demands of the urban population. The increase in the population which chooses mountainous areas for vacation and cooling purposes caused classical yayla areas to have a modern image. Yayla activity is no longer an indispensable part of economic life as it was in the past; today it turned out to be a recreational activity. The fact that a small number of yayla settlements which were able to keep their former functions were prevented from functional change process is related to their locations and to natural environmental characteristics unfavorable to the recreation demands of urban populations.

References

Aktan, O., 1996. Antalya Çevresinde ve Güney Anadolu'da Depreşen ve Dinen Konar-Göçer Asabiyeti. I. Akdeniz Yöresi Türk Toplulukları Sosyo-Kültürel Yapısı (YÖRÜKLER) Sempozyumu. 25-26 Nisan 1994, Antalya. Kültür Bakanlığı:1821, Halk Kültürlerini Araştırma ve Geliştirme Genel Müdürlüğü Yayınları:232, Seminer Kongre Bildirileri Dizisi:49, s: 1-24, Ankara.

Alagöz, C,A., 1993. Türkiye'de Yaylacılık Araştırmaları. Ankara Üniversitesi Türkiye Coğrafyası Araştırma ve Uygulama Merkezi Dergisi, Sayı:2, Ankara.

Bekdemir, Ü., Özdemir, Ü., 2002. *Doğu Karadeniz Bölümünde Gelişmekte Olan Yayla Turizm Merkezlerine Bir Örnek: Bektaş Yaylası.* Doğu Coğrafya Dergisi, Sayı:7, s:9-35, Erzurum.

Çetin, B., 2006. *Kefe Yaylasında (Serinhisar-Denizli) Fonksiyon Değişikliğiyle Ortaya Çıkan Rekreasyonel Faaliyetler.* Doğu Coğrafya Dergisi, Sayı:15, s:175-198, Erzurum.

Doğanay, H., Zaman, S., 2004. *Oba-Yayla Yerleşmelerine Tipik İki Örnek: Çambaşı ve Turnalık Obaları (Ordu İli).* Türk Coğrafya Dergisi, Sayı:43, s:1-31, İstanbul.

Doğanay, H., Zaman, S.,2002. *Antalya'da Az Tanınan Bir Kış Turizm Merkezi: Saklıkent.* Doğu Coğrafya Dergisi, Sayı:8, s:149-162, Erzurum.

Doğaner, S., 2003. Miras Turizminin Coğrafi Kaynakları ve Korunması. Ege Üniversitesi Coğrafya Bölümü Sempozyumları II, Coğrafi Çevre Koruma ve Turizm Sempozyumu, 16-18 Nisan, s:1-8, İzmir.

Emiroğlu, M., 1977. Bolu'da Yaylalar ve Yaylacılık. Ankara Üniversitesi D.T.C.F Yayınları No:272, Ankara.

Fleischer, A., Felsentein, D., 2000, Support For Rural Tourism. Annals of Tourism Research, Vol:27, No:4, pp: 1007-1024, Elsevier Science, Great Britain.

Gibson, L,A., 1993. The Potential for Tourism in Non-metropolitan Areas. In Economic Adoption: Alternatives for Non-metropolitan Areas, D, L, Barkley, ed, pp: 145-164, Boulder Co. West view.

Gönençgil, B., 2002. Kullanım Açısından Dağlık Alanların Sorunları ve Çözüm Önerileri. Ege Üniversitesi Coğrafya Bölümü Sempozyumları II, Coğrafi Çevre Koruma ve Turizm Sempozyumu, 16-18 Nisan, s:55-63, İzmir.

Güçlü, S., 2002. Kentlileşme ve Göç Sürecinde Antalya'da Kent Kültürü ve Kentlilik Bilinci. Kültür ve Turizm Bakanlığı, Kültür Eserleri Dizisi, Ankara.

Güner, İ., Ertürk, M., 2005. *Fethiye'de Yaylalar ve Yaylacılık*. Doğu Coğrafya Dergisi, Sayı:14, s:141-179, Erzurum.
Gürbüz, O., 1997. *Türkiye'de Göçebe Mesken Örneği: Çadır*. Türk Coğrafya Dergisi, Sayı:32, s:185–196, İstanbul.
Hadimli, H., Bulut, İ., 2001. *Antalya'da Sebze-Meyve Ticareti ve Antalya Toptancı Hali'ne Coğrafi Bir Yaklaşım*. Doğu Coğrafya Dergisi, Sayı:11, s:261-282, Erzurum.
Kara, H., 1994 (M.Bazin'den Çeviri). Orta Toros Yörüklerinden Sarıkeçili Aşireti. Ankara Üniversitesi Türkiye Coğrafyası Araştırma ve Uygulama Merkezi Dergisi, Sayı:3, s:323–350, Ankara
Karabacak, M., 2002. Türk Konar-Göçerliğine Yeni Bir Bakış ve Son Yörükler. Yörük ve Türkmenlerde Günlük Hayat Sempozyumu Bildirileri, s:159-166, Ankara.
Koca, H., 1995. Gözne'de Yayla Turizmi. Doğu Coğrafya Dergisi, Sayı:1, s:281–301, Erzurum.
Özgüç, N., 1977. *Sayfiye Yerleşmeleri: Gelişme ve Başlıca Özellikleri*. İstanbul Üniversitesi Coğrafya Enstitüsü Dergisi, Sayı:22, s:142-162, İstanbul.
Pekcan, N., 1993. Çakırlar (Boğaçayı) Ovası Jeomorfolojik Etüdü-Antalya. Türk Coğrafya Dergisi, Sayı:28, s:155-163, İstanbul
Somuncu, M., 2005. Aladağlar: Yaylacılık ve Dağ Göçebeliği Konusunda Bir Araştırma. Gündüz Eğitim ve Yayıncılık. Ankara.
Soykan, F., 2003. Coğrafi Çevrenin Turizm Amaçlı Değerlendirilmesinde Turizm Potansiyelini Saptamanın Önemi. Ege Üniversitesi Coğrafya Bölümü Sempozyumları II, Coğrafi Çevre Koruma ve Turizm Sempozyumu, 16-18 Nisan, s:17-24, İzmir.
Tunçdilek, N., 1986. Türkiye'de Yerleşmenin Evrimi. İst. Üniversitesi Yayınları No:3367, İ.Ü. Deniz Bilimleri ve Coğrafya Enstitüsü Yayınları No:4, İstanbul.
Tümertekin, E., 1973. Türkiye'de Şehirleşme ve Şehirsel Fonksiyonlar. İstanbul Üniversitesi Yayınları No:1840, Coğrafya Enstitüsü Yayınları No:72, İstanbul.
Yalman, A,R., Cenupta Türkmen Oymakları, II. Kültür Bakanlığı Yayınları: 256, Kültür Eserleri: 14, Ankara.
Yiğitbaşıoğlu, H., 2000. *Finike ve Yakın Çevresinin İklim Özellikleri*. Ankara Üniversitesi, Dil ve Tarih-Coğrafya Fakültesi Dergisi, C:40, S:1-2, s:107-135, Ankara.
Yücel, T., 1988. Türkiye'de Kır Yerleşme Tipleri. Türk Kültürü Araştırmaları Dergisi. C:XXVI S:1, S:61–69, Ankara.

Zaman, S., 2002. Antalya Merkez İlçenin Doğal Turistik Kaynakları. Atatürk Üniversitesi Sosyal Bilimler Enstitüsü, Yayımlanmamış Yüksek Lisans Tezi, Erzurum.

—. 2005. Kemer'in Turizm Coğrafyası. Atatürk Üniversitesi Sosyal Bilimler Enstitüsü, Yayımlanmamış Doktora Tezi, Erzurum.

CHAPTER ELEVEN

CHANGES OF GLOBAL RADIATION WITH RESPECT TO GLOBAL WARMING IN THE MEDITERRANEAN REGION

BLANKA BARTÓK

Introduction

Global radiation is determined by astronomical factors (e.g. solar declination, the length of day and night) and by actual conditions in the atmosphere (including cloudiness, aerosols and water vapour). Of these factors, cloudiness is the most active component in the solar radiation flux of the earth-atmosphere system (Salby, 1996:39), reflecting about one half of incoming solar radiation into space, thus reducing the energy-balance of the surface. Clouds also modify Outgoing Longwave Radiation (OLR) by absorption and by emission on lower temperature. Hence, in a way, OLR is a cloud-indicator as well. In sum, clouds represent a negative component of the energy budget of the planet (Harrison et al., 1990).

The present study intends to estimate how global radiation may change, parallel to the global warming. This aim is realised in a three-step combination methodology (Section 2) of regression estimations. Satellite-based data for global radiation and for OLR are used (Section 3) besides ground-based visual cloud estimates. Statistical relationships between global radiation and cloud indicators (Section 4.1) are established, in combination with another kind of regression, between cloud indicators and hemispherical temperature (Section 4.2). Finally, the combination of these regression coefficients leads to a point-wise estimation of the expected change in global radiation scaled for a 0.5 K northern hemispherical warming. This empirical methodology is generally recommended for regional scale estimation of changes in solar energy, mostly because cloud parameterisation is one of the most uncertain components in global and regional climate models (IPCC, 2001).

The targeted area of this study represents the southern part of Europe, the Mediterranean region situated between 340N and 460N latitude and 200E and 300E longitude (Fig.10-1).

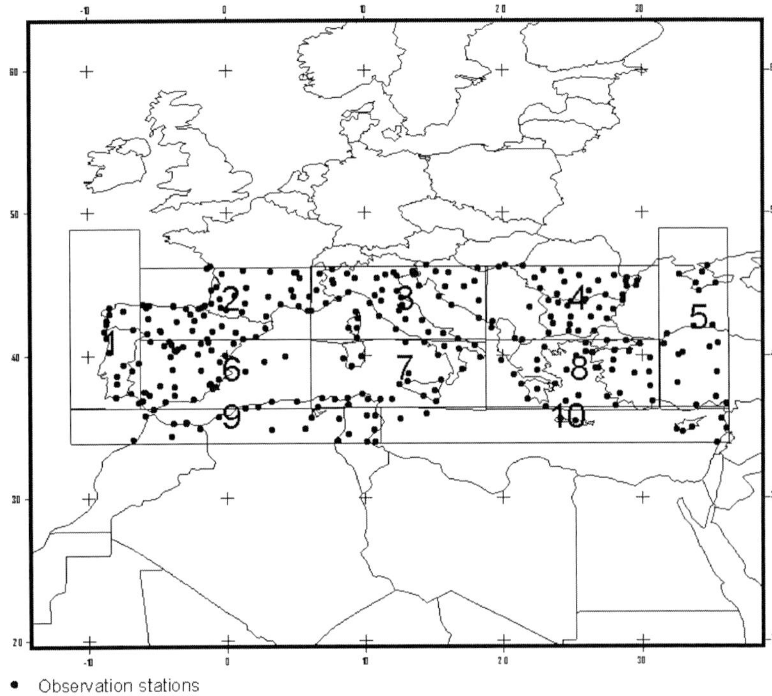

Fig. 11-1: The study area. Black points mark locations of surface–based stations; the lines delimit 2.5x2.5 degrees squares. Thick lines specify the merged regions for which changes in global radiation are finally established.

Regression Estimations

The Statistical Model: Let us have an environmental variable, W, which is an indirect function of variable X through the intermediate variable Y: W{Y(X)}. For a simplified example, biomass of a plant, W, depends on the water content of the same plant, Y, which, in turn, depends on soil moisture content, X.

If the W(Y) and Y(X) are not too complicated (i.e. not especially non-linear) and ΔX changes of X are not too large, then both functions can be approximated by linear relationships within a narrow interval around their initial conditions: W(Y) ≈ Wo+(dW/dY).Y and Y(X) ≈. Yo+(dY/dX).X. Having appropriate sets of observed or experimental data, the above derivatives can be numerically determined by statistical estimates of regression.

If we wish to know how much the given biomass would increase (ΔW) after the increase of soil moisture by additional irrigation, ΔX, we should apply the following simplified formula:

$\Delta W = (dW/dY).(dY/dX).\Delta X$, (1)

where the product of the derivatives can be estimated by the product of the corresponding regression coefficients.

In the following, we determine the regression coefficients for global radiation (in the role of W) depending on cloudiness or OLR (Y) in Section 4.1, and also for cloudiness and OLR (role of Y, in two parallel ways) being statistically dependent on the hemispherical mean temperature (role of X). A combination of the coefficients leading to global radiation estimation attributed to hemispherical temperature changes will follow in Section 4.3.

Regression estimation, however, is being performed by two different methods due to differences in the sample size and in physical background of the relationships. Using the broadly known methodology of least squares, the estimation of dW/dY is performed. These relationships between global radiation and both cloud indicators are physically plausible and the methodology is standard. Hence, here we do not go into further details.

On the other hand, however, the relationship between cloud characteristics and the hemispherical mean temperature is not obvious and sampling errors may contaminate both time series. Hence, instead of the method of least squares we apply another approach.

Method of Instrumental Variables: Let us have a dependent variable, Y, and an independent one, X, similar to Section 2.1. Let us further mark dY/dX as b, representing the regression coefficient of the linear stochastic connection, Y = Y0 + bX. One possible way to estimate the regression coefficient, b, a linear stochastic connection, Y = Y0 + bX, is the method of instrumental variables, first applied by Groisman and his colleagues (Vinnikov, 1986) in climatology. The criteria for an instrumental variable are:

non-zero correlation with observed values of the independent variable;

no correlation with the errors of the independent variable;

no correlation with the residuals of regression in the dependent variable. In case of an instrumental variable, Z, the linear regression coefficient should be estimated as the proportion of appropriate covariance values:

b=Cov(Y,Z)/Cov(X,Z) (2)

For the independent variable, X, is selected the hemispherical mean temperature (Jones, 1994 and updated: http://cdiac.esd.ornl.gov/trends/temp/jonescru/jones.html, with reference to Jones et al., 2000.). The instrumental variable for cloudiness is the sequence of years from the 24 years warming-up period, 1973-1996, exhibiting high (r = 0.796) correlation and a strong (+0.021 K/ year) warming trend and in case of OLR a sequence of years from the 22 years warming-up period, 1979-2000, characterised also by a high (r = 0.736) correlation and by a warming trend (+0.0.022 K/ year). These numbers, as well as the covariance in the denominator of (2), are collected in Table 10-1.

The applied method yields an undistorted point-wise estimation of the regression coefficient but it is rather difficult to establish significance criteria for these estimations which is a disadvantage compared to classical regression estimations that require longer time series. The present approach, developed due to the lack of very long series, is a combination of that applied by Groisman, Vinnikov and their co-workers (overview by Vinnikov, 1986) and that suggested by van Loon and Williams (1976). The former is the application of instrumental variables in estimating regression coefficients of linear connections between hemispheric and regional variables. The latter analyzes the geographic distribution of local trends in appropriate 30 year periods from the past series when the global variables exhibit definite trends.

The Applied Data

Satellite Based Global Radiation: Global radiation data are calculated by a method developed by Rimóczi-Paál et al. (1999) at the Hungarian Meteorological Service to derive all components of the radiation budget with a spatial resolution of 10x10 km using METEOSAT satellite data, empirical formulae and a radiative transfer model. Global radiation data are integrated into 2.5x2.5 degree rectangles whereas in the following analysis the results are presented for merged grids in ten (Fig. 10-1). These data of global radiation obtained from satellite images represent the whole region but they also have a disadvantage: they are available only for a relatively short period (1992-1996). This is why the statistical relationship between global radiation and hemispherical mean temperature can not be established directly by computing one immediate regression analysis. Therefore, it was necessary to include local fields of cloud coverage indicators that exhibited much longer time series.

Cloudiness and Outgoing Longwave Radiation: The surface-based cloudiness data are from the "Extended Edited Cloud Reports Archive" (EECRA: Hahn and Warren, 1999). The studied region is represented by 310 stations (Fig. 10-1.), for the period of 1973- 1996 where the monthly average amount of total cloud cover is given in percentage of the sky. In case of sporadic missing data gaps are filled by linear regression from another station that exhibited the strongest correlation with the given one.

The source of Outgoing Longwave Radiation (OLR) data is the NOAA/NCEP, elaborating and disseminating the measurements performed from the broad of the TIROS polar orbiting satellites (Tiros N, NOAA 6, 7, …, 14) by the Advanced Very High Resolution Radiometer/AVHRR (Grüber and Krüger, 1984). The resolution of the OLR at the top of atmosphere is 2.5x2.5 geographic degrees. The data are expressed in monthly averages for the period of 1979-2000 in Wh/m2.

Global Radiation Estimations for a Warmer Climate

Dependence of Global Radiation on Cloud Coverage: The statistical relationship between global radiation and cloudiness is characterised by correlation coefficients. The 310 stations with surface-based cloudiness observations are grouped by averaging into 10 squares of 2.5x2.5 geographical degree resolution, according to the availability of the OLR and global radiation data sets.

Having short periods to analyse with relatively small samples (5 years for global radiation in the period of 1992-1996), statistical relationships are established after merging the squares of 2.5x2.5 degree resolution into 10 clusters including 10 squares with 2.5x2.5 degrees one by one.

Significant correlation coefficients between monthly means of visual cloudiness and satellite based global radiation series are demonstrated in Table 10-1. Negative correlation coefficients obviously explain that less global radiation reaches the surface if cloud cover increases. Strong statistical relationships occur mostly in the warm period of the year (May – August) in all merged regions.

Table 11-1: Correlation coefficients between satellite based global radiation and visual cloudiness. Values that are non-significant at the 95% level are excluded.

Region	Winter	Spring	Summer	Autumn
1	-.80	-.91	-.98	-.88
2	-.67	-.63	-.85	-.72
3	-.62	-.45	-.78	-.74
4	-.66	-.72	-.71	-.44
5	-.67	-.76	-.88	-.78
6	-.29	-.43	-.48	
7	-.27	-.60	-.44	-.39
8	-.68	-.82	-.93	-.76
9		.57	.54	
10	-.60		-.67	

Outgoing Longwave Radiation measured at the top of the atmosphere is also taken into account as an indicator of cloud cover. The modifying role of clouds in both shortwave and longwave radiation is well known. This means that more clouds result in less longwave radiation at the top of the atmosphere emitted by the surface and the atmosphere. Thus, a positive correlation is mainly detected between global radiation and Outgoing Longwave Radiation (Table 10-2), and higher OLR values at the top of atmosphere indicate less cloud cover. Hence, there is more solar energy at the surface. In Table 10-2, the significant correlation coefficients are shown between global radiation and OLR at the level of 95 % probability. A strong relationship is characteristic during the entire year.

Table 11-2: Correlation coefficients between satellite based global radiation and Outgoing Longwave Radiation. Values that are non-significant at the 95% level are excluded.

Region	Winter	Spring	Summer	Autumn
1	.81	.95	.94	.91
2	.68	.75	.77	.78
3	.85	.91	.93	.90
4	.55	.79	.81	.78
5	.60	.60	.69	.80
6	.70	.67	.47	.54
7	.55	.86	.72	.80
8	.88	.90	.70	.85
9	.76	.67		.43
10	.70	.70	.36	.58

Late 20th Century Cloud Coverage Trends and Global Warming

In the second step of the study, past variations of local cloud cover indicators are calculated in relation to hemispherical temperature by means of instrumental variables (Section 2.2). By this method, the changes in local variables in the condition of the 0.5 K hemispherical mean temperature increases are calculated. Table 10-3 shows the behaviour of cloudiness in the 1973-1996 period relative to temperature in the four seasons of the year. The given period is characterised by a 0.21K/decade warming trend. Cloudiness changes are given in absolute percent (100 % equal to total cloudiness). These changes in the condition of 0.5 K hemispherical mean temperature show temporal and spatial differences during the year. Changes with positive or increasing and negative or decreasing signs appear as well.

Table 10-4 shows the relationship between hemispheric mean temperature and Outgoing Longwave Radiation at the top of atmosphere for the period 1979-2000, characterised by a 0.22K/decade warming trend. The regression coefficient is calculated by the same method of instrumental variables, and the changes are given in Wh/m2. During the year, positive changes of OLR, i.e. a decrease in cloudiness, are characteristic in cases of warming by 0.5K. Positive results are obtained in the spring, summer and winter seasons. In spite of the other periods of the year, autumn represents homogeneously decreasing tendencies with a maximum -6.45 Wh/m2 in OLR if the hemispherical mean temperature increases 0.5 K.

Table 11-3: Cloudiness changes given in absolute percent (100% equals total cloud cover) computed for a 0.5K hemispherical mean temperature increase.

Region	Winter	Spring	Summer	Autumn
1	1.62	-1.94	1.36	6.87
2	-5.46	-5.90	-2.58	3.56
3	-7.21	-6.42	-7.86	2.02
4	-4.49	-4.14	-9.15	-0.10
5	-1.96	-3.84	-2.79	-1.40
6	-2.44	-6.10	-3.55	6.63
7	-4.47	-6.09	-2.85	4.17
8	-6.16	-5.51	-3.20	-2.08
9	-1.46	-4.49	1.32	5.52
10	-3.57	-0.78	-0.84	1.75

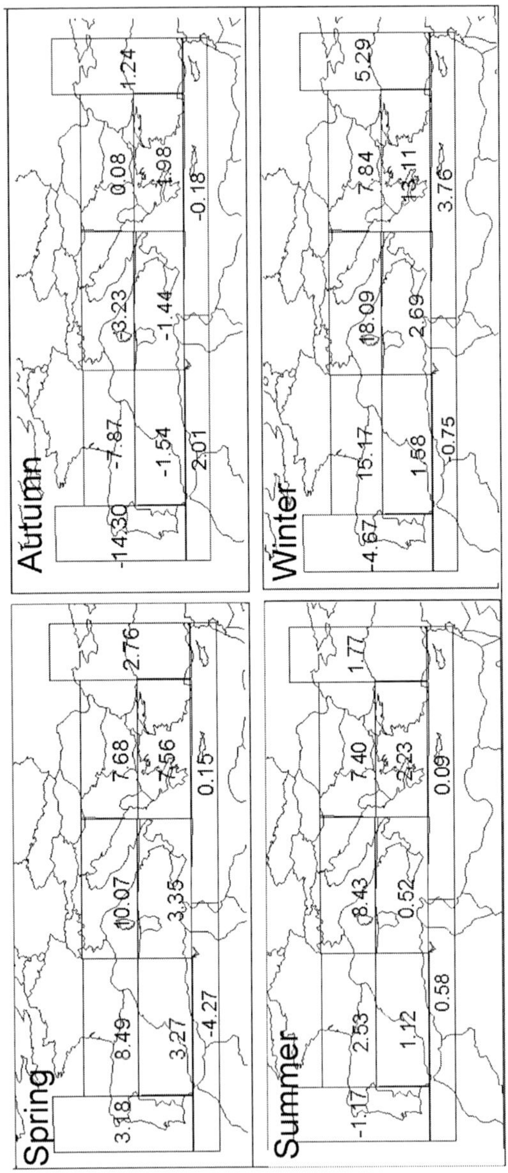

Fig. 11-2: Estimated global radiation changes from the two-step applications of point-wise regression coefficients, involving the visual cloudiness series. The values are given in relative percent (%) compared to the average values of global radiation for the four-month periods, considering a 0.5 K hemispherical mean temperature increase.

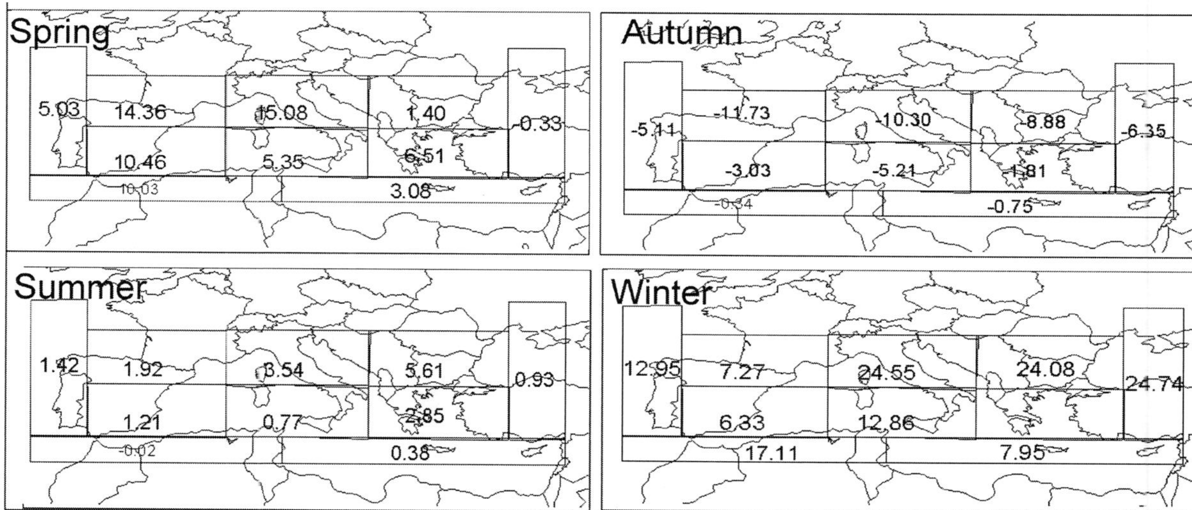

Fig. 11-3: Estimated global radiation changes from a two-step application of point-wise regression coefficients, involving the Outgoing Longwave Radiation series. The values are given in relative percent (%) compared to the average values for global radiation for the four-month periods, considering a 0.5 K hemispherical mean temperature increase.

The method of instrumental variables have a weakness, namely that confidence intervals of regression coefficients are not specified, which means that no uncertainty band can be established for global radiation changes.

Statistical Estimation of Global Radiation Based on Cloudiness and OLR Changes

The main purpose of this study is to estimate global radiation change in the case of a 0.5 K increases in hemispherical mean temperature. In the above two sections, regression coefficients of global radiation and cloudiness indicators, on one hand, and between these cloudiness indicators and hemispherical temperature, on the other hand, have been determined. In this section, the results of appropriate multiplication of these regression coefficients are presented. Significance tests could be performed for the global radiation and cloudiness (and OLR) branch of the estimation only. Hence the results below are shown without final consideration of significance. The spatial resolution of combined results is determined by the merged regions of 2.5x2.5 squares, leading to a 12.2x5 geographical resolution (Fig. 10-1).

Figure 10-2 displays global radiation changes considering a 0.5 K hemispherical warming in relative percent (compared to the 1992-1996 time average). An estimation based on cloudiness shows positive change in global radiation at a 0.5 K mean hemispherical temperature increase in the whole area. Computations applying Outgoing Longwave Radiation also indicate increased global radiation for the case of a 0.5 K warmer climate.

Numerical values of the point-wise estimates are in fair agreement with the sign and order of magnitude, i.e. a several percent increase in value. In some cases, however, there are definite differences between the two values. The reason for these differences and the selection between them needs further investigation, especially selective estimations considering types of cloudiness according to their altitude.

A strong statistical relationship exists between global radiation and cloudiness/OLR in September. The general increase of cloudiness paralleling warming in that month leads to strongly negative changes in global radiation. This feature and its contrast to the previous summer months are displayed in two parts of Fig. 10-3. The above contrastive feature is demonstrated by the data obtained from both visual observations and satellite measurements (OLR). Explanation of this contrast governed mainly by the cloudiness increase in September needs further investigation, including the available information on cloud types (Hahn and Warren, 1999).

Table 11-4: Outgoing Longwave Radiation, absolute changes in Wh/m2 computed for a 0.5K hemispherical mean temperature increase.

Region	Winter	Spring	Summer	Autumn
1	3.20	3.25	2.01	-2.48
2	2.92	8.53	1.84	-6.45
3	8.21	7.42	3.25	-5.87
4	8.09	0.64	6.64	-4.75
5	4.29	-0.26	1.73	-3.92
6	3.71	9.75	3.55	-3.98
7	9.63	6.66	1.80	-4.84
8	8.54	4.45	4.96	-1.24
9	7.23	8.28	3.74	-0.47
10	4.42	6.18	4.19	-2.08

Conclusion

The estimate of global radiation by statistical methods assuming climate change (0.5 K hemispherical mean temperature increasing) indicates an increase in solar energy available for the Mediterranean region. The existence of an energy surplus is demonstrated by the decrease of cloud cover analysed on two sets of data. In the other part of the year the relationship between global radiation and cloudiness is frequently non - significant at the level of 95% probability. This may be a result of the inaccurate interpretation of satellite data, or of visual observations in the case of snow cover, or may be due to the presence of fog.

Acknowledgements

The author hereby expresses her sincere thanks to Anikó Rimóczi-Paál and János Mika, both at the Hungarian Meteorological Service, for providing the author with the estimated global radiation data set and for the idea of the study, respectively.

References

Gruber, A. and A. F. Krueger, (1984). The status of the NOAA Outgoing Longwave Radiation data set. Bull. Amer. Meteor. Soc., 65, 958-962.

Hahn, C. J. and Warren, S. G. (1999). Extended Edited Synoptic Cloud Reports from Ships and Land Stations Over the Globe, 1952-1996. Internet publication and data. http://cdiac.esd.ornl.gov/epubs/ndp/ndp026c/ndp026c.html

Harison, E.F., P Minnis, B.R. Barkstrom, V. Ramanathan, R.D.Cess, G.G.Gibson, (1990). Seasonal variation of cloud radiative forcing derived from the earth Radiation Budget experiment. J. Geophys. res. 95, 18687-18703,

IPCC (2001): Climate Change (2001). The Scientific Basis. Contribution of Working Group I to the Third Assessment Report of the Intergovernmental Panel on Climate Change (IPCC), Cambridge University Press, Cambridge, UK,

Jones, P.D. (1994). Hemispheric surface air temperature variations: a reanalysis and an update to 1993. J. Climate, 7, 1794-1802

Mika J. and Bálint G. (2000). Rainfall scenarios for the Upper-Danube catchment. In: Proc. XXth Conf. Danubian Countries, Bratislava, Slovakia, 4-8 Sep., 2000. CD pp. 990-995

Rimóczi-Paál, Kerényi J., Mika, J., Randriamampianina R., Dobi, I, Imecs Z. and Szentimrey, T. (1999). Mapping daily and monthly radiation components using METEOSAT data Adv. Space Res., Vol. 24 No. 7, pp. 967-970

Salby L. M. (1996). Fundamentals of Atmospheric Physics, Academic Press (International Geophysics Series: v 61), San Diego

Vinnikov, K. Ya. (1986). Sensitivity of climate. Gidrometeoizdat. 219 p. (In Russian)

CHAPTER TWELVE

THE MUNICIPAL ENERGY PLANS AS TOOLS FOR URBAN PLANNING IN THE VENICE PROVINCE, ITALY

ALESSANDRO CALZAVARA, LORIS LOVO, PAOLA MINOIA AND GABRIELE ZANETTO

Introduction: The State of the Italian Legislation and the Role of Local Authorities

In recent years, Italy's emission of greenhouse gases (GHGs) has accounted for nearly 2% of the world's emissions, and approximately 13% of the European Union's actual. In 2003, the most important GHG in Italy was carbon dioxide (CO_2), contributing 85.5 % to the total national GHG emissions expressed in CO_2 equivalents, followed by nitrous oxide (N_2O), 7.4 % and methane (CH_4), 6.1 %. The country's policies on climate change over the years have shifted from voluntary commitments by industry and individuals to obligatory GHG emissions reduction. In fact, following the ratification of the Kyoto Protocol, Italy had a legally binding commitment to reduce GHGs emissions by 6.5% below 1990 levels (equal to 519.75 million tons –Mt- of $C0_2$eq.), over the period of 2008-2012 (Parliamentary Act Number. 120 of 1 June 2002, in the Official Journal of the Italian Republic Number 146 of 19 June 2001). The main sector to be included in domestic policies and subject to measures in order to reduce emissions is the energy sector, accounting for 83.7 % of total national GHG emissions, followed by industrial processes (6.9 %), agriculture (6.8 %) and waste (2.2 %).

The National Action Plan for 2003-2010 for the Reduction of GHG Emissions is based on European legislation and international agreements. However, its application suffers from significant delays, which are officially explained by the opposition of local authorities to approving industrial and infrastructural projects at the basis of these policies (Italian Third National Communication, Policy and Measures, 2002). However, in same cases, as this paper explores, the local commitment towards GHG emissions reduction has increased, as climate change has undergone increased research, and local officials have come to have a deeper understanding of the consequences of local actions on global climate change. The "Piano di Assetto del Territorio" (Plan of Territorial Management) of Martellago – a municipality located near Venice -- is just one of the "virtuous" example of local initiatives to seriously tackle climate change, aiming at introducing the energy factor as a fundamental aspect of the entire urban planning and management process.

In general terms, regional governments maintain the functions of establishing guidelines and coordinating all activities, while planning and management functions are transferred to the next lowest administrative level (see also Regional Law 13 April 2001, n. 11 "Conferimento di funzioni e compiti amministrativi delle autonomie locali in attuazione del decreto legislativo 31 marzo 1998, n. 112"). The Regions have to adopt, in agreement with local entities and industries, the regional energy plan on the use of renewable energy sources (Article 5 (2) of Law 10/1991). The Veneto Regional Law Number 25 of 27 December 2000 envisages norms for regional energy planning, incentives for energy savings, and the development of renewable energy sources. Law Number 10 of 1991 also prescribes an Energy Plan for municipalities with over 50,000 citizens (see Article 5 (5) of Law 10/1991). Its aim is to develop an energy system where renewable energy sources and energy savings are priority instruments adopted to achieve improvement in safeguards to the physical environment.

It is in this scenario that Italian municipalities can define their climate change adaptation and mitigation measures in the framework of an integrated sustainable urban development policy.

This approach assumes that only a definite synergy between local authorities, the regions and the Central State can guarantee the design and concrete implementation of GHG emissions reduction measures.

Such a promising outcome will offer greater efficiency of efforts and a more coherent harmonization of adaptation efforts at the national and local levels.

In the case of Martellago, the changing environment has promoted a beneficial reaction by stimulating voluntary measures to limit GHG emissions. Martellago is a small town of about 20,000 inhabitants, far less than the minimum numbers required by law for the implementation of energy plans. The environmental awareness shown by the municipal officers of this town is pointing out sustainable solutions for the developmental issues that this area is facing, such as a growing urbanisation trend, the marginalisation of agricultural activities, and intensification off road traffic in a land characterised by high vulnerability due to the proximity to the Venice Lagoon.

The Municipality of Martellago: Description of the Area and Planning Needs

The area of Martellago is part of the drainage basin flowing into the Venice Lagoon and, therefore, is integrated in a wide and relevant environmental system, managed by the planning tool named PALAV (Plan of the lagoon and Venetian areas).

During the elaboration of the Plan of Territorial Order (PAT), a planning scheme was foreseen by Regional Urban Law 11/2004 for the territorial government of the municipality. The local administration of Martellago considered the option of an appropriate tool to be used for territorial planning, in line with the need to ameliorate the urban, territorial and environmental quality as well as responding to sustainable development principles (Oliva F., 2005). The methodological need was to identify a quantitative and qualitative parameter that could perform as a synthetic quality index for the interventions that are foreseen in the municipal area, to support planning choices in a shared and consolidated manner.

The technicians considered GHG emissions as the main indicator for the intrinsic quality of the Plan. Two strategic goals were set: the first was to respect the main targets of the Kyoto protocol (reduction of the total GHG emissions, equivalent to 6.5% of the 1990 levels by the period 2008–2012) for the local share; while the second was to implement a housing development plan with "zero impact", in order to foster urban growth without negatively impacting environmental conditions, by systematically ensuring efficiency in the planning exercise.

These priorities have been articulated in a series of sectoral objectives, as follows:

a. reduction of emissions from the residential system through greater energy efficiency in new buildings as well as in buildings under restoration;
b. offsetting emissions from all new GHG sources by the establishment of new green areas, according to typologies and specifications of the Plan;

c. a reduction in emissions from road traffic through the optimization of street networks and the incorporation of intensive vegetation belts along streets.

In addition to these developments, an additional planning objective was added, which was connected with GHG limitation and directly focused on the reduction and control of pollution from PM10 (Jaecker-Voirol and Pelt, 2000).

One stipulation was that the program had to be to be implemented and updated, and adequate to the planning and control needs of small-scale municipalities like Martellago.

The urban plan reflected in the PAT of Martellago had to consider two factors: the town's territorial position, which can favour a low-density expansion and the reallocation of tertiary functions from the Venice area; and the positioning around the Lagoon, which poses environmental challenges, and for which there is a need for ecological preservation (by a re-naturalisation of the river banks with an orientation east-west, and by a of the non-built territories) through the enhancement of the natural systems converging towards Laghetti Park. Laghetti Park is an area that is recognised as an European Site of Community Importance (SCI) and is located in the central area of the municipal territory. These systems present a North-South trend, and this fact allows an action of complementation and interconnection with the river network. The proposed structure is based on a "wider green-belt conception," one designed to protect the Lagoon area. This protection is multifunctional, as it includes hydraulic regimen control, the reduction of flowing nutrients, the consideration of the articulation of landscape and environmental systems, and the prevention of conditions of local climate alteration.

Emission Control from the Residential Sector

The first step was to assess current GHG emissions, comparing them with 1990 levels and then reducing them by 6.5%. This would allow estimation of the CO_2 excess, which would need to be made up for specific adaptation measures. The first approach to the quantitative evaluation had been accomplished by a geo-referenced location of the buildings in the municipal area; each one was then identified by a geo-code and a specific file. By such coding, each property could be categorized by classes of age and building typologies (Table 12-1).

Table 12-1: Age classes of buildings in Martellago

Construction periods	Number of buildings
Before 1800	111
1801-1900	47
1901-1945	174
1946-1961	435
1962-1971	1017
1972-1995	1328
After 1995	331

This data could be interlinked with the utilization (in percentage) of each energy vector, in order to estimate the current consumption relations.

Considering just two years of reference, 1990 and 2006, for PAT purposes, the relation between consumed energy sources are reported in the table 12-2.

Table 12-2: Repartition of energy vectors in 1990 and 2006.
Source: Data from Autorità per l'Energia Elettrica ed il Gas, 2004.

Energy vector	% 1990	% 2006
Gpl	0,03	0,05
Gasoli	0,06	0,02
Natural gas	0,78	0,78
Electricity	0,13	0,15

Each age class of building showed a different pattern of energy transmittance and yearly thermal need. This estimation had been realized through the attribution of transmittance values and of annual energy need values to the different residential building typologies.

The assessment relieved the physiological amelioration of the energy characteristics of the building, both by considering the house covering, and by considering the adoption of appropriate heating systems. The application of provisional models allowed an evaluation, for the year 2012, the levels of CO_2 emissions as 10.307 tons, while the data for the reference year 1990 showed emissions of 11.807 tons (Della Ragione L. et alii 2005). Therefore the forecast, only attributable to the technological amelioration, is equal to 12.7%, but it is even more significant, considering a population growth from 18,506 inhabitants in 1990 to estimated 21,055 inhabitants in 2012 (+13,8%). In this perspective, the evaluated CO_2 emissions per capita will drop from 0.64 t/yr to 0.49 t/yr.

Such a result will not only be ensured by the change of energy values and by improved heating technologies, but also by a larger application of new building regulations, requiring an energy certification for each building. This will allow evaluation of the energy performances of new buildings together with the restructured pre-existing buildings.

A system has been implemented to evaluate the efficiency of buildings in accordance with the principle of environmental sustainability. The plan distinguishes between compulsory requirements and volunteer compliance. Compulsory requirements depend on current regulations in the fields of environment, energy savings and security. Volunteer compliance are in some cases, further restrictions with respect to the legal threshold values, or building advice for reducing the environmental impact of the new construction.

As a complement to this strategy, some collateral tools are being studied, in order to stimulate the compliance of qualitative methods, by transforming the administrative impositions into economic opportunities. For instance, the achievement of energy efficiency levels can allow the benefit of a volumetric bonus. This option will permit the community to share the added costs deriving from efforts towards the objectives of satisfying social interests, including the reduction of GHG emissions.

The Vegetation Component

A further objective of the Plan is to decrease the local CO_2 emissions caused by population growth, by adaptation measures that can sustain local community development without causing deterioration of environmental conditions. While in the building sector the strategy is to adopt technological innovation in order to increase energy efficiency, mainly by sharing collectively the intervention costs, this sector needs to identify some adaptation measures that can compensate for the sector's ongoing emissions. The main possibility is the addition of adequate vegetation, which has the capacity to balance consistent CO_2 emissions (Lorenzini G., 1983).

To this aim, the planning process for Martellago undertook a quantitative and qualitative project for a green component that could counterbalance emissions of CO_2 produced by heating plants, transport means and new real estates. This was expected to be done by fixing them in vegetation biomass and organic substances to the soil.

The planning exercise was preceded by two parallel analyses: the first one to calculate the impacts of new settlements generated by the population increase, in terms of added emission; the second one was to strengthen the capacity of the local vegetation to absorb the CO_2 emissions (McPherson G., 1994 - McPherson G., 1998).

The first analytic phase was the realization of a system to calculate the added emissions that would foresee the following inputs:

1. primary energy need (kwh/m^2 year) of the new building;
2. number of equivalent inhabitants of the new building;
3. superficies (in m^2) of the new building;
4. energy vectors utilized for heating purposes and their CO_2 contribution; and
5. relocations generated by the new settlements and caused by added number of population.

The second analytic phase is aimed to assess the available green areas within the municipal area, both in order to evaluate its compensation capacity of the current emissions, and particularly to draw strategic indications to utilize for the design of new interventions. This analysis has lead to the following articulation of green structures in the territory of Martellago:

i) Forest areas of great dimension, considered as "green lungs" of high efficiency and effectiveness in respect to the proposed functions, serve in the strategy of territorial naturalization, are difficult to achieve in peri-urban contexts (Paletto A. et al 2006).

ii) Trees outside forest areas: all trees or groups of trees that are located in areas that do not belong to the categories "forests" or "other wood systems", as they do not reach the minimum thresholds for extension, coverage, width and height at maturity that are fixed for these categories. Wood systems like e.g. hedges, hedgerows, plantations, can easily be diffused in the territory, and increase the environmental and landscape complexity of the sites. If positioned next to the CO_2 emission targets, they favour absorption and also contribute to the creation of specific micro-climates (APAT, 2003; Borin and Maccatrozzo 2005).

iii) Private green areas: these are present in areas of low inhabited density, and represent a resource that needs to be maintained, in order to reproduce the soil, to guarantee the maintenance of non-impermeable superficies, and to limit emissions that, without these areas, would necessitate very large areas specifically devoted to forests or woodlands.

The third phase of this study focused on the evaluation of the CO_2 absorption capacity of vegetation systems. The absorption capacity from the atmosphere is essentially dependant on various factors, such as the types of species of plants which are utilized in the proposed vegetation systems, their treatment and maintenance, the particular environment in which they are placed, and, of course, the age of the plants. A simple evaluation model was implemented to assess the atmospheric CO_2 fixation capacity as a function of abaci plants linked to the urban planning for the expansion areas.

In this way it was possible to establish the size and magnitude of green areas to assign to the municipal administration and to give indications for the realization of private green areas that would be required to achieve the "zero emissions" objective.

The Transport Component

It is well known that a relevant part of the GHGs emissions are derived from the transport sector. Therefore, it is essential to include this aspect of the equation in any emissions control plan. The approach of the plan was to quantify and fix the CO_2 emissions derived from the equivalent inhabitants by means of transportation, and to set the objective to reduce their emissions into the atmosphere. In this case a model was established to evaluate existing transportation conditions, in terms of circulating vehicles, traffic conditions and driving styles, and to foresee future trends (Berta G.L. et al. 2000).

The data on emissions from this sector was obtained comparing the population increase with some economic coefficients recorded by the Venice Province Chamber of Commerce for the years 1991 – 2001 (Sistema Informativo Camerale, 2001). These parameters were used to calculate emissions for the years 1990 and 2000. For the analysis of future trend for the year 2010, the reference model is based only on a correlation between the number of vehicles and the population.

Based on the number of inhabitants in the Martellago municipal area, the CO_2 emissions have been estimated for the years 1990 and 2000. It is possible to estimate future trends of CO_2 emissions from the traffic increase due to projected demographic growth. Trends were evaluated for the different vehicle classes in use in Italy as a whole. A scenario has been developed based on a GDP growth of 2% and a light decrease in the growth trend of the overall car stock (Berta G.L. et al, 2000). Considering the same classes of vehicles, the emissions scenarios have been built, both for Venice province and for the Martellago municipality, for the year 2010.

In order to evaluate the per capita emissions and demographic development, planners excluded relative values for heavy vehicles (3.5 t or more). These must be calculated within their respective productive sectors. The obtained value was then divided by the 2010 estimated population. The result is that, for the municipality of Martellago, the emission charge is about 1050 kg CO_2eq per year per inhabitant.

This reference value has been compared with a series of alternative viability solutions, capable of reducing greenhouse gases and having different social, economic and landscape impacts. The solution is expected to allow a significant decrease in greenhouse gas emissions by 2010, from 8819 t/year (the "do nothing" scenario) to 4856 t/year.

These values only consider the emissions from the primary transport network within the Municipality. Total emissions from transportation includes a much larger networks, such as motorways and other roads within the same municipal area, which is expected to reach 54,000 t/year by 2012.

Finally, the proposed interventions, based on a development and utilization of vegetation, were also considered as functional when capturing the suspended particulates, in particular PM10, which is dependent on suspended dusts, organic material dispersed by vegetations, inorganic material produced by natural agents, industrial processes, combustion in general, and, of course, vehicles traffic. Vegetation does accomplish the fundamental functions of containing the atmospheric pollution induced by vehicle traffic: vegetation has the effect of purifying and filtering various gases and particulate pollutants. The purifying effect of vegetation is certainly linked to the volume of the foliage, but also to other intrinsic and extrinsic characteristics of the plant. Vegetation can reduce pollution in various ways: by absorption of polluting agents, by modification of atmospheric characteristics, and by the separation of the source of emissions from elements that are most negatively affected.

Plants, within certain limits, can act as air filters by fixation of the atmospheric dust particles. The interception of solid particulate by the leaves can occur by sedimentation, by impact due to air currents, and by deposition through precipitation. Moreover, vegetation can also extract from the atmosphere relevant quantities of pollutants. In light of these functions, the urban plan called for establishing vegetation corridors in areas that are most conducive to protecting residential areas.

Concluding Remarks

According to national Law 10/91, the urban plans for city areas of more than 50,000 inhabitants must define a specific means by which municipalities will use renewable sources of energy. This is the Municipal Energy Plan (Piano Energetico Comunale - PEC). The general aim of the PEC is the integration of energy factors in territorial planning, to promote strategic choices for the amelioration of the environmental state of the urban area and for the promotion of the rational use of resources, in line with sustainable development principles. Some smaller towns, including Martellago, are voluntarily complying with this regulation.

This approach, which is still rarely applied in Italy, significantly changes regional and urban planning, as it requires innovation in the method of decision making, and in the response of planners to residential settlement, the physical environment and to transport needs.

The case presented in this paper is one of the first experiences of coupling urban development needs with the environmental needs, by mainstreaming the energy-climate change policies in the whole planning process. Three components mainly design the planning process undertaken by the urban planners for the municipality of Martellago: 1) the analysis of the energy needs of the final users, in order to produce a municipal energy balance; 2) the analysis of traffic mobility and other major local GHG producers, in order to address the needs of settlements development from an energy point of view; and 3) the development and implementation of mechanisms for dealing with the municipal PAT, including building regulation; the identification of action that is needed to meet specified objectives, and the transfer of responsibility to the relevant technical administration offices.

What has been proposed is not an experiment, but a true planning tool presenting various environmental goals: a reduction of CO_2 emissions; control of the impact of PM10, either by transport policies and also by planning, reforestation and management techniques; an increase of use of renewable resources, particularly biomass from a re-conversion of marginal vegetation and from a proactive management of the vegetation cover; and the realization of energy saving by utilizing new technological options, coupled with sustainable building and urban regulations.

Dissemination activities are also proposed in order to stimulate other local authorities to follow the sustainable energy planning option, in order to unify the efforts towards to reduction of GHGs which need the broadest cooperation at all governmental levels.

References

APAT 2003. Le biomasse legnose. Un'indagine sulle potenzialità del settore forestale italiano nell'offerta di fonti di energia, APAT Rome.

ARPA. 2007 http://www.arpa.emr.it

Autorità per l'Energia Elettrica ed il Gas, Relazione Annuale sullo stato dei servizi e sull'attività svolta, Rapporti alle Camere, Roma, anni 1998 - 2004.

Berta G.L., Casoli P., Ferrari G., 2000. Sui fattori di emissione di diversi tipi di veicoli in ambito urbano. Giornata di studio Inquinamento e traffico nei centri urbani: situazione, soluzioni tecniche e prospettive – ATA Associazione Tecnica dell'Automobile, Genova 26 ottobre 2000. (http://www.ingegneria.unige.it/associazioni/ata/berta%5Cindex.htm)

Borin M., Maccatrozzo G., 2005. Immobilizzazione della CO_2 da parte delle siepi campestri. SISEF

Della Ragione L., Meccariello G., Prati M. V., Rapone M. (2005), Un modello statistico per lo sviluppo di fattori di emissione, Environment Information and Observation NETwork, Expert Panel Trasporti, CNR, Bologna

Italian Report on Demonstrable Progress under Article 3.2 of the Kyoto Protocol, 26 January 2007

Italian Third National Communication, Policy and Measures, 2002 (http://www2.minambiente.it/Sito/settori_azione/pia/att/PIA_accordi_internazionali_menu.asp., at 92)

Jaecker-Voirol and Pelt, 2000. Osservatorio Pm_{10} 2003.

Lorenzini G., 1983. "Le piante e l'inquinamento dell'aria". Edagricole, Bologna

McPherson G., 1994. Using urban forests for energy efficiency and carbon storage. Journal of Forestry 92(10):36-38,40-41.

—. 1998. Atmospheric carbon dioxide reduction by Sacramento's urban forest. Journal of Arboriculture. 24(4) :215-223

Oliva F., 2004. La qualità ambientale del progetto urbanistico. Proceedings of the National Conference "Pianificare con l'ambiente - Ambiente, urbanistica, territorio: idee e strumenti delle Agende 21 locali per una pianificazione sostenibile", Associazione Nazionale Coordinamento Agende 21 Locali, Modena.

Paletto A, De Natale F, Gasparini P, Morelli S, Tosi V, 2006. L'Inventario degli Alberi Fuori Foresta (IAFF) come strumento di analisi del paesaggio e supporto alle scelte di pianificazione territoriale. Forest@ 3 (2): 253-266. (http://www.sisef.it/)

Parliamentary Act n. 120 of 1 June 2002, in Official Journal of the Italian Republic n. 146 of 19 June 2001

Sistema Informativo Camerale STARNET http://www.starnet.unioncamere.it/

Veneto Region, Law of 27 December 2000, n. 25 published in BUR n.114/2000, available on line. http://www.consiglioveneto.it/crvportal//leggi/2000/00lr0025.html

CHAPTER THIRTEEN

THE ROLE OF ECOTOURISM IN LOCAL TOURISM DEVELOPMENT: THE VIEW OF BULGARIAN MUNICIPAL ADMINISTRATIONS

ELKA DOGRAMADJIEVA AND MARIANA ASSENOVA

Introduction

Over the past decade, ecotourism[1] has obtained considerable popularity in Bulgaria. Ecotourism is highly appreciated for its potential contribution to both the preservation of resources and the prosperity of local communities. As one of the most rapidly growing tourism sectors in the world, and an industry with significant potential in Bulgaria, there are high expectations for ecotourism. In Bulgaria, as in some other countries, ecotourism is expected to help balance regional differences, stimulate local economies, and promote sustainable development in socially and economically-depressed areas. The development of ecotourism in Bulgaria is associated mainly with protected[2] and rural[3] areas. However, ecotourism

---

[1] In the current paper ecotourism is discussed from the point of view of the definition provided by WTO (Vereczi, 2002).

[2] According to the national legislation protected areas in Bulgaria occupy about 5% of the country's territory. However, areas that meet ecological criteria of the European network of protected zones NATURA 2000 are much wider and cover about 35% of the area.

[3] Rural municipalities in Bulgaria are determined as ones without settlements of more than 30 000 dwellers and population density lower than 150 people / km². They comprise 87 % of all 265 municipalities in the country.

also supplies opportunities for traditional destinations to diversify their main traditional offerings, such as sea recreation, skiing and spa-tourism.

Well-preserved nature and cultural heritage create great potential for ecotourism development in Bulgaria. With a relatively small territory of 111,000 km², a large variety of relief forms, climatic zones, and landscapes can be observed, from the Black Sea coast to the highest mountain ranges. Mountains take up twenty eight percent (28%) of the country's area but there are numerous additional natural sites of interest, including in valleys and especially in wetlands along the Black Sea coast and the Danube River. Bulgaria is also a country of great biodiversity, with more than 30,000 plant and animal species. Almost one-third (1/3) of the country's territory is covered with forests. Vast territories preserve deciduous and coniferous woods 250-300 years of age and there are some of the most significant habitats for hundreds of endemic, rare and endangered spices in Europe.

For example, bird life in Bulgaria is rich in the variety of its species. There are 405 bird species that make up more than the half of the total number in Europe, including about 35 birds-of-prey. Preserved natural environment and an abundance of living and nonliving nature are the reason for the establishment of a network of protected areas, which is expected to widen in the years to come. Srebarna Lake and Pirin National Park are Bulgaria's natural sites included in the UNESCO List of Natural Heritage Sites.

The potential for ecotourism development in Bulgaria is not limited to the country's rich natural landscapes. One of the important features of the country is the dense network of small picturesque villages with preserved traditions, and the availability of rich cultural and historical resources that carry the mark of the crossroad between European and Middle Eastern cultures. These conditions create favorable preconditions for the development of various forms of alternative tourism and the successful combination of ecotourism with other types of tourism.

Since the beginning of the 1990s, Bulgaria has made noticeable advances in the field of ecotourism. This process has been demonstrated through the following developments:

The establishment of numerous NGOs focused on fostering and promoting ecotourism. These include local and regional tourism organizations, national product associations and ecological organizations. Some of these organizations are very active and are tightly related to the main initiatives concerning ecotourism at the local, regional and national levels.

The development of ecotourism-oriented businesses. Included are firms providing accommodation and catering to specific destinations, and travel agencies and tour-operators offering specific services or complete ecotourism packages. There is no official statistical data on the number of companies operating in the sphere of ecotourism in Bulgaria, but they are mainly small enterprises with from 2-3 to 10-15 employees.

The implementation of a range of ecotourism projects at the local and regional levels.[4] These are generally aimed at the construction of specialized infrastructures, providing support to local ecotourism businesses, the development and promotion of ecotourism destinations, and of new products and services. As a result, ecotourism attractions have been recently created and many destinations have gradually become recognizable at the market level by their resources and supply.

Despite these efforts, however, Bulgarian public policy on ecotourism suffers serious shortages and is still far from being highly effective. For the past years it has been strongly dependent on foreign donors[5], while the initiatives that have already been undertaken have often shown to be unsustainable after the expiration of external funding. This problem can be explained by a number of factors, including:

The implementation of isolated small-scale projects within separate municipalities that do not take advantage of economies-of-scale and do not bring tangible economic benefits to local communities;

Evidence of "ready-made" decisions and the implementation of a "copy-paste" approach to solutions without enough consistency with local potential and needs;

Insufficient capacity of local authorities and NGOs for tourism planning and destination management under the conditions of market economy and decentralization of tourism policy;

Problematic coordination and cooperation between the different initiatives and stakeholders involved (e.g. the public and private sector; the local, regional and national levels; NGOs and official authorities; within the NGO-sector and the business, etc.).

An attempt was made for overcoming some of these problems through the elaboration of the National Ecotourism Strategy in 2004. This was the first planning document in the field of tourism at the national level that was presented to public audiences, with great enthusiasm, and was

---

[4] Usually these projects are implemented in collaboration among NGOs, local or regional authorities.

[5] UK Know-How Fund, USAID, Swiss Agency for Development and Cooperation, Swiss Program for Protection of Biodiversity, GETZ, JICA, UNDP, PHARE, etc.

expected to strongly push ahead ecotourism development in the country. Unfortunately, serious concerns about the accomplishments[6] of the Strategy have appeared. Probably for this reason, the Strategy has not yet been officially adopted by the Bulgarian government.

As ecotourism takes place mainly at the local and regional scale, the role of local planning and management capacity is essential for its further development. Municipalities in Bulgaria have gradually come to occupy the leading position in regards to ecotourism initiatives and their implementation, after a long period of time when they had been considered as institutions without possibilities to influence tourism policy.

According to the national legislation[7], local authorities in Bulgaria play an important role in the coordination, fostering and control of tourism development through tourism planning, land-use planning and a number of administrative procedures as well as through the maintenance of dialogue and collaboration with business and with different kinds of NGOs. The specific functions of municipal authorities that are closely related to ecotourism include: local infrastructure development; grading of low standard accommodation establishments and restaurants; the collection of a tourist tax; support for the promotion of local tourism offerings; the provision of information to visitors (e.g. development of tourism information centers); and the preservation and protection of natural and cultural heritage sites within their territories.

Institutions of local self-government are in charge of preparing, adopting and implementing development strategies and annual tourism development programs, containing measures and actions aimed at fulfilling the functions described above. They bear the main responsibility and dispose of various mechanisms to provide financing to tourism-supporting activities, either by their own resources (utilizing the tourist tax fund) or by attaining external funding. Currently, the role of municipalities is getting even more significant, as they are expected to have a leading position in the absorption of EU Structural and Cohesion Funds. Hence, the assessment by local officials of the potential for ecotourism and the possibility for its further adoption is indicative of future trends in local tourism policy, and is critical to particular projects that are to be developed and implemented.

---

[6] The problem is discussed by Marinov and Dogramadjieva (2005)

[7] Tourism Act (2002), Regional Development Act (2004), Spatial Development Act (2001), Local Self-Government and Local Administration Act (1991), etc.

Data and methods

This paper presents results of a wider questionnaire-based survey of Bulgarian municipalities[8] that was carried out in 2006 in order to support the elaboration of the Strategic Framework for Tourism Planning and Development and the Main Directions of the Strategy for Development of Bulgarian Tourism for 2006-2009[9].

A total of 124 local administrations participated in the survey. This accounts for nearly 50% of the entire number of municipalities in Bulgaria (265) and covers more than half of the country's territory. Areas of different tourism potential and types of tourism development were represented, including seaside areas, mountain regions of various altitudes, regions with developed centres of spas, cultural and city tourism, as well as areas in which the tourism industry is almost absent. Thus, comparable information about the current development of tourism, and its potential and main problems at the local level, were provided from the point of view of municipal administrations, as these are important stakeholders in Bulgarian tourism development.

The current analysis is focused on several questions (out of 44 included in the questionnaire) concerning fostering and limiting factors of tourism development in municipalities, the most important activities that have to be undertaken, the local assessment of current and potential tourism products, the undeveloped resources and necessary measures for their development, and guidelines for tourism promotion, etc. Conclusions are drawn from the information which was gathered, focusing mainly on the strategic position of ecotourism in the future development of tourism as a whole in Bulgaria.

Results and discussions

Self assessment of tourism potential in municipalities

In general, the potential of tourism resources within Bulgarian municipalities was highly evaluated by local administrations. Nearly half (1/2) of the valid respondents estimated it as "very good," one third (1/3)

---

[8] The Report on Municipalities Tourism Survey (2006) was prepared by the authors of the current paper and transmitted to Bulgarian State Tourism Agency. It has not been officially published but has supported the elaboration of the Strategic Framework for Tourism Development (2006).

[9] The document has been submitted to the Bulgarian State Tourism Agency and its official adoption is expected soon.

of them as "good," and fourteen percent (14%) as "average." There were few negative assessments (4%). The greatest majority of municipalities that have estimated their potential as good and very good are located in mountainous regions and near water basins. In most of the cases, the natural environment within their territories is well-preserved and resources for ecotourism are available.

Respondents were also asked to evaluate the protection and quality of environment within their municipalities, and environmental aspects that had to be improved. Generally, the quality of nature was more highly evaluated than its protection (Fig. 13-1).

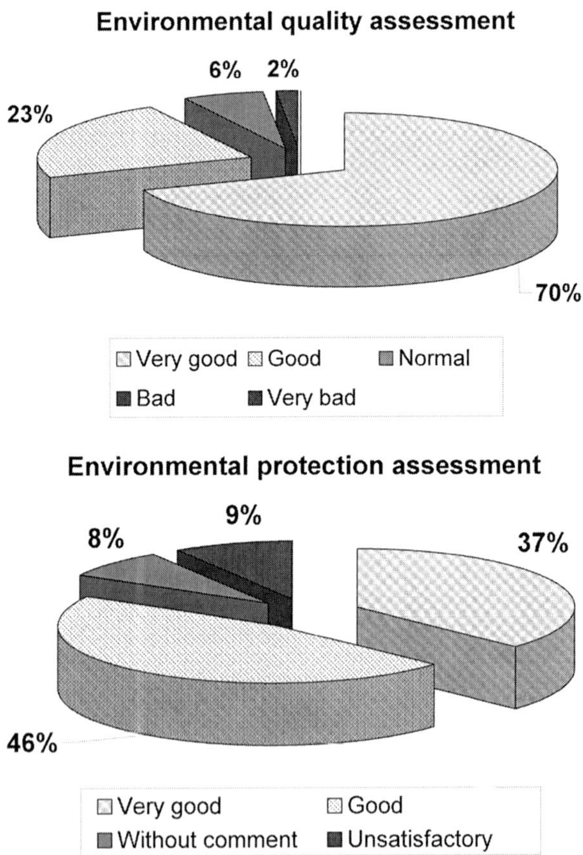

Fig. 13-1: Assessment of environmental quality and protection

According to the results, seventy percent (70%) of the municipalities are characterized as the environment as "very good quality" and about one fifth (1/5) of them as "good quality." Negative assessments referring to this part of the question were almost absent. On the other hand, municipal administrations were more critical of protection policies. Protection was estimated as very good by only thirty-seven percent (37%) of the respondents, while negative estimations of protection made up nearly ten percent (10%). The fact that eight percent (8%) of the respondents left this question without comment indicates that they probably do not consider themselves involved in such policies.

Most of the responding municipalities did not believe that many of the environmental factors listed in the questionnaire had to be improved (Fig, 13-2). The only element of local environmental policy that was widely estimated as needing improvement was solid waste collection (81% of the valid respondents). High support was also demonstrated for the better protection of woods (42%) and for solving of drinking water problems (32%). Other problems that were pointed out by a considerable number of municipalities included poor sanitation (25%), pollution from farming (23%), and pollution of lakes and rivers (20%). Few municipalities considered quality of air, uncontrolled hunting and fishing, and noise pollution to be problematic in their area.

Fostering and limiting factors in the development of local tourism

Thirteen factors that could foster future tourism development were enumerated in the questionnaire. Respondents had to check those recognized as important for their area and to arrange them according to their priority. Free answers could be added as well.

Results show that all of the enumerated factors were estimated as significant, but the natural environment occupied the first position. The natural environment was listed as the main driving force of tourism development by forty percent (40%) of the respondents and was mentioned as one of the three priority factors by sixty-eight percent (68%) of those responding (Fig. 13-3). Other factors tightly related to ecotourism development were considered as very important by a great share of respondents: local customs and traditions, historical sites and other cultural attractions, mountain environments, possibilities for rural tourism.

Another question in the survey was dedicated to the limiting factors of tourism development in destinations that reduce tourist satisfaction. Respondents had to check the main reasons for visitors' complaints and to arrange them according to their importance. Comparatively low support was demonstrated to most of the enumerated shortages (Fig, 13-4) which indicates low self-criticism and certain inclination of municipal administrations to underestimate limitations of local tourism development.

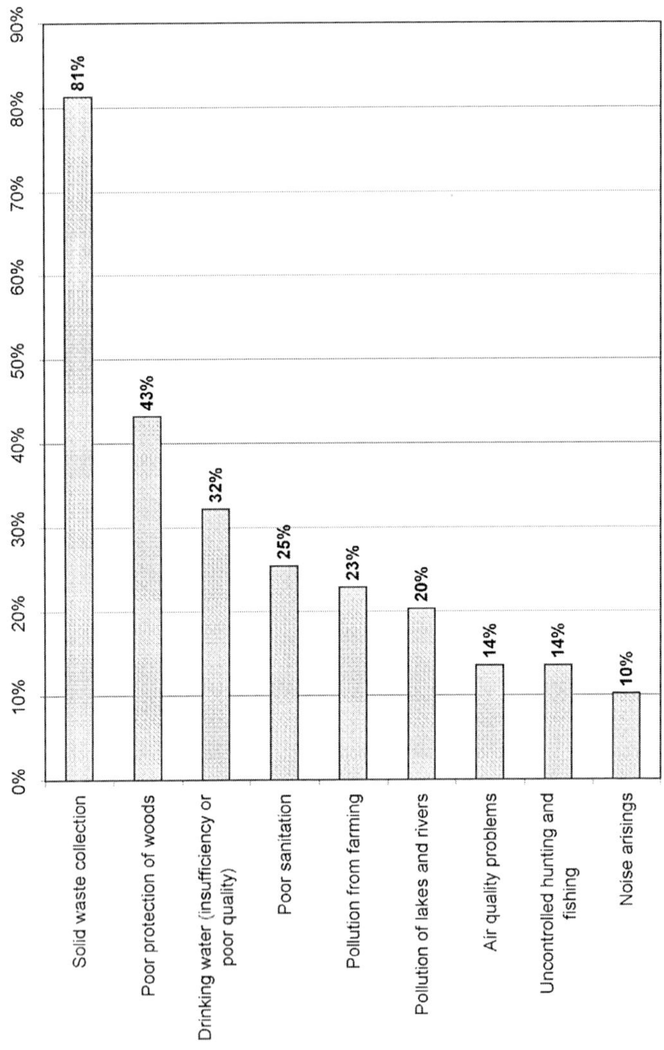

Fig. 13-2: Environmental aspects that have to be improved (share of valid respondents)

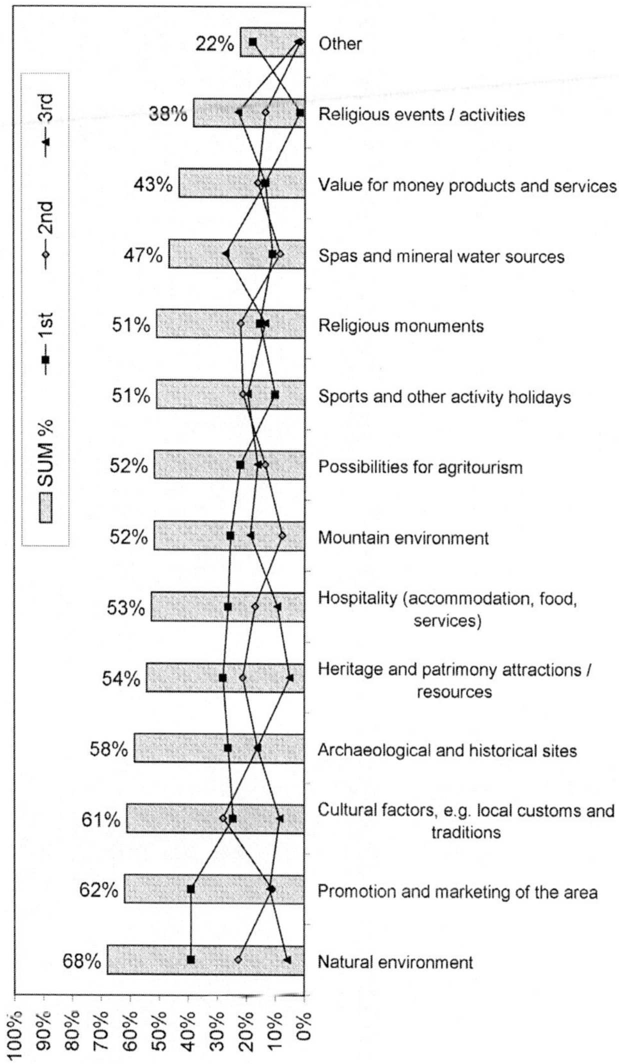

Fig. 13-3: Fostering factors of future tourism development (% of valid respondents)

Based on the answers of the respondents, a wide group of factors that reduce the satisfaction of customers and create obstacles to the future of tourism development can be outlined. Most of these directly refer to ecotourism. Insufficient development of common infrastructure generally, and of tourism infrastructure specifically (especially the bad condition of roads and the scarcity of tourism facilities) appeared as the most widespread problem. This was pointed out by fifty-four percent (54%) of valid respondents. Other important shortages that reduce the possibility for the development of ecotourism in a great share of municipalities include the lack of-or insufficient-tourism information (38%), low level of service (40%) and difficult communication between locals and foreigners (38%).

Local tourism products and services

Analysis of local products and services in the field of ecotourism is based on respondents' answers to several questions concerning current and potential tourism activities, the most famous local attractions and their competitors, as well as undeveloped resources and measures necessary for their adoption.

Results show a great variety of nature-oriented and heritage-based tourism activities were listed as currently being offered in most of the municipalities. However, the majority of these were characterized by low and medium levels of development (Fig. 13-5). Hiking and hunting together with visiting museums and monasteries seem to be most advanced, but they were mentioned as highly developed by only one-fifth (1/5) of the respondents. A few activities appeared to be of comparatively limited territorial distribution: canyoning, scuba diving and river rafting. Yet, according to the answers, these are currently present in about forty per cent (40%) of the municipalities, though with very low levels of development.

All the enumerated activities were considered as having potential for future development and most of them were estimated as very promising by an overwhelming majority of municipalities (Fig. 13-6). The majority of nature-orientated and heritage-based activities were checked by more than sixty percent (60%) of the respondents and ecotourism was most preferred among them. These were pointed out as highly promising by seventy-eight percent (78%) of the valid respondents.

Rural tourism, hiking, educational tours and participation in cultural events were also considered to be of good potential in more than two thirds (2/3) of the municipalities. Highly specialized activities like bird, flora and animal watching appeared of relatively lower potential from the point of view of local authorities. Lower support was demonstrated for some adventure activities, like mountain biking, paragliding and water sports.

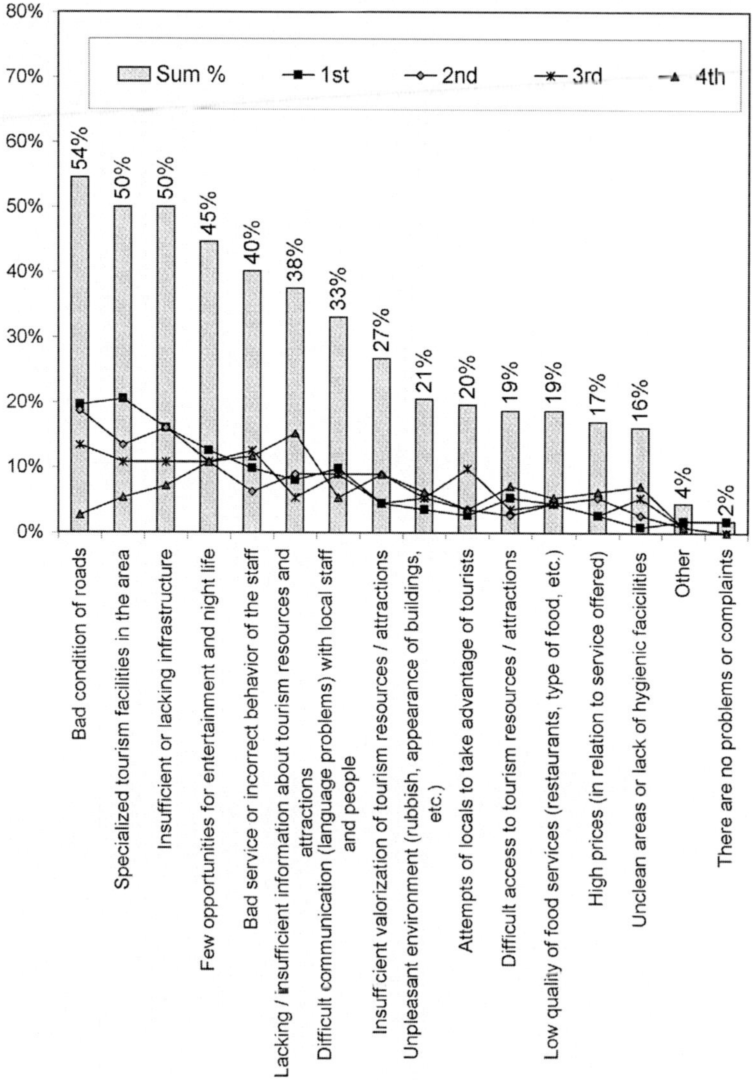

Fig. 13-4: Limiting factors of local tourism development (% of valid respondents)

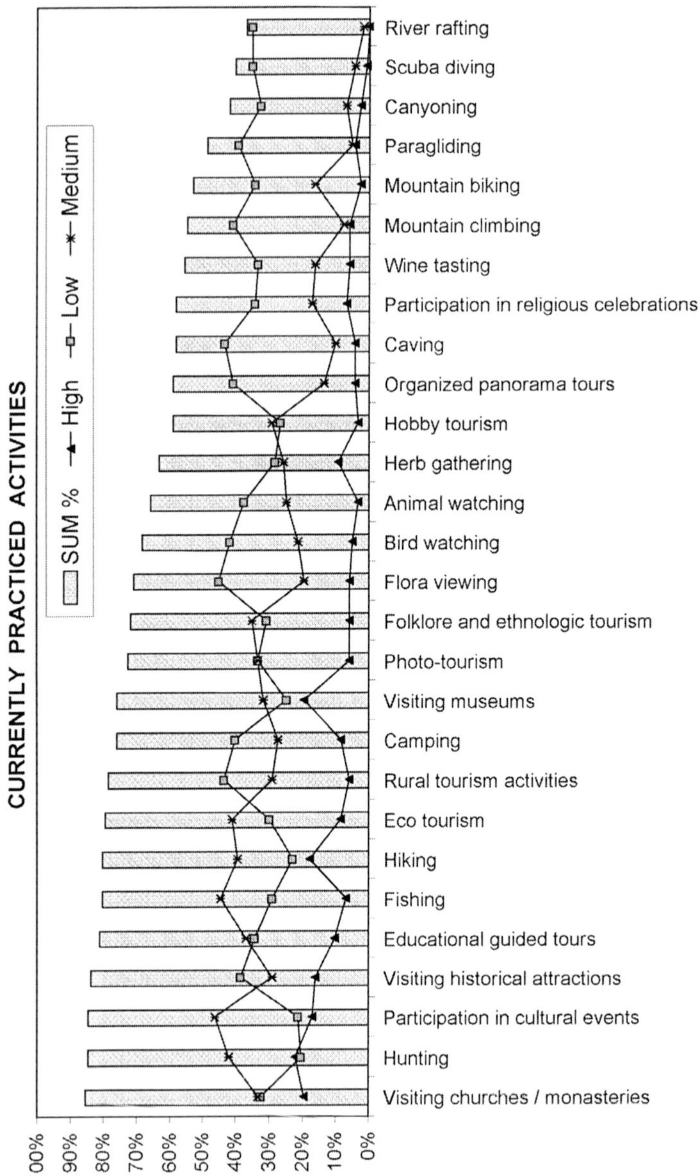

Fig. 13-5: Currently practiced nature-based and ecotourism activities and their level of development (share of valid respondents)

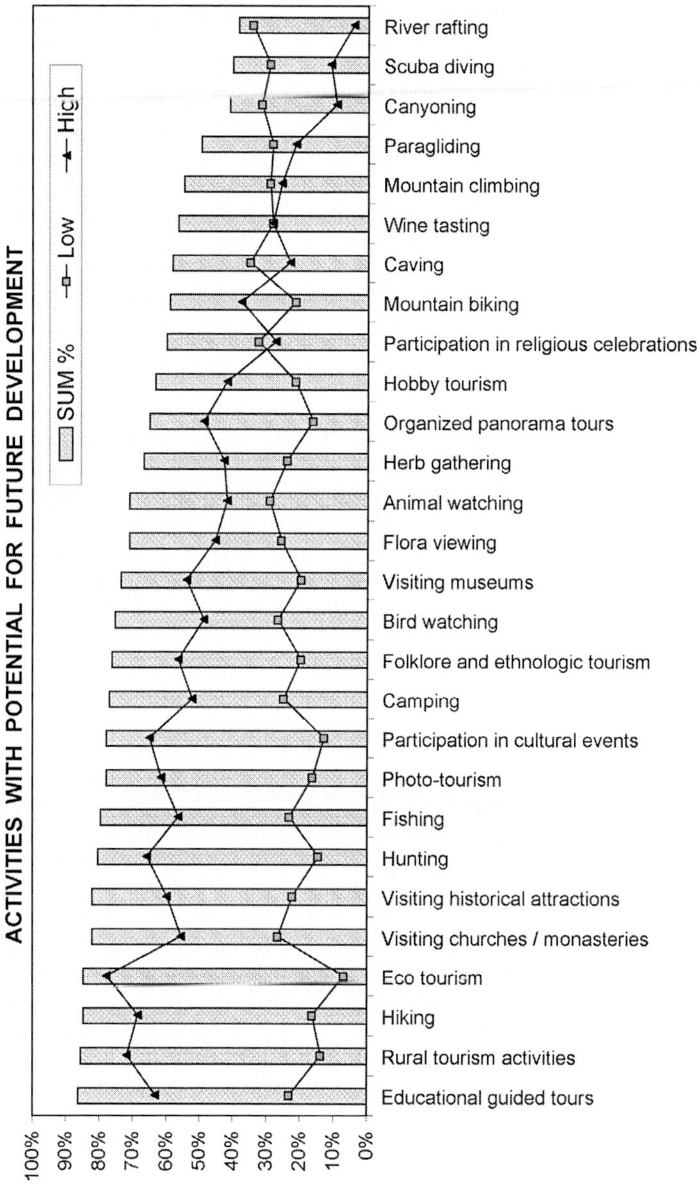

Fig. 13-6: Nature-based and ecotourism activities with potential for future development (share of valid respondents)

In another question, respondents were asked to outline the main tourism attractions in their area and to estimate whether they had competitors. Additionally, the respondents were also asked to identify their competitors. Generalized results show that, on average, respondents mentioned 2.8 attractions in each municipality (Table 13-1). About half of the municipalities were able to point out four attractions while seventeen percent (17%) of them relied on just one or two attractions. There was also a considerable share of respondents (10%) that mentioned no famous attractions in their area. Thus, the current situation in the country is revealed, showing that many of the available resources in municipalities are not transformed into real attractions and are not included in complete tourism products, despite their high cognitive and aesthetic value.

Table 13-1: The arrangement of municipalities by the number of attractions

	Number of municipalities	Share of municipalities
4 attractions	65	52%
3 attractions	26	21%
2 attractions	10	8%
1 attraction	11	9%
No attractions	12	10%
Average number of mentioned attractions	2,8	

A total of 341 attractions were enumerated by the respondents in free answers. They could be arranged in eight categories (Fig. 13-7). Natural sites, tourist routes and eco-trails made up 34% of all answers. Historical and architectural sites, religious sites, local traditions, and events that are usually tied to ecotourism activities, comprised forty percent (40%) of all answers. Thus, ecotourism attractions of both natural and cultural origin were outlined as the leading ones in Bulgaria from the point of view of local administrations.

Answers to the second part of the question relating to the estimation of local sites revealed to what extend local authorities take into consideration the competitive factor in the development of tourism. About half of the municipalities which listed at least one attraction did not answer the question. This means that the majority of respondents did not even think about competitors, and ignored the importance of competitive advantage for successful positioning within the dynamic tourism market. A few municipalities stated their attractions had no competitors. In fact, 9 attractions were estimated as unique in Bulgaria, and most of these referred to ecotourism: 4 natural sites, 2 historical sites, 1 religious site, 1 museum and a gallery.

Participants in the survey were also asked to enumerate potential tourism sites and to identify necessary activities for their utilization. A total of 287 undeveloped tourism sites were enumerated and their average number in responding municipalities was 2.4 (Table 13-2). Results show that more than two thirds (2/3) of the municipalities have 3 undeveloped resources, which confirm the above-made conclusions about the existing high undeveloped tourism potential and the low level of transformation of resources in real attractions.

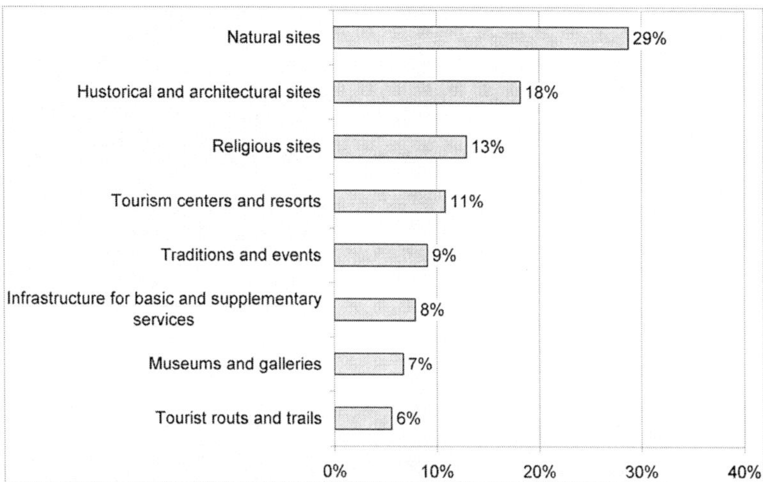

Fig. 13-7: Main types of attractions (share of received free answers)

Table 13-2: Grouping of municipalities by the number of undeveloped tourism sites

	Number of municipalities	Share of municipalities
1 undeveloped resource	20	16%
2 undeveloped resources	12	10%
3 undeveloped resources	84	68%
Without answer	8	6%
Average number of undeveloped resources	2,4	

Figure 13-8 illustrates the arrangement of respondents' answers with reference to types of undeveloped resources. Obviously undeveloped resources for rural and ecotourism predominated (35%), followed by those for cultural tourism (24%). According to local authorities, undeveloped resources for adventure and spa tourism were much smaller in share (10-11%). Some municipalities stated the necessity to adopt resources for hunting, fishing and ski tourism, but their relative weight was too small (3-5%). Practically there were no outlined undeveloped resources for sea recreation. There was another small group of answers referring to the necessity of specialized infrastructural development and to more rational use of existing facilities. It is worth mentioning that only three respondents stressed the unused possibilities of favorable location and cross-border collaboration.

Respondents pointed out numerous measures for development of the above-mentioned resources which could be roughly united into several groups:
-The development of common and specialized infrastructures to improve accessibility to sites of interest;
-The restoration and conservation of historical sites;
-Education and training of local people and tourism staff;
-Obtaining financial resources for marketing and promotion;
-Improvement of tourism information at points of destination.

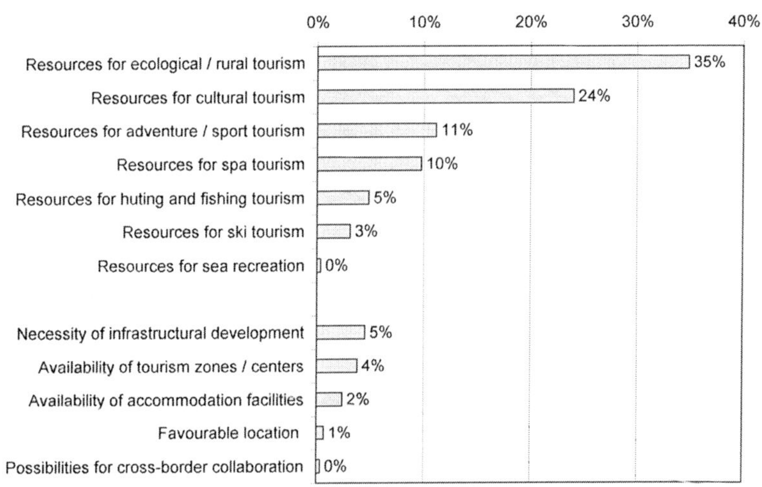

Fig. 13-8: Grouping of undeveloped tourism resources (share of free answers)

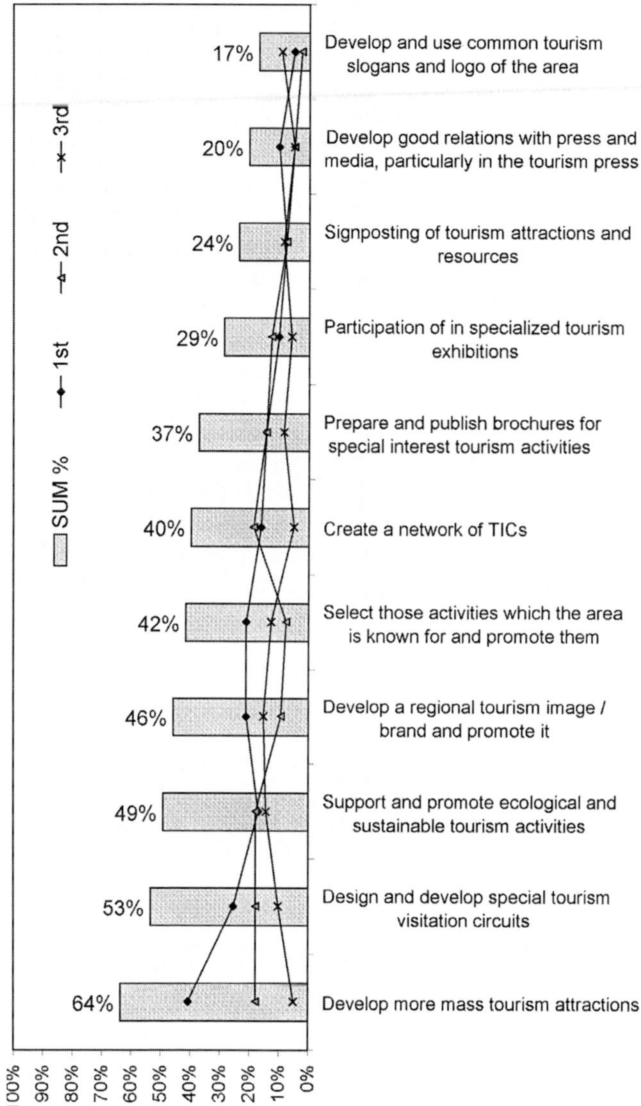

Fig. 13-9: Priorities of marketing and promotional strategy (share of valid respondents)

Guidelines for development of tourism product and promotion

Since tourism promotion was outlined as a leading factor of local tourism development, it is necessary to analyze how municipal administrations arranged their priorities in this central field of activity. Analysis is based on answers by respondents to the question concerning the most important guidelines for their promotional strategy (Fig. 13-9).

Elaboration of specialized tourism circuits and promotion of ecological and sustainable tourism activities appeared as a significant trend of future development, immediately after the creation of more mass tourism attractions. These were supported by half of the responding municipalities and were pointed out as a first priority by twenty to twenty-five percent (20%-25%) of them.

Next by cumulative result came the development of a regional tourism image, together with the selection and promotion of those activities which the area is known for (46%-42%). These were estimated as highest priority activities by 21% of the valid respondents. Relatively strong support was demonstrated also for the creation of a network of Tourism Information Centers (40%) and to preparation of brochures for special interest tourism (37%).

Based on the answers of the question, a conclusion could be made that local authorities stress the *creation* of attractions more than on the *promotion* of attractions. This could be interpreted as a misunderstanding of the significance of tourism promotion. But it also indicates the crucial need for transformation of favorable resources in real tourism attractions which has not been achieved yet in the greatest majority of Bulgarian municipalities.

Another observation regarding "pure" tourism promotion shows that a regional tourism image is accepted as hypothetically important. However, it is still far from a real implementation priority in the marketing policy of most municipalities. This is proved by the lowest support demonstrated for the development and use of common tourism slogans and logos of the area that should be the main instrument to accomplish such a goal.

Conclusions

According to Bulgarian municipal administrations, currently nature-oriented tourism activities are widely spread over the territory of the country, but are of low and medium levels of development. They include mainly hiking, ecotourism, fishing and hunting, as well as rural tourism and herb gathering. The results of the survey correspond to the Directional Policy Matrix (Strategic Framework…, 2006) developed to evaluate the strategic scenarios for tourism development in Bulgaria. Bulgarian ecotourism at the present moment is assumed to have negligent market share and extremely low revenue growth. This product is not well perceived and is often misinterpreted; there is no product branding.

At the same time, in spite of its internationally small market segment, the growth prospects for ecotourism are promising and it is attractive to invest in. Municipal administrations in Bulgaria also regard nature-based tourism as having potential for future development and are very positive in terms of their view of its potential impact for the future. In fact, municipalities demonstrate the greatest support to ecotourism and other nature-based activities for the future, followed by educational guided tours, agro-tourism and various activities, mostly related to cultural tourism.

Obviously, the country is in the ecotourism *product development* phase, since most municipalities put the emphasis on creation of attractions, elaboration of visitation circuits, and the establishment of a network of Tourism Information Centers, while *branding and promotion* are considered as hypothetically important but not as urgent. The limiting factors, which have to be tackled, include the improvement of infrastructures, especially of solid waste treatment, and of roads. Lack of adequate infrastructures has a negative impact on the accessibility of attractive tourism sites.

As ecotourism is regarded as one of the most promising types of tourism in Bulgaria, reworking and actualizing of the already developed National Ecotourism Strategy is necessary. Clearly, the role of municipal administrations as main stakeholders in the ecotourism industry should be recognized. Finally, the state should be resolutely engaged in the implementation of the Ecotourism Strategy.

References

Marinov, V., E. Dogramadjieva. 2005. Evaluation of tourism development strategies: the case of the National Strategy for Development of Ecotourism in Bulgaria. *Annular of the University of Sofia*, Book 2–Geography, vol. 97, pp. 257:275

National Ecotourism Strategy and Action Plan. 2004. USAID, Sofia, Bulgaria. http://www.ecotourism.bulgariatravel.org

Strategic Framework for Tourism Planning and Development in Bulgaria and Main Directions of Strategy for the Development of Bulgarian

Tourism for 2006–2009. 2006. Bulgarian State Tourism Agency, PHARE, Sofia, Bulgaria. http://www.tourism.government.bg

Report on Municipalities Tourism Survey. 2006. Bulgarian State Tourism Agency, PHARE, Sofia, Bulgaria.

Vereczi, G. 2002. Summary of the WTO Research Programme on Ecotourism Generating Markets, Sofia, Bulgaria http://www.ecotourism.bulgariatravel.org

Local Self-Government and Local Administration Act. 1991. Promulgated, State Gazette, SG No. 77/17.09.1991, last amended SG No. 30/11.04.2006.

Tourism Act. 2002. Promulgated, State Gazette No. 56/7.06.2002 (effective 1.10.2002), last amended SG No. 80/3.10.2006.

Regional Development Act. 2004. Promulgated, SG. 14/20.02.2004, amended and supplemented SG No. 32/12.04.2005, SG No. 88/4.11.2005, supplemented SG No. 24/21.03.2006.

Spatial Development Act. 2001. Promulgated, State Gazette No. 1/2.01.2001 (effective 31.03.2001), last amended and supplemented SG No. 79/29.09.2006.

CHAPTER FOURTEEN

CULTURAL ASPECT OF GEOGRAPHICAL STUDIES: THE MEDITERRANEAN REGION

ELZBIETA ORLOWSKA

Introduction

Geographically-specific human culture is the most important factor that differentiates people from other living things and shapes the way we live. Culture serves as a guide for how we act and interpret the world around us. This paper focuses on cultural aspects of the Mediterranean Region and its modern multicultural space, a space which is constantly undergoing transformation due to globalization. The effects of this transformation are most evident in the regions which form the frontiers of cultural boundaries. In the process of widening the Mediterranean *ecumene*, particular human communities have developed different lifestyles. In addition, there have been active forms of human adaptation to the environment through the hereditary transfer of knowledge and models of behaviour suited to specific cultural levels and specific existential conditions in particular locations. These are components of culture, which in turn are products of individual societies creating and shaping images of their own regions.

When studying how cultural characteristics of place develop, geographers often focus on the region as the key concept of classification and analysis. In fact, the "region" is *the* basic building block of geographic analysis. In every geographic region, we discover certain patterns which are related to culture and serve as a reflection of culture. To understand these patterns, it is first necessary to understand their cultural context. Indeed, questions about the essence of human geography are also questions about human life and cultural identity in the multicultural space of Earth.

Study Area

The Mediterranean Region is defined as the region that includes countries bordering the Mediterranean Sea (plus Portugal) between about 27° N to 47° N and 10° W to 37° E.

The term Mediterranean derives from the Latin *mediterraneus* ("medius" or "middle" plus "terra," which means "land," or "earth"). To the ancient Romans, the Mediterranean was the center of the Earth as they knew it.

Throughout history, the Mediterranean Sea has been known by a number of alternative names. It was, for example, commonly called *Mare Nostrum* (Latin, "Our Sea"), and occasionally, *Mare Internum* by the Romans. The Greeks called the Mediterranean *Mesogeios*, meaning "inland," or "interior." In the Old Testament, it was on the "west coast" of the Holy Land, and therefore behind a person facing the east, it was called the "Hinder Sea," sometimes translated as "Western Sea" and also the "Sea of the Philistines," as ancient Philistines occupied a large portion of its shores near the Israelites. Mostly, however, it was the "Great Sea" or simply "The Sea." In Hebrew, it is called *HaYam HaTikhon* – the "Middle Sea," a literal adaptation of the German equivalent *Mittelmeer*. In Turkish, it is *Akdeniz* – the "White Sea." In Arabic, it is *Al-Baħr Al-Abyad Al-Muttawasit* – the "Middle White Sea" (www.wiktionary.org).

The Mediterranean, which literally means the "sea between lands," offers favorable environmental conditions for significant human habitation, such as climate, biological diversity and natural resources. Since earliest times, a variety of people from faraway lands have chosen its shores to settle. Indeed, the region is the birthplace of some of the oldest, most deeply rooted cultures and civilizations on our planet (Fig. 14-1 to 14-6) and of three of the most influential global religions: Christianity, Islam and Judaism.

The Mediterranean Sea is the largest of the semi-enclosed European seas. It is surrounded by 22 countries and has shores on three continents (Europe, Africa and Asia) with a combined population of 130 million people in the catchment draining into the sea, and sharing a coastline of 46,000 kilometer. It is one of the leading tourist areas in the world, hosting 100 million visitors every year. This influx of people increases the waste discharge from domestic and industrial sources.

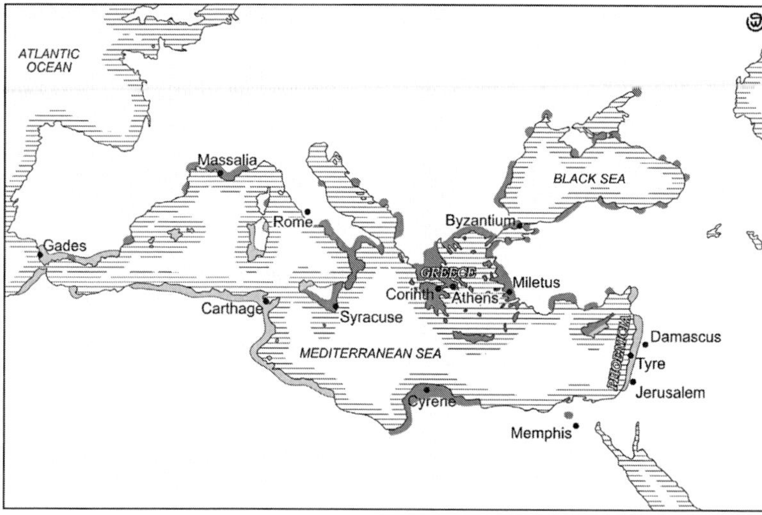

Fig. 14-1: Greek and Phoenician Colonies in Fourth Century BC

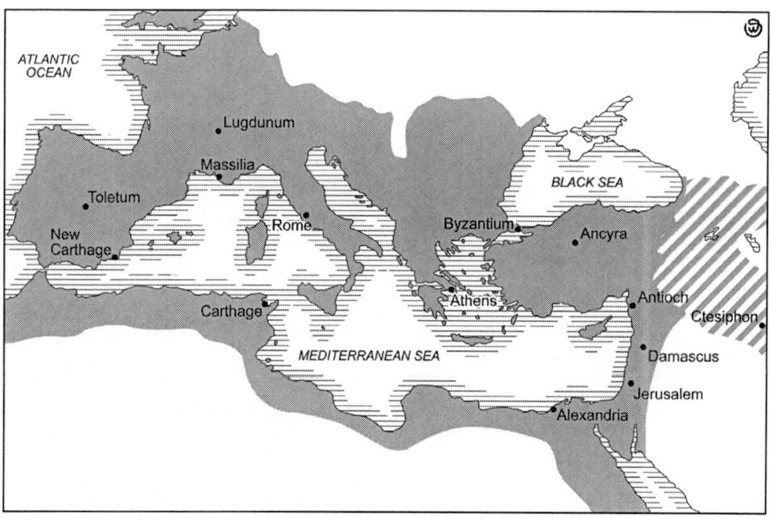

Fig. 14-2: The Roman Empire in the Second Century AD

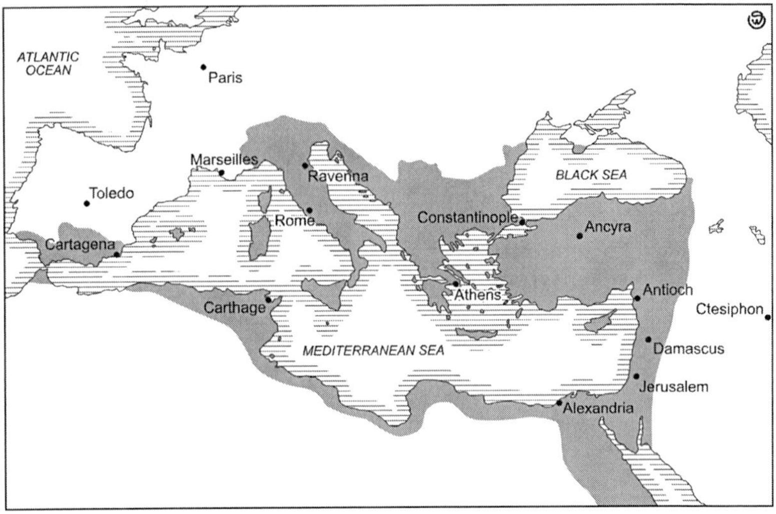

Fig. 14-3: The Byzantine Empire in the Sixth century.

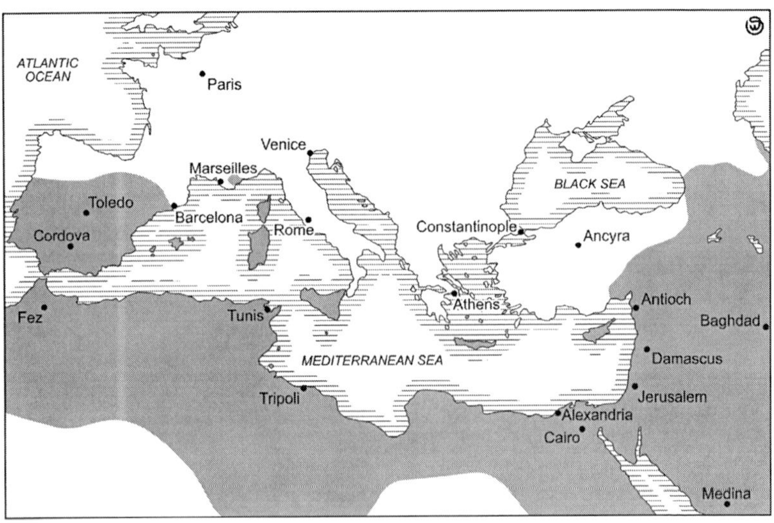

Fig. 14-4: Muslim-Controlled Areas at the End of Tenth Century .

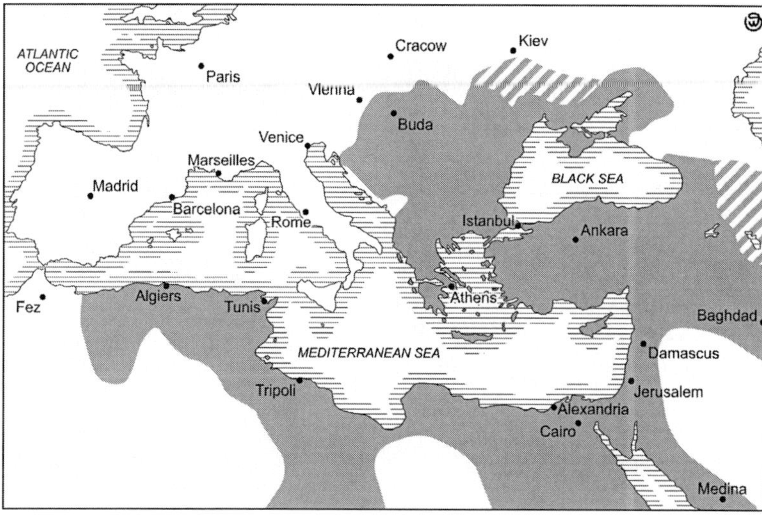

Fig. 14-5: The Ottoman Empire in 17th century

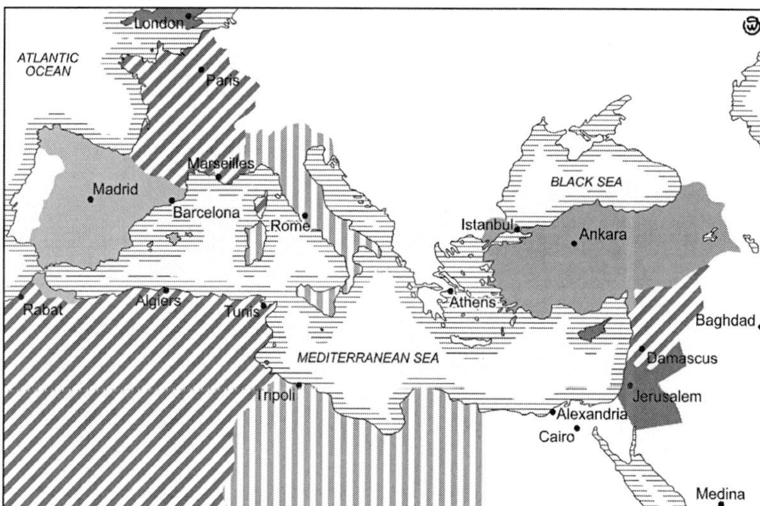

Fig. 14-6: Colonialism in the Mediterranean World in 1938 . 1 France, 2 Great Britain, 3 Italy, 4 Spain, 5 Turkey.

The region of the northern Mediterranean includes Albania, Bosnia-Herzegovina, Croatia, France, Greece, Italy, Malta, Monaco, Montenegro, Slovenia, Spain, while the south-eastern Mediterranean includes Algeria, Cyprus, Egypt, Israel, Lebanon, Morocco, Libya, the Palestinian Authority, Syria, Tunisia, and Turkey. Collectively, the countries of the Mediterranean region cover 8,759 million km^2 and presently hold 427 million people. It is estimated that population of the northern-rim states will grow from 192 million in 2000 to 196 million in 2025, while the population of the southern and eastern-rim states will grow faster – from 235 million to 327 million (www.wiktionary.org).

Because it is a crossroad of civilizations and interests, the Mediterranean Sea region is also an area of transit; a zone of circulation and passage. Due to this position, every kind of concern be they political, cultural, ideological and/or economic and military, have transited, and continue to transit, via this region. It was an important route for merchants and travelers during ancient times, allowing for trade and cultural exchange between emergent peoples of the region: Mesopotamian, Egyptian, Semitic, Persian, Phoenician, Carthaginian, Greek, Turkish and Roman.

The history of the Mediterranean is crucial to understanding the origins and development of many modern societies. The key indicator of culture in a region is its cultural landscape (Sauer 1931; Kenzer 1985; Myga-Piatek 2001). Landscape is more of an intellectual than a material notion. In every social environment, one can distinguish distinct cultural features (including religious ones), which influence the formation of the model way of thinking commonly observed within a particular environment. Thus, an interpreted cultural landscape makes a certain visual whole and physiognomic unity, which expresses the sense of space and its organization (sense of order, form and colour). Interesting results are provided by the observation of landscape in the cultural borderlands on both sides of the Mediterranean Sea, separating various states and nations.

Methods and Materials

At present there are at least two research trends in geography: one dealing with the essence of the cultural and humanistic point of view (Claval 1992, Buttimer 1993, Mitchell 2000, Rembowska 2002a, 2002b, 2007, Jedrzejczyk 2001, Wilczynski & Wardenga 1998, Orlowska 2002-2005, Czepczynski 2007) and the second concerned with the spatial aspect of social, political and religious structures and with relations among territorial systems and cultural processes (Huntington, Braudel, Campione, papers presented on the websites of the South East European Studies at Oxford & the RAMSES 2 and *European Journal of Turkish Studies*).

The new theoretical topics were discussed in Poland, including in two venues. One was a conference called the "Cultural Aspect of Geographical Research: Theoretical and Regional Studies" held in Wroclaw in 2002 (works edited by Orlowska) and another was the Cultural Landscapes Commission's studies (edited by Myga-Piatek), based on a conference held in Sosnowiec, Poland in 2001.

It is very difficult to define any method used in humanistic geographical research as "geographical" sensu stricto (since spatial character of a study does not appear sufficient). However, theoretical development in other scientific disciplnes (e.g. sociology, economy, psychology, cultural anthropology) largely determines the conceptual state and content priorties of geography (Madurowicz 2007).

In the opinion of Czepczynski (2007), since the 1990s, traditional, modernist cultural geography has undergone severe methodological and thematic transformation (Mitchell 2000; Philo 2000). New, post-Berkley cultural geography is focused on synthesizing the spatial features of broadly understood culture to reconstruct identities and distinctiveness. One of the most effective and valuable ways to conceptualize space in post-modern thought is to rely on a holistic and interpretative synthesis based on cultural approaches.

These methodologies, often anchored in French philosophical schools of the late Twentieth Century, can be most beneficial in any holistic and structural studies on the urbanized world. New cultural geography investigations can employ an assortment of research paradigms, including the structuralist, the iconographical (Daniels, Cosgrove 2004;) and the textual, while the most frequently used incorporate the social constructivist approach and theory of representation. The diversity and often mixture of research methods and approaches, frequently based on positionality, can be most prolific in the process of narrating post-modern and fast developing spaces, places and milieux (Foucault 2001; Tuan 1978).

Cultural geography lacks a conceptually rigorous and empirically meaningful understanding of human behavior to facilitate analyses of human landscape creation. A possible solution is employment of the concepts and principles of behavior analysis, as developed in psychology, and as applied in sociology, anthropology, and economics. During the period of evolution of the social science disciplines, positivism was seen not as one philosophy of science but rather as the one philosophy; accordingly, questions of meaning and interpretation were of little importance in both the physical and social sciences. This situation changed dramatically in the mid-twentieth century, following the waning of positivism as a social science philosophy.

Today, there are numerous philosophies to which the social scientist can turn, one of which is positivism, as is most clearly articulated in psychological behaviorism and as applied in sociology and anthropology. A consideration of recent ideas about place creation suggests that it is possible that cultural geography might benefit from a judicious use of behaviorist concepts in such analyses. For cultural geographers, this suggestion is at variance with most other recent conceptual contributions that emphasize alternative social theorizations, such as those associated with humanism, or critical science.

Results

The Mediterranean Region is characterized by a significant variety of traditionally distinct ethnic groups and state formations, border fluctuations, and human migrations in times of the great empires and in the countries formed after their break-ups. The modern cultural landscape of the region is the effect of the overlapping of the cultural heritage of different civilizations and religious circles. The dialogue and co-existence of nationally and ethnically diversified societies of the Mediterranean Basin is the process of building its specific multicultural regional identity. Mediterranean cultures vary, not only in regards to values, but also in regards to the kinds of social activity pursued by these societies.

In the Mediterranean, with its typical diversity and regional contrasts, everything tends to have its own local color. Braudel (1994) spoke about complexities which exist in the sphere of culture, whose history cannot be separated from the earthly environment which surrounds it. It is a rather complicated jigsaw puzzle of ethnic groups and countries, of populations and religions with antinomies which divide the two sides of the Mediterranean both from a structural and a symbolic point of view: North-South, West-East, Christian-Islamic world, the European-Arab world.

This is even if the territorial contiguity, the geopolitical reasons, the tight net of commercial and cultural relationships which have existed for time immemorial and which have deeply influenced and mutually contaminated both sides of the Mediterranean basin, still seem to have a role in trying to amalgamate the different realities in an integrated context. Without this complexity it would be impossible to understand the unitary meaning of the area we are examining (Mainardi 1994).

In a world with a freer bipolar management, there are new considerations of belonging and identity. New processes of territorial redistribution have begun, that are often accompanied by enclosures and conflicts (Huntington 1993). Therefore, territory is again at the center of our attention and interest, at the very moment when the outcome of the world economy seems to want to involve the network of relationships and motivations, new local convergences, in a globalizing logic.

In the case of the great Mediterranean region, the logic of decomposition and of a renewed cohesion might require the opening of relationships, with a high level of interaction, preventing both neo-colonizing temptations and refusals based on prejudice (Campione, 1998). This process seems to have developed in Europe, at least from a conceptual point of view, in order to first of all avoid possible negative effects from the consequences of a failed integration.

Notwithstanding everything that has been said, however, a Mediterranean policy of identity has yet to developed, and the aspect of an "informal" regionalization have only been touched on. Had it developed, this regionalization might have contributed to resolving the problems of tension and violence. The fact that the Mediterranean is the focus of many conflicts – from the beginning of this century there have been over 50 wars, revolutions or armed conflicts, about 20 of which relate to the Arab-Israel conflict (Kliot, 1993) – has not been enough to change the logic of European priorities. In this focus of conflicts, and with identities which are becoming increasingly stronger, the Mediterranean has become a geopolitically ignored area of recent times.

There are fundamental differences between oral and written cultures, especially for religious creeds and practices, because cultural systems also depend on the modalities of communication. It is the written word which strongly marks identity and differences above all in the Mediterranean characterized, as previously said, by regional contrasts. In this sea of "infinite variety", in a set of creeds, ideas and icons (Gottmann, 1966), especially on a local scale, the influence of the three great monotheistic religions on history, culture and identity of places, particularly in Europe and the Mediterranean, must be considerable (Campione, 1995).

On the religious level also, diversity is a current talking point, whether within Islam or vis à vis the other religions. Egypt for example retains a large Coptic Christian community, whose existence predates the Arab conquest by centuries, and which is no less attached to the country. The minorities of Islam itself are also well integrated and accepted, since the Sunni majority respects the Shiites as well as mystical movements which emanate from Sufism. Religious tensions still exist today, however, and the Israeli-Palestinian conflict does not cease to illustrate the strength of the examples of radicalism present in the region. At times, the source of conflicts, the very diversity of this region, is also an asset, for it provides its cultural and historic wealth and its creative dynamism.

Discussion

In Search of New Directions and Research Perspectives in Cultural and Humanistic Geography – Example of Mediterranean Studies

Among the basic problems mentioned in this study, some are worth considering very carefully by geographers exploring cultural differences in environmental perceptions and experiences through the prism of Mediterranean history.

There is culture as an organiser of geographical space; the problems of the historical heritage (the time of the great empires and European colonisation); the influence of the great monotheistic religions (Judaism, Christianity, Islam) on the regional cultures and the identities of places; the influence of globalisation (economic segregation and separation etc.) and cultural landscape as a regional indicator of culture.

Cultural geography at the beginning of 21st century is looking for a change and for a new paradigms and research questions. Some concepts are centered in the problems: national, religious and cultural minorities in the Mediterranean; culture and geopolitics in the Mediterranean countries; geographical education for inter-ethnic and inter-religious dialogue in the multicultural Mediterranean society (Table 14-1) and questions of geographical and historical causes of antagonism between the North and the South coasts of the Mediterranean Sea.

Fernand Braudel (1994), describing his approach to research, suggests that "Everything must be recaptured and relocated in the general framework of history, so that despite the difficulties, the fundamental paradoxes and contradictions, we may respect the unity of history which is also the unity of life."

The richness of European and Mediterranean societies lies in cultural diversity expressed also by the number of religious, ethnic and cultural groups and communities present across the region. Cultural and ethnic diversity have in fact been a characteristic of European and Mediterranean societies for many centuries, even if sometimes events seem to give the impression of the opposite.

Table 14-1a: A Summary of The Ethnic Groups of the Mediterranean Region*

Country	Ethnic groups (%)
Albania	Albanian 95%, Greek 3%, other 2% (Vlach, Roma (Gypsy), Serb, Macedonian, Bulgarian) (1989 est.) note: in 1989, other estimates of the Greek population ranged from 1% (official Albanian statistics) to 12% (from a Greek organization)
Algeria	Arab-Berber 99%, European less than 1% note: almost all Algerians are Berber in origin, not Arab; the minority who identify themselves as Berber live mostly in the mountainous region of Kabylie east of Algiers; the Berbers are also Muslim but identify with their Berber rather than Arab cultural heritage; Berbers have long agitated, sometimes violently, for autonomy; the government is unlikely to grant autonomy but has offered to begin sponsoring teaching Berber language in schools
Bosnia and Herzegovina	Bosniak 48%, Serb 37.1%, Croat 14.3%, other 0.6% (2000) note: Bosniak has replaced Muslim as an ethnic term in part to avoid confusion with the religious term Muslim – an adherent of Islam
Croatia	Croat 89.6%, Serb 4.5%, other 5.9% (including Bosniak, Hungarian, Slovene, Czech, and Roma) (2001 census)
Cyprus	Greek 77%, Turkish 18%, other 5% (2001)
Egypt	Egyptian 98%, Berber, Nubian, Bedouin, and Beja 1%, Greek, Armenian, other European (primarily Italian and French) 1%
France	A majority comprised mainly of a Latin-Teutonic-Celtic blend with Slavic, North African, Indochinese, Basque minorities and admixtures. Overseas departments: black, white, mulatto, East Indian, Chinese, Amerindian
Gibraltar	Spanish, Italian, English, Maltese, Portuguese, German, North Africans
Greece	population: Greek 93%, other (foreign citizens) 7% (2001 census) note: percents represent citizenship, since Greece does not collect data on ethnicity
Israel & the area of Palestinian National Authority with Gaza Strip	Jewish 76.4% (of which Israel-born 67.1%, Europe/America-born 22.6%, Africa-born 5.9%, Asia-born 4.2%), non-Jewish 23.6% (mostly Arab) (2004)

Italy	Italian (includes small clusters of German-, French-, and Slovene-Italians in the north and Albanian-Italians and Greek-Italians in the south)
Lebanon	Arab 95%, Armenian 4%, other 1% note: many Christian Lebanese do not identify themselves as Arab but rather as descendents of the ancient Canaanites and prefer to be called Phoenicians

Table 14-1b: A Summary of The Ethnic Groups of the Mediterranean Region*

Libya	Berber and Arab 97%, other 3% (includes Greeks, Maltese, Italians, Egyptians, Pakistanis, Turks, Indians, and Tunisians)
Malta	Maltese (descendants of ancient Carthaginians and Phoenicians, with strong elements of Italian and other Mediterranean stock)
Monaco	French 47%, Monegasque 16%, Italian 16%, other 21%
Montenegro	Montenegrin 43%, Serbian 32%, Bosniak 8%, Albanian 5%, other (Muslims, Croats, Roma (Gypsy) 12%
Morocco	Arab-Berber 99.1%, other 0.7%, Jewish 0.2%
Slovenia	Slovene 83.1%, Serb 2%, Croat 1.8%, Bosniak 1.1%, other or unspecified 12% (2002 census)
Spain	Composite of Mediterranean and Nordic types
Syria	Arab 90.3%, Kurds, Armenians, and other 9.7%
Tunisia	Arab 98%, European 1%, Jewish and other 1%
Turkey	Turkish 86%, Kurdish 8% others 6% (estimated)

* (Source: Data from the [U.S.] Central Intelligence Agency. The World Fact Book: www.cia.gov/library/ publications/the-world-factbook/fields/2075.html). 'Ethnologue Data from Languages of the World' http://www.ethnologue.com/

In this context, the role of intercultural dialogue appears as an essential approach to counter and overcome mutual prejudices and the self-fulfilling prophecy of Huntington's (1993, 1998) "clash of civilizations", with its purpose of "learning to live together peacefully and constructively in a multicultural world and developing a sense of community and belonging." It is clear that this result can be achieved only if it involves everyone, young people included.

As a global microcosm, the Mediterranean is a manifestation of the interaction of the great monotheistic religion with all the trappings of the social dilemmas and stereotype, and delimitations we are witnessing worldwide. As in the World globally, the Mediterranean manifests all the problems found world wide, and peoples on opposite sides of the

Mediterranean still feel each other to be distant and foreign. Indeed, people from opposite shores of the Mediterranean perceive each other too much through the dangerous prism of stereotypes.

Conclusions

The humanization of geography by viewing geographical space through a cultural prism has become a necessity. Humanism contained in culture is a kind of criticism helping to overcome the excessive abstractness of the scientific method applied so far in regional studies.

Space diversity implies the necessity to use the research methods of humanist geography drawing on other disciplines. The complexity of the problems of space and its internal relations, as well as the wide variety of research methods were used by geographers (a specific methodological eclecticism), calls for the integration of specialists dealing with regional studies. In the opinion of some Polish geographers (Rembowska, 2002c; Orlowska, 2004; Jedrzejczyk, 2001; Wilczynski, 1991), a "cultural turn" in geography has created a new perspective of seeing traditional and often isolated spheres of examined reality and has uncovered their social meanings and cultural basis of their variety.

Acknowledgement

The author hereby expresses her sincere thanks to Dr. Waldemar Spallek at the, Institute of Geography and Regional Development University of Wroclaw for drawing drawing the figures.

References

Andrews, P.A. 2007. 'Ethnologue Data from Languages of the World'

Bonnemaison, J. 2005. Culture and space. Conceiving a new cultural geography. I.B. Tauris, London & New York.

Braudel, F. 1994. History of Civilization. Allen Lane-Penquin Press, New York.

Buttimer, A. 1993. Geography and the human spirit. The Johns Hopkins University Press, Baltimore and London.

Campione, G. 1995. Il governo delle giraffe – la comunita internazionale e le grandi conferenze sui valori. Geotema 3.

—. 1998. Monotheism and conflict in the Mediterranean Region. In: W. Wilczynski & U. Wardenga (eds.) Religion, ideology and geographical thought. WSP Kielce Studies in Geography. Vol. 3, pp. 65:72.

cia.gov/library/publications/the world factbook/fields/2075.html.

Claval, P. 1992. Champs et perspectives de la géographie culturelle. Géographie et Cultures, 1 (1), pp. 7:38.

Crang, M. 2004. Cultural Geography. Routledge, London-New York.

Czepczynski, M. 2007. Podejscie badawcze w nowej geografii kultury. In: W. Maik, K. Rembowska, A. Suliborski. Podstawowe idee i koncepcje w geografii. Bydgoszcz. Vol. 3, pp. 203:211.

Daniels, S., Cosgrove, D. 2004. Introduction. In: D. Cosgrove, S. Daniels (eds.) The iconography of landscape. Essays on the symbolic representation, design and use of past environments. Cambridge University Press, Cambridge, pp. 1:10.

Doni, E. 1995. I guerrieri di Dio. Milano: Rizzoli.

European Journal of Turkish Studies. February 2005. Archives. URL.

Foucault, M. 2001. Space, Power and Knowledge. In: S. During (ed.) The Cultural Studies Reader. Routledge, London-New York, pp. 134:141.

Gottmann, J. 1966. Geographie politique, in Encyclopedie de la Pleiade. Geographie Generale. Gallimard. Paris.

http://www.ejts.org/document547.html.

http://en.wikipedia.org/wiki/Mediterranean_Sea.

http://www.wiktionary.org (a multilingual, Web free content dictionary).

Huntington, S. P. 1993. The Clash of Civilizations? Foreign Affairs 72, 22:49.

—. 1998. Zderzenie cywilizacji i nowy ksztalt ladu światowego. WWL Muza S.A. Warszawa.

Jedrzejczyk, D. 2001. Wprowadzenie do geografii humanistycznej. Wyd. Uniwersytetu Warszawskiego. Warszawa.

Kenzer, M.S. 1985. Milieu and the “Intellectual Landscape”. Carl O. Sauer’s Undergraduate Heritage. In: Annals of the Association of American Geographers.

Kliot, N. 1993. La dove esplodono i conflitti. Sistema Terra anno II,1. Laterza. Roma.

Madurowicz, M. 2007. Szkielko i wiara w geografii czlowieka. In: W. Maik, K. Rembowska, A. Suliborski (eds.) Podstawowe idee i koncepcje w geografii. Bydgoszcz. Vol. 3, pp. 231:242.

Mainardi, R. 1994. Geografia regionale. Nuova Italia Scientifica. Roma.

Mitchell, D. 2000. Cultural Geography. A Critical Introduction, Blackwells, Oxford.

Myga-Piatek, U. (ed.). 2000. Krajobraz kulturowy, idee, problemy, wyzwania. Wydzial Nauk o Ziemi Uniwersytetu Slaskiego. Oddzial Katowicki PTG. Sosnowiec.

Nalkowski, W. 1911. Co to jest geografia (What is geography)? Paprocki. Warszawa.

Orlowska, E. (ed.) 2002a. Kultura jako przedmiot badan geograficznych, Vol 1, Oddzial Wroclawski Polskiego Towarzystwa Geograficznego. Uniwersytet Wroclawski, Wroclaw.

—. (ed.) 2002b. Kultura jako przedmiot badan geograficznych. Studia teoretyczne i regionalne. Vol. 2. Oddzial Wroclawski PTG. Uniwersytet Wroclawski. Wroclaw.

Orlowska, E. Klementowski, J. (eds.). 2003. Kulturowy aspekt badan geograficznych. Studia teoretyczne i regionalne. Vol 3. Oddzial Wroclawski PTG. Instytut Geografii i Rozwoju Regionalnego Uniwersytetu Wroclawskiego. Wroclaw.

Orlowska, E. (ed.). 2004. Kulturowy aspekt badan geograficznych. Studia teoretyczne i regionalne. Vol 4. Oddzial Wroclawski PTG. Instytut Geografii i Rozwoju Regionalnego Uniwersytetu Wroclawskiego. Wroclaw.

Orlowska, E. 2005. Kulturowy aspekt badan geograficznych. In: U. Myga-Piatek. Krajobraz kulturowy. Aspekty teoretyczne i metodologiczne. Prace Komisji Krajobrazu Kulturowego PTG. Vol 4. Sosnowiec.

Philo, Ch. 2002. Foucault geography In: M. Crang, N. Thrit (eds.) Thinking Space, Routledge. London-New York, pp. 205:238.

Rembowska, K. 2002a. Kultura w badaniach geograficznych. In: E. Orlowska (ed.) Kultura w badaniach geograficznych. Oddzial Wroclawski Polskiego Towarzystwa Geograficznego. Uniwersytet Wroclawski. Wroclaw Vol 1. pp. 9:28.

—. 2002b. Perspektywy badan geograficznych w świetle dialektycznych koncepcji rzeczywistości spoleczno-ekonomicznej. In: E. Orlowska (ed.) Kultura jako przedmiot badan geograficznych. Studia teoretyczne i regionalne. Oddzial Wroclawski Polskiego Towarzystwa Geograficznego. Uniwersytet Wroclawski. Wroclaw. Vol 2. pp. 9:18.

—. 2002c. Kultura w tradycji i we wspolczesnych nurtach badan geograficznych. Wydawnictwo Uniwersytetu Lodzkiego. Lodz.

—. 2007. Kulturowy zwrot w geografii. In: W. Maik, K. Rembowska, A. Suliborski. Podstawowe idee i koncepcje w geografii. Bydgoszcz. Vol. 3, pp.189:202.

Sauer, C. 1931. Cultural Geography. In: P.L. Wagner, M.W. Mikessel (eds.) Readings in Cultural Geography. University of Chicago Press. Chicago.

Tuan, Y.F. 1978. Sign and Methapor. In: Annals of the Association of American Geographes. Vol. 68.

Websites of the South East European Studies at Oxford & the RAMSES 2.

Wilczynski, W. and E. 1991. Waclaw Nalkowski 1851-1911. In: G.J. Martin (ed.) Geographers. London: Mansell. Vol. 13. pp. 45:52.

Wilczynski, W. Wardenga, U. (eds.). 1998. Religion, ideology and geographical thought. WSP Kielce Studies in Geography. Vol. 3. pp. 7:11.

CHAPTER FIFTEEN

THE MANAGEMENT OF FOREST FIRES IN PROTECTED AREAS IN PORTUGAL

FANTINA TEDIM

Introduction

The designation of protected areas by a number of governments around the world is meant to safeguard and promote ecological diversity and to preserve natural and cultural resources through legal instruments and other regulations. Forest fire hazards which may affect such areas contribute to the degradation of the landscape, the destruction of specific ecosystems, and change in the areas affected in terms of their ecological and economic value. In addition to these consequences, forest fires contribute to global warming due to greenhouse gas emissions.

In Portugal, despite prevention and emergency measures which have been implemented over the last few years, recent data suggest that there has been an increase in fire severity in protected areas. The consequences, namely in terms of biodiversity loss, are not always assessed, but they are certainly important, since in 2005, for instance, 35% of the area burned was identified as *Área Prioritária Para a Conservação da Natureza* (Priority Area for Nature Conservation).

Study Focus and Methodology

Through a questionnaire survey sent out by e-mail to all protected areas in Portugal, the author of the present study attempted to identify the main characteristics of forest fire management policies as far as protected areas are concerned and to analyze the integration of these policies in the overall national policy of forest protection against fires established in 2006 (Resolution of the Council of Ministers 65/2006 of the 26th of May).

Table 15-1: Identification of Protected Areas in Portugal Returning Questionnaire
* Protected areas not returning questionnaire

Protected area	Area (ha)	Establishment date
Peneda-Gerês National Park	69,592.50	Law Decree n. 187/71
Sapal de Castro Marim and Vila Real de Santo António Natural Reserve*	2,312	Law Decree n. 162/75
Serra da Estrela Natural Park	99,478.05	Law Decree n. 557/76
Estuário do Tejo Natural Reserve*	9,846.48	Law Decree n. 565/76
Arrábida Natural Park*	11,054.74	Law Decree n. 622/76
Ria Formosa Natural Park	17,900.92	Law Decree n. 45/78
Dunas de São Jacinto Natural Reserve*	733.52	Law Decree n. 41/79
Montesinho Natural Park	74,544.62	Law Decree n. 355/79
Serras d'Aire e Candeeiros Natural Park	38,392.53	Law Decree n. 118/79
Paul do Boquilobo Natural Reserve	817.62	Law Decree n. 198/80
Estuário do Sado Natural Reserve*	17,206.06	Law Decree n. 430/80
Sintra-Cascais Natural Park*	14,450.85	Law Decree n. 292/81
Serra da Malcata Natural Reserve*	16,158.66	Law Decree n. 294/81
Serra do Açor Protected Landscape*	373.40	Law Decree n. 67/82
Alvão Natural Park*	7,202.66	Law Decree n. 237/83
Arriba Fóssil da Costa da Caparica Protected Landscape	1,551.50	Law Decree n. 164/84
Litoral Norte Natural Park*	1,316.64	Law Decree n. 357/87
Paul de Arzila Natural Reserve*	586.76	Law Decree n. 219/88
Sudoeste Alentejano e Costa Vicentina Natural Park*	60,485.05	Law Decree n. 241/88
Serra de São Mamede Natural Park	56,058.87	Law Decree n. 121/89
Vale do Guadiana Natural Park	69,669.30	Reg. Decree n. 28/95
Douro Internacional Natural Park	87,011.26	Law Decree n. 8/98
Albufeira do Azibo Protected Landscape*	4,897	Reg. Decree n. 13/99
Corno do Bico Protected Landscape	2,175	Reg. Decree n. 21/99
Serra de Montejunto Protected Landscape	4,5	Reg. Decree n. 11/99
Tejo Internacional Natural Park	23,728.05	Reg. Decree n. 9/00
Lagoas dc Bertiandos e São Pedro dos Arcos Protected Landscape	346	Reg. Decree n. 19/00
Lagoas de Santo André e da Sancha Natural Reserve	3,123.39	Reg. Decree n. 10/00

The questionnaire, which contained a total of 50 questions, was meant to elicit information on the following issues related to national policy:

- dimension and causes of forest fires;
- fire prevention measures;
- fire surveillance;
- fire suppression and recovery of burned areas.

A total of 28 questionnaires were sent to all parks, protected landscapes and natural reserves (Table 15-1). We excluded from the study the remaining protected areas, including natural monuments and private protected areas, which are commonly named "sites of biological interest," since these areas are, in principle, less vulnerable to forest fires. Only 15 of the 28 questionnaires were returned, representing a coverage ratio of 54%. Still, the returned questionnaires represent 83% of the number of forest fires and 94% of total burned areas. Thus we can consider the sample as significant.

The Severity of Forest Fires in Portugal

In Portugal, the number of forest fires in protected areas-which cover 638,617 hectares, or 7.6% of the Portuguese territory-has increased considerably since 1992 (Fig.15-1). In 1998 and 2000, an unprecedented number of occurrences were recorded. In 2003 and 2005, the largest burned areas were recorded, 28,272.90 hectares, and 20,432.44 hectares, respectively (MAOTDR, 2006:9). Between 1992 and 1999 the annual average area burned was 6,298 hectares, and in the period 2000-2005 about 15,904 hectares were burned. Of the total area of the country that was burned after 1992, approximately 59% burned after the year 2000.

The results of the questionnaire show us that in most protected areas, the most severe fire occurred after 2000 (Table 15-2).

Protected areas in Portugal are characterised by different bio-geographical, topographic, and demographic features, and therefore have very different levels of fire risk. By analysing the area burned between 1992 and 2005, it is possible to conclude that in Serra da Estrela Natural Park, the burned area corresponds to 59% of the total area of the park. Other problematic protected areas include Montesinho Natural Park, Serra de S. Mamede Natural Park, Peneda-Gerês National Park and Douro International Natural Park with burned area rates of more than 12% of the total area of each park.

It should be pointed out that in the Montesinho, Douro Internacional and Serra da Estrela natural parks the annual average area burned, between

1992 and 2005, was over 1,000 hectares. Emphasis should also be placed on the fact that some of the fires that hit protected areas were ignited outside their perimeter and when the areas were hit the fires were already out of control. Although the concern with the protection of natural heritage against fire arose effectively with the establishment of protected areas, this does not mean that there were adequate human and technical resources and a legal framework supporting the implementation of an effective prevention and suppression fire policy.

Table 15-2: Largest Fires per Year of Occurrence

Protected area	Year (1)	Area (1) (ha)	Ann, Averag. Area Burned (1992-2005)(2)
Serras d'Aire e Candeeiros Natural Park	2006	2,997	879
Peneda-Gerês National Park	2006	3,945	763
Lagoas de Santo André e da Sancha Natural Reserve	2006	0.006	0.5
Corno do Bico Protected Landscape	2006	150	No data*
Serra da Estrela Natural Park	2005	4,772	4,178
Ria Formosa Natural Park	2004	173	31
Serra de Montejunto Protected Landscape	2003	2,23	No data*
Douro Internacional Natural Park	2003	1,3	1,309
Parque Natural do Vale do Guadiana	2003	1000	318
Paul do Boquilobo Natural Reserve	2003	12,8	1
Serra de S. Mamede Natural Park	2003	13,35	815
Tejo Internacional Natural Park	2002	293.2	48
Montesinho Natural Park	1998	1,185.51	1,529
Arriba Fóssil da Costa da Caparica Protected Landscape	1996	65	9
Lagoas de Bertiandos e São Pedro dos Arcos Protected Landscape	-	0	0

The data available concerning the causes of fires in protected areas is inconclusive, since the great majority of occurrences were not investigated or the results of the investigations conducted were inconclusive. However, it is clear that the causes of forest fires are mainly human, such as accidents, negligence and arson.

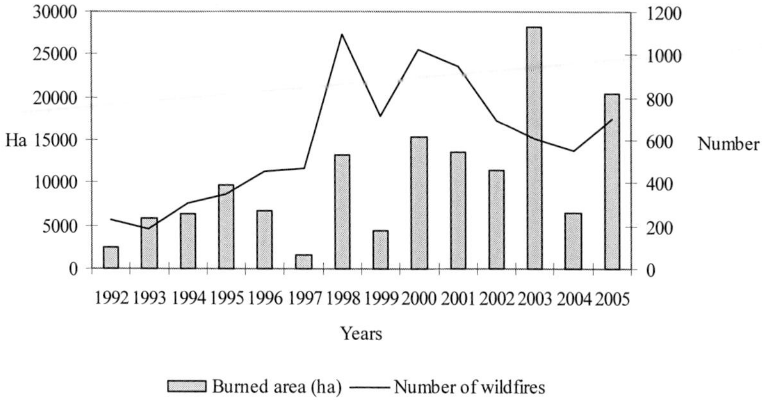

Fig. 15-1: Evolution in the Number of Forest Fires and Burned Territory in Protected Areas
Source: Institute for Nature Conservation, 2006

Forest Fire Prevention

From the answers to the questionnaire, it becomes clear that respondents from all protected areas acknowledge the existence of a policy of forest protection against fire, which includes prevention and surveillance measures as their main priorities. The active participation of the protected areas in fire suppression is not so evident, since only nine of the areas have properly-equipped initial attack teams. Their participation is even less significant in terms of the recovery of burned areas, since the current policy of forest protection does not effectively promote it.

The management of protected areas in Portugal is the responsibility of the Institute for Nature Conservation and Biodiversity (ICNB). The Institute established in its report for 2005 the following topics as the main priorities in terms of forest protection against fire:

- to decrease burned area, especially in priority areas for nature conservation as the most valuable ecological areas;
- to reduce the number of ignitions;
- to reduce the vulnerability of these areas;

and finally

- to monitor and restore burned areas.

These objectives are effectively acknowledged as priorities by the management of protected areas in Portugal, even by the areas of regional interest which are not under management of the Institute for Nature Conservation and Biodiversity (Table 15-3), aiming at improving suppression efficacy and promoting a more effective organizational process.

Table 15-3: Objectives of Forest Fire Policy in Portugal (Source: Survey, 2007)

Objectives	Number	%
Decrease burned area	11	73
Reduce number of ignitions	10	66
Reduce vulnerability	9	60
Increase resilience of forest	9	60
Improve efficacy of suppression	9	60
Promote a more effective organization process	8	53
Restore burned areas	6	40
Did not answer	1	7

In 2006, with the implementation of a new policy of forest protection against fires, each municipality assumed responsibility for preparing a Municipal Fire Plan, which included a strategy to ensure forest protection and fire prevention within their territorial areas. In this way, actions carried out by those responsible for protected areas, which sometimes extend along several bordering municipalities, have become part of the above-mentioned plans. However, there are protected areas where those responsible for protection promote other complementary actions.

There is a generalized belief that there have been improvements in forest fire prevention over the last few years, especially due to improved surveillance capacities and other detection technologies. In fact, it should be emphasised that, in recent years, there has been a greater awareness and commitment of public and private institutions to promoting greater surveillance efforts, especially through mobile teams. Additionally, teams have been increasingly provided with better training and equipment, resulting in actions that have gradually removed possible causes of forest fires (Table 15-4). There is also greater collaboration between the entities involved in fire prevention, and consequently, better coordination of the way in which material and human resources are used. This collaboration involves not only a number of public institutions and civil protection agents, but also many private entities. Among the latter, we should point out entities such as scout associations, hunting associations, fishermen's associations, foresters associations, farmer associations, environmental associations, common land managers, and off-road clubs. Thus, it is possible to conclude that there is great participation of community based organizations in forest fire prevention, but this participation varies with the local context.

Table 15-4: Prevention Measures (Source: Survey, 2007)

Prevention measures	Number	%
Land Mobile Surveillance	15	100
Risk Areas Identification	14	93
Fuel Removal	9	60
Education and Awareness Campaigns	9	60
Construction of water points	8	53
Track Construction and Improvement	8	53
Forest Management	7	47
Definition and Maintenance of Buffer Zones	6	40
Preventive silviculture	6	40
Land Fixed Surveillance	6	40
Use of Prescribed Burning	4	27
Clearance for using Prescribed Burning	3	20
Fire Hazard Information	2	13
Information Signs	1	7
Restricted Access by cars to high risk areas	1	7
Removal of barbecues from picnic parks	1	7
Purchase of communication equipment	1	7
Contact with shepherds	1	7

Other factors, such as changes in legislation, namely as far as the control of offenses is concerned, have also been considered positive. The respondents have also indicated two other factors which are considered important, but which are not sufficiently effective. The first is forest management, both in terms of fuel removal and the choice for fire resistant plant species, which is a measure of particularly difficult implementation. In this context, the use of prescribed fire was only mentioned by the Montesinho, Peneda-Gerês, Vale do Guadiana and Serra d'Aires e Candeeiros Natural Parks. This practice is designed to meet specific objectives, especially pasture renovation, with a limited focus on fuel removal. The second is awareness of local communities to the need for fuel removal and a reduction of potential fire risk behaviour. This is still not a widespread concern among Portuguese protected areas.

Currently, public education relies heavily on making information available to people. However simply providing people with information is not sufficient to encourage preparedness, because people do not necessarily accept this information at face value. Rather, it is how people interpret their own risk and the information available that determines whether they prepare for forest fires or not. While some people appear predisposed to prepare, but need to be guided in this endeavour, others decide not to prepare. This means that, when planning public education, it is not enough to know that levels of preparedness are low. It becomes necessary to find out if levels are low because people have decided not to prepare, or if levels are low because people need guidance to know what to do. Education campaigns should be directed at the general audience but also at specific groups such as shepherds and farmers.

The Portuguese fire prevention system is described by the majority of managers responsible for protected areas as good, but the Douro Internacional Natural Park, the Vale do Guadiana Natural Park, the Tejo Internacional Natural Park, the Paul do Boquilobo Natural Reserve and the Ria Formosa Natural Park believe the system could still be considerably improved.

Fire Detection and Suppression

Surveillance plays a vital role in the reduction of ignitions and burned area. There is a general consensus about recent improvements in surveillance in Portugal, since, in addition to fixed surveillance (in lookout towers), which still remains in use in six of the protected areas where the questionnaire was answered, mobile surveillance has evolved tremendously. There has been an increase in terms of the human resources dedicated to mobile surveillance. This type of surveillance is highly effective in dissuading possible criminal actions, but is also efficient, as it alerts the population to risk behavior. Surveillance teams are made up of staff from the protected areas, members of the Portuguese Republican National Guard, and fire-fighters and brigades of "sapadores florestais" (forest sappers). In the Peneda-Gerês National Park, for instance, twelve teams of "sapadores florestais" were established by common lands management and foresters associations and are in charge of the surveillance of the park. Only the Serra de Montejunto Protected Landscape is equipped with a video surveillance system.

All of the effort put into the development, equipment and training of surveillance teams explains the opinion of nine of the respondents from protected areas that their surveillance system is effective. Those responsible for the remaining areas believe their system to be "acceptable," although they believe it is still possible to improve it, in particular because they do not have a number of surveillance teams adequate to the fire risk of the area, or because it is necessary to better train and equip the brigades. Since the decrease of burned area is intimately related with a fast, effective initial attack, surveillance teams in some protected areas are also equipped with a first intervention kit.

While some of the protected areas can count on initial attack teams equipped with various resources such as booster units or even backhoes, other protected areas depend completely on external means to suppress the fires. These initial attack teams are generally very effective because they are highly trained, extremely professional experts, who keep regularly updated on the latest techniques; in addition, most of the staff already possesses practical experience. Nevertheless, respondents of some protected areas indicated that their teams are made up of seasonal staff, which significantly reduces the efficacy of the team; working with full-time staff members would be desirable in order to promote greater stability as well as regular training actions.

With respect to fire suppression itself, respondents from most protected areas believe suppression has improved over the last few years, thanks to a more effective action by initial attack teams, to better coordination, to the availability of more resources, and to better fire-fighter training. However, there are still negative aspects that need to be overcome, such as the lack of knowledge of the territory and existing infrastructures (tracks, water points), excessive use of water or the poor choice of other suppression techniques by the staff member in charge of the operations. It has been also mentioned that mop-up is sometimes deficiently handled, promoting re-ignitions.

Recovery of Burned Areas: Current Situation

The first legislation (Law Decree n. 488/70 of 21 October 1970) established the need to reconstitute forest settlements hit by forest fires. In 1988 Law Decree n. 139/88 of 22 April orders the reforestation of areas affected by forest fires to be carried out within 2 years and establishes the need to request authorization to the Portuguese Forest Service to introduce any changes to the type of settlement. In 1989, Law Decree n. 180/89 of 30 May establishes the legal framework to ensure the management of burned areas through specific rules in protected areas. However, the need to implement this policy has been so far disregarded.

The monitoring and recovery of burned areas is one of the objectives set out in the policy implemented by the Institute for Nature Conservation and Biodiversity. However, the concrete actions that are actually carried out are still scarce.

According to the records that have been so far analyzed, the first intervention was carried out in the Arriba Fóssil da Costa da Caparica Protected Landscape, which was established in May 1984. This intervention followed a fire that burned 115 hectares of state-owned forest land. At that time, a recovery project was prepared by the protected area. New plans to be implemented on the burned area were drawn, changes were introduced to the orientation and width of buffer zones, and stone pines were planted. This protected area would again be hit by a fire in 1996, which burned 65 hectares of private forest land. The land owner prepared a reforestation project, which was presented to the protected area. The intervention of the protected area management was limited to delivering its opinion, identifying a set of constraints to minimize the impacts of the intervention on the dune ecosystem. The project would be later approved by IFADAP - Instituto de Financiamento e Apoio ao Desenvolvimento da Agricultura e das Pescas (Financing and Development Aid Institute for Agriculture and Fisheries) and only some of the issues raised by the protected landscape were taken into consideration.

Concerning the recovery of burned areas, it is possible to summarize the different procedures adopted by those individuals and authorities responsible for protected areas in the following statements:

- Five of the protected areas claimed to have implemented the recovery procedures right after a large fire, although with different motivations. The motivations included to recover the forest or areas of high ecological value, to minimize the effects of erosion, and to provide shelter and food to several fauna groups. The largest fires of the last few years created the opportunity to implement new recovery actions like the ones that are being carried out in the Douro Internacional Natural Park, the Serra de São Mamede Natural Park and in the Peneda-Gerês National Park. The fires that hit these protected areas burned 1,300 hectares, 13,350 hectares, and 3,945 hectares, respectively, destroying areas of high ecological value.
- Four of the protected areas already have experience in the recovery of burned areas resulting from an intervention carried out by the Portuguese Forest_Service. This intervention was conducted in properties located in the protected area and owned by the Portuguese

state and in common lands which are managed by the Portuguese Forest Service.

- Four of the protected areas declared that they did not have any type of intervention planned for the recovery of burned areas. One of them shows no interest in carrying out recovery actions.
- The Lagoas de Bertiandos e S. Pedro de Arcos Protected Area and the Lagoas de Santo André e da Sancha Natural Reserve have been little affected by forest fires, thus they do not have experience in the recovery of burned areas.

With the exception of the Peneda-Gerês National Park, which coordinates the recovery of burned areas, the intervention capacity of the remaining protected areas is much more limited, since they can only actively work in lands under their management. Often they are limited to delivering their opinion about the projects that are prepared and their recommendations are not always taken into consideration.

The intervention in burned areas aims fundamentally at controlling erosion (four answers), restoring the forest (four answers), preventive silviculture (three answers) and harvesting burned timber (two answers).

By mentioning the recovery of burned areas, what is being advocated is not the compulsory artificial recovery of these areas, but the possibility of, after a fire, monitoring the areas affected and taking advantage of the opportunity to improve their resilience to forest fires, ensuring its main aim: economic, ecological, environmental (for instance, carbon sequestration) and recreational value. The option of reforestation should only be made in cases clearly identified and based on a strategy precisely defined and only when natural regeneration is not doable or when the need to perform changes in vegetation in order to promote its restoration is obvious.

There is widespread consensus regarding the advantages of natural regeneration in the recovery of burned areas (Table 15-5) and a great concern with the removal of invasive species. Reforestation actions using species of higher environmental value are more prevalent than reforestation using species of high economic value.

Currently, the experiences that are being carried out in Portugal vary according to circumstances as well as the particular authority responsible for management and strategy. For example, in the Douro Internacional Natural Park the recovery of 360 hectares of public forest through natural regeneration and environmental promotion is currently underway with extremely positive results. The director of the park is of the opinion that in this type of intervention the focus should not be entirely on environmental promotion, but on the economic development of the burned area. In the

Serra de São Mamede Natural Park the experience has been based on natural regeneration and on reforestation using species of high environmental value, while still promoting measures of protection against fire. All planned measures have been carried out, although it is still not possible at this time to assess their results. In the Peneda-Gerês National Park the recovery began after the large fire that hit the park in 2006 in which 3,945 hectares were destroyed. The selected procedures are natural regeneration, removal of invasive species and improvement of the natural regeneration using species of high environmental value. In the areas which are being reforested, the main concern is with the environmental promotion of these spaces. The implementation of recovery measures is funded either by public budget or by the European Union.

There are several difficulties of structural, political, organisational, technical and scientific nature involved in the recovery of burned areas.

One of the main difficulties stems from the fact that most land is private and s divided into small parcels. To this we can add the lack of organization of Portuguese foresters and the incapacity of many landowners to properly promote the social, environmental and economic value of the rural world and in particular of their own lands. The insufficiencies of the Portuguese forest land register complicate contact and negotiation with private landowners.

Table 15-5: Recovery techniques (Survey 2007)

Techniques	Number
Natural regeneration	7
Control of natural regeneration removing invasive species	4
Improve natural regeneration using species of high economic value	1
Improve natural regeneration using species of high environmental value	4
Reforestation using species of high economic value	2
Reforestation using species of high environmental value	5
Establishment of permanent pastures	1

On private lands, the recovery of burned areas is the responsibility of land owners. On public and common lands, the recovery falls under the jurisdiction of the National Forest Service and in some cases of the Institute for Nature Conservation and Biodiversity. When protected areas are not responsible for direct management of common and public lands they are only responsible for issuing an opinion with regard to the actions which the entities that elaborate or approve the project intend to carry out. With regard to this we should point out the lack of coordination between the entities with the objective of envisaging the best strategy of intervention in the burned areas.

Other difficulties concern the lack of human and financial resources, extensive cattle breeding, and the lack of bush and tree species on the Portuguese market. The recovery of burned areas should be understood as an opportunity to re-organize the rural landscape, increasing its social, economic and environmental value, but always bearing in mind the strategies of forest protection against fire.

Conclusion

In Portugal, forest fires constitute a serious problem which worsened in the first years of the 2000s. To promote a better understanding of the problem of forest fires in its several dimensions, it is necessary to have accurate data (numeric and cartographic databases), since there are still errors in the delimitation and georeferentiation of burned areas.

The causes of most fires remain unknown, either because no investigation was conducted or because the result of the investigation was inconclusive. In order to achieve more effective fire prevention and surveillance, it is essential to promote a better understanding of the causes of forest fires.

There have been improvements in forest fire prevention over the last few years, especially due to the availability of a greater number of surveillance and detection means and not so much due to fuel removal.

Little attention has been paid to educating and raising awareness among the population, especially when it comes to natural and humanized landscapes with great environmental and cultural value.

Interventions in terms of the recovery of burned areas in protected areas are not effectively considered a priority, resulting in repercussions on nature conservation. In fact, some of the respondent in protected areas indicated that they did not have any financial aid to carry out actions regarding the recovery of burned areas. It is the responsibility of the Conselho Nacional de Reflorestação (National Reforestation Council) to set out the general strategic guidelines for the recovery of burned areas, respecting the objectives and prevailing functions of forest spaces. The guidelines of the National Reforestation Council specifically for the protected areas include assessing the capacity of natural regeneration of the areas hit by fires, followed, complementarily and whenever necessary, by the restoration of native plant species and the control of invasive exotic plants. (Conselho Nacional de Reflorestação, 2005:9).

The National Fire Plan (Council of Ministers n. 65/2006 of 26 May) includes the recovery and restoration of ecosystems as one of its priorities. Its core objective is the assessment and mitigation of the impacts caused by fire and the implementation of long-term restoration strategies. It is clearly stated in the Plan that the recovery of burned areas should aim at a future increase of resilience. It is the Portuguese Government's intention, in accordance with the National Fire Plan, to publish by the end of 2007 a Code of Good Practices on Recovery of Burned Areas, under the authority of the Portuguese Forest Service and prepared jointly with other public entities including the Institute for Nature Conservation and Biodiversity.

Experiments in the recovery of burned areas described above are very recent. Thus, it is still early to assess the interventions in order to define and validate the principles and practices of recovery of burned areas as well as to assess its impacts on rural landscape and on the promotion of fire resilience. This could aid the elaboration of a code of good practices on the recovery of burned areas, which is highly recommended by some managers of protected areas.

One conclusion that can be drawn from the experience of the Peneda-Gerês National Park is that the gradual replacement of coniferous trees by native non-coniferous species is advisable.

The rural world is changing and processes like depopulation, reforestation of agricultural lands with coniferous trees or eucalyptus and tourist pressure are simple examples of factors that will have an effect on land use and consequently on the future dimension of forest fires.

It should be pointed out that the recovery of burned areas is an important opportunity for social, economic and environmental enhancement of the rural landscape and also an opportunity to make these areas more resilient to forest fires.

Acknowledgments

The author would like to thank the managers of Portuguese protected areas who returned the questionnaires and who have therefore made this study possible.

References

Conselho Nacional de Reflorestação 2005. Orientações estratégicas para a recuperação das áreas ardidas em 2003 e 2004. Ministério da Agricultura, do Desenvolvimento Rural e das Pescas. Lisboa.

Dale L. 2006.Wildfire Policy and Fire Use on Public Lands in the United States. Society and Natural Resources, 19, pp.275:284,.

Field D. R., Jensen D. A. 2005. Humans, Fire, and Forests: Expanding thc Domain of Wildfire Research. Society and Natural Resources. 18, pp.355:362.

Ministério do Ambiente, do Ordenamento do Território e do Desenvolvimento Regional (MAOTDR) 2006. Relatório sobre Incêndios Rurais na Rede Nacional de Áreas Protegidas e na Rede Natura 2000. Instituto da Conservação da Natureza. Lisboa.

Pedrosa F. T., Herrmann M. L. 2007. Forest Fires in Protected Areas in Portugal and in the State of Santa Catarina (Brazil): Comparative Analysis of Policies of Prevention and Recovery of Burned Areas. Proceedings of Wildfire2007- 4th International Wildland Fire Conference.Seville. Spain.

Resolução do Conselho de Ministros n.º 65/2006 de 26 de Maio de 2006 Diário da República nº 102,I Série-B 3511.

CONTRIBUTING AUTHORS

Ayhan Akış
Assist. Prof.
Selcuk University, Faculty of Education, Department of Geography Education, Konya, Turkey
E-mail: ay_body@yahoo.com

Türkan Bayer Altın
Assist. Prof.
Nigde University, Faculty of Education, Nigde, Turkey.
E-mail: turkanaltin@yahoo.com

Ibrahim Atalay
Professor Dr. h.c.
Buca Faculty of Education, Dokuz Eylul University, Izmir, Turkey
E-mail: ibrahim.atalay@deu.edu.tr

Mehmet T. Aritan
Professor
University of Wisconsin – La Crosse, La Crosse, WI 54601 USA
E-mail: aritan.mehm@uwlax.edu

Mariana Assenova
University of Sofia "St. Kliment Ohridsky", Faculty of Geography and Geology. Sofia, Bulgaria
Email: mariana.assenova@gmail.com

Cem Aydın
Southwest Anatolia Forest Research Institute-Antalya, Turkey
E-mail: acemfb77@hotmail.com

Antonio Ceballos Barbancho
Department of Geography of Universidad de Salamanca, Salamanca, Spain.

Blanka Bartók
Meteorological Department, University of Debrecen, Debrecen, Hungary
E-mail: bartokblanka@yahoo.com

Sultan Baysan
Assist. Prof.
Adnan Menderes University, Faculty of Education, Aydın, Turkey
E-mail: sbaysan@adu.edu.tr, sbaysan02@hotmail.com

Sihem Benabdallah
LGPCE, Centre de Recherches et de Technologies des Eaux BP 273, 8020 Soliman, Tunisie,
E-mail: sihem.benabdallah@certe.rnrt.tn

Sandu Boengiu
Lecturer
University of Craiova, Geography Department, Craiova, Dolj, Romania
E-mail: sboengiu@central.ucv.ro

Rached M. Boussema
LTSIRS, Ecole National d'Ingénieurs de Tunis, Tunis Belvédère, Tunisie
E-mail: Rached.boussema@enit.rnu.tn

Shpejtim Bulliqi
Lecturer Dr.
University of Pristina, Department of Geography, Geomorphology, Natural Resources, Pristina, Kosovo
E-mail: shpejtim.bulliqi@ks.gov.net

Alessandro Calzavara
Agriteco s.c., via Mezzacapo 15, 30175 Marghera – Venice, Italy
E-Mail: calzavara@agriteco.com

Salih Ceylan
Assist. Prof.
Mehmet Akif Ersoy University, Faculty of Education, Burdur, Turkey.
E-Mail: sceylan@mehmetakif.edu.tr

Ufuk Coşgun
PhD.
Southwest Anatolia Forest Research Institute-Antalya, Turkey
E-mail: ufuk-cosgun@gmail.com

Georges G. Cravins
Professor
Department of Geography and Earth Science, University of Wisconsin-La Crosse.
Wisconsin, USA
E-Mail: cravins.geor@uwlax.edu

Gheorghe Curcan
Lecturer
University of Craiova, Geography Department, Craiova, Dolj, Romania
E-mail: gilcurcan@central.ucv.ro

Isa Cürebal
Assist. Prof.
Department of Geography, Balikesir University, Balikesir, Turkey
E-mail: curebal@balikesir.edu.tr

Hilmi Demirkayą
Assist. Prof.
Mehmet Akif Ersoy University, Faculty of Education, Burdur, Turkey
E-mail: hilmi.demirkaya@gmail.com

Elka Dogramadjieva,
University of Sofia “St. Kliment Ohridsky”, Faculty of Geography and Geology., Sofia, Bulgaria
E-mail: elidogramadjieva@gmail.com

Recep Efe
Professor
Department of Geography, Balikesir University, Balikesir, Turkey
E-mail: recepefe@hotmail.com

Gözde Emekli
Ege University Faculty of Letters, Department of Geography
Izmir, Türkiye
E-mail: gozde.emekli@ege.edu.tr

Ecmel Erlat
Ege University Faculty of Letters, Department of Geography
Izmir, Türkiye
E-mail: ecmel.erlat@ege.edu.tr

Selma Akay Ertürk
Istanbul University Faculty of Letters Geography Department
Istanbul, Turkey
E-mail: akays@istanbul.edu.tr

Andrea Adél Gál
Babeş-Bolyai University, Faculty of Geography, Department of Physical Geography, Clinicilor, Cluj-Napoca, Romania,
E-mail: gal_andi@yahoo.com

Gani Gashi
University of Pristina,Department of Geography and Hydrometeorological Institute , Pristina, Kosovo
E-mail: gani.gashi@ks.gov.net,

Çiğdem Görk
Assist.Prof. Dr.
Mugla University, Biology Department, Mugla, Turkey
E-Mail: cgork@mu.edu.tr

Güven Görk
Prof.Dr.
Mugla University, Biology Department, Mugla, Turkey
E-Mail : guven66@yahoo.com

Salih Gucel
Near East University, The Institute of Earth, Ocean, Atmosphere and Environmental Sciences, Nicosia, Northern Cyprus

Francesca Guglielmi
University Ca' Foscari of Venice, IDEAS/CESD, Italy
E-mail: francesca_guglielmi@yahoo.it

Cumhur Güngöroğlu
Southwest Anatolia Forest Research Institute, Antalya, Turkey
E-mail: cgungoroglu@yahoo.com

Mehmet Gürbüz
Assist. Prof. Dr.
Sutcu Imam University, Faculty of Arts and Science. Kahraman Maraş, Turkey
E-mail: mgurbuz@ksu.edu.tr

Ion Ionita
Professor
Department of Geography, University "Al. I. Cuza" of Iasi, Romania.
E-mail: iionita49@yahoo.com

Florim Isufi
Lecturer Dr.
University of Pristina, Department of Geography, Pristina, Kosovo
E-mail: florim_isufi@yahoo.com,

Murat Karabulut
Assist. Prof. Dr.
Sutcu Imam University, Faculty of Arts and Science. Department of Geography, Kahramanmaras, Turkey.
E-mail: mkarabulut@ksu.edu.tr

Ali Kavgacı
PhD.
Southwest Anatolia Forestry Research Institute, Antalya, Turkey
E-mail: alikavgaci1977@yahoo.com

Basturk Kaya
Assist. Prof. Dr.
Selçuk University Faculty of Education, Department of Biology Education, Konya, Turkey
E-mail: basturk@selcuk.edu.tr

Nurdan Keser
Assistant Prof.
Dumlupinar University, Education Faculty, Kütahya, Turkey
E-mail: kesernurdan@yahoo.com.tr

Muhterem Küçükönder
Kahramanmaraş Sutcu Imam University, Faculty Engineering and Architecture, Geological Engineering, Kahramanmaras, Turkey
E-mail: muhti40@gmail.com

Melánia Lakatos
College of Modern Business Sciences, Department of EU Sciences
Harghita, Romania
E-mail: lakmeli_24@yahoo.com

Loris Lovo
Agriteco s.c., via Mezzacapo 15, 30175, Marghera – Venice, Italy

Emil Marinescu
Lecturer,
University of Craiova, Geography Department, Craiova, Dolj, Romania
E-mail: marinescu.emil@personal.ro

Paola Minoia
Professor
University Ca' Foscari, IDEAS/CESD, Venice- Italy
E-mail: minoia@unive.it

Manel Mosbahi
LTSIRS, Ecole National d'Ingénieurs de Tunis, Tunis Belvédère, Tunisie,
E-mail: manel241079@yahoo.fr

Elzbieta Orlowska
Institute of Geography and Regional Development
University of Wroclaw, Wroclaw, Poland
E-mail: orlowska@uni.wroc.pl

Münir Öztürk
Professor
Ege University, Botany Department, Bornova-Izmir, Turkey
E-Mail: munirozturk@gmail.com

Ali Ozel
Assist. Prof.
Dumlupınar University, Education Faculty, Kütahya, Turkey
E-mail: aliozel2002@yahoo.com

Ibrahim Ramadani
Assistant Professor
University of Pristina,Department of Geography and Hydrometeorological Institute, Pristina, Kosovo. E-mail: ibrahimramadani@yahoo.com

Rumi Sabuncu
Provincial Directorate Environment and Forestry, Antalya, Turkey
E-mail: rumisabuncu@hotmail.com

Serdar Sakçalı
Assist. Prof. Dr.
Fatih University, Biology Department, Istanbul, Turkey
sakcali@fatih.edu.tr

Cemali Sarı
Department of Geography, University of Akdeniz, Antalya, Turkey
E-mail: cemsari@yahoo.com

Rehman Seferov
PhD.
Azerbaijan National Academy of Sciences (ANAS), Baku, Azerbaijan.
E-mail: rahmansel@yahoo.com

Abdullah Soykan
Assoc. Prof.
Department of Geography, Balikesir University, Balikesir, Turkey
E-mail: asoykan@balikesir.edu.tr

Füsun Soykan
Professor
Ege University Faculty of Letters Department of Geography
Izmir-Türkiye
E-mail: fusun.soykan@ege.edu.tr

Fantina Tedim
Professor
Faculty of Arts of the University of Porto, Porto, Portugal
E-mail: fpedrosa@letras.up.pt

Enrique Morán-Tejeda
Department of Geography, Universidad de Salamanca, Salamanca, Spain.
Email: emoran@usal.es

Ülkü Eser Ünaldı
Assoc. Professor
Gazi University, Faculty of Education, Department of Geography, Ankara, Turkey
E-mail: unaldi@gazi.edu.tr

Alina Vlăduţ
Lecturer
University of Craiova, Geography Department, Craiova, Dolj, Romania
E-mail: vladut_alina2005@yahoo.com

Aylin Yaman
Research Assistant
Istanbul University, Faculty of Letters,
Department of Geography, Istanbul-Turkey
E-mail: yaman@istanbul.edu.tr

Celal Yarcı,
Professor.
Marmara University, Biology Department, Goztepe- Istanbul, Turkey
E-Mail: celalyarci@marmara.edu.tr

Serhat Zaman
Assist. Prof.
Atatürk University, Department of Geography Education,
Erzurum, Turkey
E-mail: serhatzaman@hotmail.com

Gabriele Zanetto
Professor
University Ca'Foscari, IDEAS/CESD, Venice, Italy
E-mail: gzanetto@unive.it